FROM
BACTERIA
TO BACH AND
BACK

ALSO BY DANIEL C. DENNETT

FROM BACTERIA TO BACH AND BACK

THE EVOLUTION OF MINDS

Daniel C. Dennett

W. W. NORTON & COMPANY

Independent Publishers Since 1923

New York | London

For information about special discounts for bulk purchases, please contact
W. W. Norton Special Sales at specialsales@wwnorton.com or 800-233-4830

Manufacturing by LSC Communications, Harrisonburg, VA
Book design by Chris Welch
Production manager: Anna Oler

ISBN 978-0-393-24207-2

W. W. Norton & Company, Inc.
500 Fifth Avenue, New York, N.Y. 10110
www.wwnorton.com

W. W. Norton & Company Ltd.
15 Carlisle Street, London W1D 3BS

1 2 3 4 5 6 7 8 9 0

TO BRANDON, SAMUEL, ABIGAIL, AND ARIA

CONTENTS

Part I
TURNING OUR WORLD UPSIDE DOWN

Part II
FROM EVOLUTION TO
INTELLIGENT DESIGN

Part III
TURNING OUR MINDS INSIDE OUT

14. Consciousness as an Evolved User-Illusion

15. The Age of Post-Intelligent Design

LIST OF ILLUSTRATIONS

Color insert following page 238

PREFACE

I started trying to think seriously about the evolution of the human mind when I was a graduate student in philosophy in Oxford in 1963 and knew almost nothing about either evolution or the human mind. In those days philosophers weren't expected to know about science, and even the most illustrious philosophers of mind were largely ignorant of work in psychology, neuroanatomy, and neurophysiology (the terms *cognitive science* and *neuroscience* would not be coined for more than a decade). The fledgling enterprise dubbed Artificial Intelligence by John McCarthy in 1956 was attracting attention, but few philosophers had ever touched a computer, whirring mysteriously in an air-conditioned prison guarded by technicians. So it was the perfect time for an utterly untrained amateur like me to get an education in all these fields. A philosopher who asked good questions about what they were doing (instead of telling them why, in principle, their projects were impossible) was apparently such a refreshing novelty that a sterling cadre of pioneering researchers took me in, gave me informal tutorials, and sent me alerts about whom to take seriously and what to read, all the while being more forgiving of my naïve misunderstandings than they would have been had I been one of their colleagues or graduate students.

Today there are dozens, hundreds, of young philosophers who do have solid interdisciplinary training in cognitive science, neuroscience, and computer science, and they are rightly held to much higher standards than I was. Some of them are my students, and even grandstudents, but other philosophers of my generation jumped into the deep end (often with more training than I) and have their own distinguished flocks of students making progress on the cutting edge, either as interdisciplinary philosophers or as philosophically trained scientists with labs of their own. They are professionals, and I am still an amateur, but by now a well-informed amateur, who gets invited to give lectures and participate in workshops and visit labs all over the world, where I continue my education, having more fun than I ever imagined an academic life could provide.

I consider this book to be, among other things, my grateful attempt to pay my tuition for all that instruction. This is what I think I've learned—a lot of it is still very conjectural, philosophical, out on a limb. I claim that it is the sketch, the backbone, of the best scientific theory to date of how our minds came into existence, how our brains work all their wonders, and, especially, how to think about minds and brains without falling into alluring philosophical traps. That is a controversial claim, of course, and I am eagerly looking forward to engaging with the reactions of both scientists and philosophers, and the amateurs who often have the most perceptive comments of all.

Many have helped me with my books, but I will concentrate here on thanking the people who have specifically helped me with the ideas in *this* book, and who, of course, are not responsible for the errors that they were unable to talk me out of. These include the participants in the Santa Fe Institute working group I organized on cultural evolution in May of 2014: Sue Blackmore, Rob Boyd, Nicolas Claidière, Joe Henrich, Olivier Morin, Pete Richerson, Peter Godfrey-Smith, Dan Sperber, and Kim Sterelny, as well as others at SFI: especially Chris Wood, Tanmoy Bhattacharya, David Wolpert, Cris Moore, Murray Gell-Mann, and David Krakauer. I would also like to express my gratitude to Louis Godbout of the Sybilla Hesse Foundation, for supporting the workshop.

Then there are my Tufts students and auditors who participated in a seminar in the spring of 2015 that went through most of the chapters in early drafts: Alicia Armijo, Edward Beuchert, David Blass, Michael Dale, Yufei Du, Brendan Fleig-Goldstein, Laura Friedman, Elyssa Harris, Justis Koon, Runeko Lovell, Robert Mathai, Jonathan Moore, Savannah Pearlman, Nikolai Renedo, Tomas Ryan, Hao Wan, Chip Williams, Oliver Yang, and Daniel Cloud, who visited the seminar to discuss his new book. And Joan Vergés-Gifra, Eric Schliesser, Pepa Toribio, and Mario Santos Sousa and the rest of the happy group who gathered at the University of Girona where I spent an intensive week as Guest Lecturer for the Ferrater Mora Chair of Contemporary Thought in May. Another test-bed was provided by Anthony Grayling and the faculty and students he has convened at the New College of the Humanities in London, where I have been trying out versions of these ideas for the last four years.

Others who wrestled with my drafts, changed my mind, noted my errors, and urged me to try for greater clarity include Sue Stafford, Murray Smith, Paul Oppenheim, Dale Peterson, Felipe de Brigard, Bryce Huebner, Enoch Lambert, Amber Ross, Justin Junge, Rosa Cao, Charles Rathkopf, Ronald Planer, Gill Shen, Dillon Bowen, and Shawn Simpson. Further good advice has come from Steve Pinker, Ray Jackendoff, David Haig, Nick Humphrey, Paul Seabright, Matt Ridley, Michael Levin, Jody Azzouni, Maarten Boudry, Krys Dolega, Frances Arnold, and John Sullivan.

As with my previous book, *Intuition Pumps and Other Tools for Thinking*, editors Drake McFeely and Brendan Curry at Norton challenged me to clarify, simplify, compress, expand, explain, and sometimes expunge, making the finished book a much more unified and effective reading experience than it would have been without their expert advice. John Brockman and Katinka Matson, as always, have been the perfect literary agents, advising, encouraging, entertaining—and, of course, selling—the author at home and abroad. Teresa Salvato, Program Coordinator at the Center for Cognitive

Studies, has handled all the logistics of my academic life for years, releasing thousands of prime-time hours for writing and researching, and played a more direct helping role for this book, tracking down books and articles in libraries and organizing the references.

Finally, my wife Susan, who has been my mainstay, advisor, critic, and best friend for more than half a century, has kept just the right amount of heat on the stove to keep the pot happily simmering, through all the ups and downs, and deserves accolades for her contributions to our joint enterprise.

Daniel Dennett
North Andover, MA
March 28, 2016

Part I

TURNING OUR WORLD UPSIDE DOWN

1

Introduction

Welcome to the jungle

How come there are minds? And how is it possible for minds to ask and answer this question? The short answer is that minds evolved and created thinking tools that eventually enabled minds to know how minds evolved, and even to know how these tools enabled them to know what minds are. What thinking tools? The simplest, on which all the others depend in various ways, are spoken words, followed by reading, writing, and arithmetic, followed by navigation and mapmaking, apprenticeship practices, and all the concrete devices for extracting and manipulating information that we have invented: compass, telescope, microscope, camera, computer, the Internet, and so on. These in turn fill our lives with technology and science, permitting us to know many things not known by any other species. We know there are bacteria; dogs don't; dolphins don't; chimpanzees don't. Even bacteria don't know there are bacteria. Our minds are different. It takes thinking tools to understand what bacteria are, and we're the only species (so far) endowed with an elaborate kit of thinking tools.

That's the short answer, and stripped down to the bare generalities it shouldn't be controversial, but lurking in the details are some surprising, even shocking, implications that aren't yet well understood or appreciated. There is a winding path leading through a jungle of science and philosophy, from the initial bland assumption that we people are physical objects, obeying the laws of physics, to an understanding of our conscious minds. The path is strewn with difficulties, both empirical and conceptual, and there are plenty of experts who vigorously disagree on how to handle these problems. I have been struggling through these thickets and quagmires for over fifty years, and I have found a path that takes us all the way to a satisfactory—and satisfying—account of how the "magic" of our minds is accomplished without any magic, but it is neither straight nor easy. It is not the only path on offer, but it is the best, most promising to date, as I hope to show. It does require anyone who makes the trip to abandon some precious intuitions, but I think that I have at last found ways of making the act of jettisoning these "obvious truths" not just bearable but even delightful: it turns your head inside out, in a way, yielding some striking new perspectives on what is going on. But you do have to let go of some ideas that are dear to many.

There are distinguished thinkers who have disagreed with my proposals over the years, and I expect some will continue to find my new forays as outrageous as my earlier efforts, but now I'm beginning to find good company along my path, new support for my proposed landmarks, and new themes for motivating the various strange inversions of reasoning I will invite you to execute. Some of these will be familiar to those who have read my earlier work, but these ideas have been repaired, strengthened, and redesigned somewhat to do heavier lifting than heretofore. The new ones are just as counterintuitive, at first, as the old ones, and trying to appreciate them without following my convoluted path is likely to be forlorn, as I know from many years of trying, and failing, to persuade people piecemeal. Here is a warning list of some of the hazards (to comfortable thinking) you will meet on my path, and I don't expect you to "get" all of them on first encounter:

1. Darwin's strange inversion of reasoning
2. Reasons without reasoners
3. Competence without comprehension
4. Turing's strange inversion of reasoning

5. Information as design worth stealing
6. Darwinism about Darwinism
7. Feral neurons
8. Words striving to reproduce
9. The evolution of the evolution of culture
10. Hume's strange inversion of reasoning
11. Consciousness as a user-illusion
12. The age of post-intelligent design

"Information as design worth stealing? Don't you know about Shannon's mathematical theory of information?" "Feral neurons? As contrasted with what, domesticated neurons?" "Are you serious? Consciousness as an illusion? Are you kidding?"

If it weren't for the growing ranks of like-minded theorists, well-informed scientists and philosophers who agree with at least large portions of my view and have deeply contributed to it, I'd no doubt lose my nerve and decide that I was the one who's terminally confused, and of course it's possible that our bold community of enthusiasts are deluding each other, but let's find out how it goes before we chance a verdict.

I know how easy, how tempting, it is to ignore these strange ideas or dismiss them without a hearing when first encountered because I have often done so myself. They remind me of those puzzles that have a *retrospectively* obvious solution that you initially dismiss with the snap judgment: "It can't be that," or don't even consider, it is so unpromising.[1] For someone who has often accused others of mistaking failures of imagination for insights into necessity, it is embarrassing to recognize my own lapses in this regard, but having stumbled

---

1   One of my favorites: Four people come to a river in the night. There is a narrow bridge, and it can only hold two people at a time. They have one torch and, because it's night, the torch has to be used when crossing the bridge. Person A can cross the bridge in one minute, B in two minutes, C in five minutes, and D in eight minutes. When two people cross the bridge together, they must move at the slower person's pace. The question is, can they all get across the bridge in fifteen minutes or less?

upon (or been patiently shown) new ways of couching the issues, I am eager to pass on my newfound solutions to the big puzzles about the mind. All twelve of these ideas, and the background to render them palatable, will be presented, in *roughly* the order shown above. Roughly, because I have found that some of them defy straightforward defense: you can't appreciate them until you see what they can get you, but you can't use them until you appreciate them, so you have to start with partial expositions that sketch the idea and then circle back once you've seen it in action, to drive home the point.

The book's argument is composed of three strenuous exercises of imagination:

> turning our world upside down, following Darwin and Turing;
> then evolving evolution into intelligent design;
> and finally turning our minds inside out.

The foundation must be carefully secured, in the first five chapters, if it is to hold our imaginations in place for the second feat. The next eight chapters delve into the empirical details of the evolution of minds and language *as they appear from our inverted perspective.* This allows us to frame new questions and sketch new answers, which then sets the stage for the hardest inversion of all: seeing what consciousness looks like from the new perspective.

It's a challenging route, but there are stretches where I review familiar material to make sure everybody is on the same page. Those who know these topics better than I do can jump ahead if they wish, or they can use my treatments of them to gauge how much they should trust me on the topics they don't know much about. Let's get started.

A bird's-eye view of the journey

Life has been evolving on this planet for close to four billion years. The first two billion years (roughly) were spent optimizing the basic machinery for self-maintenance, energy acquisition and reproduction, and the only living things were *relatively* simple single-celled entities—bacteria, or their cousins, archaea: the *prokaryotes*. Then an amazing thing happened: two *different* prokaryotes, each with its own competences and habits due to its billions of years of independent evolution, collided. Collisions of this sort presumably happened countless numbers of times, but on (at least) one occasion, one cell engulfed the other, and instead of destroying the other and using the parts as fuel or building materials (eating it, in other words), it let it go on living, and, by dumb luck, found itself fitter—more competent in some ways that mattered—than it had been as an unencumbered soloist.

This was perhaps the first successful instance of *technology transfer*, a case of two different sets of competences, honed over eons of independent R&D (research and development), being united into something bigger and better. We read almost every day of Google or Amazon or General Motors gobbling up some little start-up company to get its hands on their technological innovations and savvy, advances in R&D that are easier to grow in cramped quarters than in giant corporations, but the original exploitation of this tactic gave evolution its first great boost. Random mergers don't always work out that way. In fact, they almost never work out that way, but evolution is a process that depends on amplifying things that almost never happen. For instance, mutation in DNA almost never occurs—not once in a billion copyings—but evolution depends on it. Moreover, the vast majority of mutations are either deleterious or neutral; a fortuitously "good" mutation almost never happens. But evolution depends on those rarest of rare events.

Speciation, the process in which a new species is generated when some members get isolated from their "parent" population and wander away in genetic space to form a new gene pool, is an exceedingly rare event, but the millions or billions of species that have existed on this planet each got their start with an event of speciation. Every birth in every lineage is a potential initiation of a speciation event, but speciation almost never happens, not once in a million births.

In the case we are considering, the rare improvement that resulted from the fortuitous collision of a bacterium and an archaeon had a life-changing sequel. Being fitter, this conjoined duo reproduced more successfully than the competition, and every time it divided in two (the bacterial way of reproducing) both daughter cells included an offspring of the original guest. Henceforth their fates were joined—symbiosis—in one of the most productive episodes in the history of evolution. This was *endosymbiosis* because one of the partners was literally inside the other, unlike the side-by-side ectosymbiosis of clownfish and sea anemone or fungus and algae in lichens. Thus was born the *eukaryotic* cell, which, having more working parts, was more versatile than its ancestors, simple *prokaryotic* cells, such as bacteria.[2] Over time these eukaryotes grew much larger, more complex, more competent, *better* (the "eu" in "eukaryotic" is like the "eu" in euphonious, eulogy, and eugenics—it means *good*). Eukaryotes were the key ingredient to make possible multicellular life forms of all varieties. To a first approximation, every living thing big enough to be visible to the naked eye is a multicellular eukaryote. We are eukaryotes, and so are sharks, birds, trees, mushrooms, insects, worms, and all the other plants and animals, all direct descendants of the original eukaryotic cell.

This Eukaryotic Revolution paved the way for another great transition, the Cambrian "Explosion" more than half a billion years ago, which saw the "sudden" arrival of a bounty of new life forms. Then came what I call the MacCready Explosion, after the late great Paul MacCready, visionary engineer (and creator of the Gossamer Albatross, among other green marvels). Unlike the Cambrian diversification, which occurred over several million years about 530 million years ago (Gould 1989), the MacCready Explosion occurred in only about 10,000 years, or 500 human generations. According to MacCready's calculations (1999), at the dawn of human agriculture

2   Lane (2015) has a fascinating update (and revision) of the story of the endo-symbiotic origin of eukaryotes that I have been conveying for the last twenty years or so. It now is quite secure that a bacterium and an archaeon were the Adam and Eve, not two different bacteria, as I had often said.

10,000 years ago, the worldwide human population plus their live-stock and pets was only ~0.1% of the terrestrial vertebrate biomass. (We're leaving out insects, other invertebrates, and all marine animals.) Today, by his estimation, it is 98%! (Most of that is cattle.) His reflections on this amazing development are worth quoting:

> Over billions of years, on a unique sphere, chance has painted a thin covering of life—complex, improbable, wonderful and fragile. Suddenly we humans . . . have grown in population, technology, and intelligence to a position of terrible power: we now wield the paintbrush. (1999, p. 19)

There have been other *relatively* sudden changes on our planet, mass extinctions such as the Cretaceous-Paleogene extinction about sixty-six million years ago that doomed the dinosaurs, but the MacCready Explosion is certainly one of the fastest major biological changes ever to occur on Earth. It is still going on and picking up speed. We can save the planet or extinguish all life on the planet, something no other species can even imagine. It might seem obvious that the order of MacCready's three factors—population, technology, and intelligence—should be reversed: first our human *intelligence* created the *technology* (including agriculture) that then permitted the *population* boom, but as we shall see, evolution is typically an interwoven fabric of coevolutionary loops and twists: in surprising ways, our so-called native intelligence depends on both our technology and our numbers.

Our human minds are strikingly different from the minds of all other species, many times more powerful and more versatile. The long answer of how we came to have such remarkable minds is beginning to come into focus. The British biologist D'Arcy Thompson (1917) famously said, "Everything is the way it is because it got that way." Many of the puzzles (or "mysteries" or "paradoxes") of human consciousness evaporate once you ask how they could possibly have arisen—and actually try to answer the question! I mention that because some people marvel at the question and then "answer" it by saying, "It's an impenetrable mystery!" or "God did it!" They may in the end be right, of course, but given the fabulous bounty of thinking tools recently put at our disposal and hardly used yet, this is a strikingly premature surrender. It may not be defeatist; it may be defensive. Some people would like to persuade the curious to keep their hands off the beloved mysteries, not realizing that a mystery solved is even more ravishing than the ignorant fantasies it replaces. There are some people who have looked hard at scientific explanations and disagree: to their taste, ancient myths of fiery chariots, warring gods, worlds hatching from serpent eggs, evil spells, and enchanted gardens are more delightful and worthy of attention than any rigorous, predictive scientific story. You can't please everybody.

This love of mystery is just one of the potent imagination-blockers standing in our way as we attempt to answer the question of how come there are minds, and, as I already warned, our path will have to circle back several times, returning to postponed questions that couldn't be answered until we had a background that couldn't be provided until we had the tools, which couldn't be counted on until we knew where they came from, a cycle that gradually fills in the details of a sketch that won't be convincing until we can reach a vantage point from which we can look back and see how all the parts fit together.

Douglas Hofstadter's book, *I Am a Strange Loop* (2007), describes a mind composing itself in cycles of processing that loop around, twisting and feeding on themselves, creating exuberant reactions to reflections to reminders to reevaluations that generate novel structures: ideas, fantasies, theories, and, yes, thinking tools to create still more. Read it; it will take your imagination on a roller-coaster ride, and you will learn a lot of surprising truths. My story in this book is of the larger strange looping process (composed of processes composed of processes) that generated minds like Hofstadter's (and Bach's and Darwin's) out of nothing but molecules (made of atoms made of . . .). Since the task is cyclical, we have to begin somewhere in the middle and go around several times. The task is made difficult by a feature it doesn't share with other scientific investigations of processes (in cosmology, geology, biology, and history, for instance): people care so deeply what the answers are that they have a very hard time making themselves actually *consider* the candidate answers objectively.

For instance, some readers may already be silently shaking their heads over a claim I just made: Our human minds are strikingly different from the minds of all other species, many times more powerful and more versatile. Am I really that *prejudiced*? Am I a "species chauvinist" who actually thinks human minds are that much more wonderful than the minds of dolphins and elephants and crows and bonobos and the other clever species whose cognitive talents have recently been discovered and celebrated? Isn't this a barefaced example of the *fallacy* of "human exceptionalism"? Some readers may be ready to throw the book across the room, and others may just be unsettled by my politically incorrect lapse. It's amusing (to me, at least) that human exceptionalism provokes equal outrage in opposite directions. Some scientists and many animal lovers deplore it as an intellectual sin of the worst kind, scientifically ill-informed, an ignoble vestige of the bad old days when people routinely thought that all "dumb" animals were put on this planet for our use and amusement. Our brains are made of the same neurons as bird brains, they note, and some animal brains are just as large (and just as smart, in their own species-specific ways) as ours. The more you study the actual circumstances and behaviors of animals in the wild, the more you appreciate their brilliance. Other thinkers, particularly in the arts and humanities and social sciences, consider the *denial* of human exceptionalism to be myopic, doctrinaire, *scientism* at its worst: *Of course* our minds are orders of magnitude more powerful than the cleverest animal mind! No animal creates art, writes poetry, devises scientific theories, builds spaceships, navigates the oceans, or even tames fire. This provokes the retort: What about the elegantly decorated bowers built by bowerbirds, the political subtlety of chimpanzees, the navigational prowess of whales and elephants and migrating birds, the virtuoso song of the nightingale, the language of the vervet monkeys, and even the honey bees? Which invites the response that these animal marvels are paltry accomplishments when compared with the genius of human artists, engineers, and scientists. Some years ago,[3] I coined the terms *romantic* and *killjoy* to refer to the sides of this intense duel over animal minds, and one of my favorite memories of this bipolar reaction to claims about animal intelligence occurred at an international scientific workshop on animal intelligence where one distinguished researcher managed to play both romantic and killjoy roles with equal passion: "Hah! You think insects are so stupid! I'll show you how smart they are. Consider this result. . . . !" Followed later on the same day by, "So you think bees are so clever? Let me show you how stupid they *really* are! They're mindless little robots!"

Peace! We will see that both sides are right about some things and wrong about others. We're not the Godlike geniuses we sometimes think we are, but animals are not so smart either, and yet both humans and (other) animals are admirably equipped to deal "brilliantly" with many of the challenges thrown at them by a difficult, if not always cruel, world. And our human minds are uniquely powerful in ways that we can begin to understand once we see how they got that way.

Why do we care so much? That is one of the many hanging questions that needs an answer, but not right now, except in briefest outline: While the processes that gave rise to this caring go back thousands of years, and in some regards millions or even billions of years, they first became a *topic*—an object to think about and care

---

3   Since this book is the culmination of a half century of work on these topics, normal academic citation practices would sprinkle its pages with dozens of interruptions of the "(Dennett 1971, 1991, 2013)" variety, but such voluminous self-citation would send the wrong message. My thinking has been shaped by hundreds of thinkers, and I have tried to credit the key sources on all the ideas discussed as they arise, while sequestering most of the information about where I myself have expanded on these points to the Appendix: The Background (p. 415), for the convenience of anyone who is curious to see how the arguments developed.

about—at the birth of modern science in the seventeenth century, so that is where I will break into the ring and start this version of the story.

The Cartesian wound

Si, abbiamo un anima. Ma é fatta di tanti piccoli robot!
(Yes, we have a soul, but it's made of lots of tiny robots!)
—Headline for an interview with me by Giulio Giorello
in *Corriere della Serra*, Milan, 1997

René Descartes, the seventeenth-century French scientist and phi-losopher, was very impressed with his own mind, for good reason. He called it his *res cogitans*, or thinking thing, and it struck him, on reflection, as a thing of miraculous competence. If anybody had the right to be in awe of his own mind, Descartes did. He was undoubtedly one of the greatest scientists of all time, with major work in mathematics, optics, physics, and physiology; and the inventor of one of the most valuable thinking tools of all time, the system of "Cartesian coordinates" that enables us to translate between algebra and geometry, paving the way for calculus and letting us plot almost anything we want to investigate, from aard-vark growth to zinc futures. Descartes propounded the original TOE (theory of everything), a prototypical Grand Unified Theory, which he published under the immodest title *Le Monde* (*The World*). It purported to explain everything from the orbits of the planets and the nature of light to the tides, from volcanoes to magnets, why water forms into spherical drops, how fire is struck from flint, and much, much more. His theory was almost all dead wrong, but it held together surprisingly well and is strangely plausible even in today's hindsight. It took Sir Isaac Newton to come up with a better physics, in his famous *Principia*, an explicit refutation of Descartes's theory.

Descartes didn't think it was just his mind that was wonderful; he thought that all normal human minds were wonderful, capable of feats that no mere animal could match, feats that were beyond the reach of any imaginable *mechanism*, however elaborate and complicated. So he concluded that minds like his (and yours) were not material entities, like lungs or brains, but made of some *second* kind of stuff that didn't have to obey the laws of physics—articulating the view known as *dualism*, and, often, *Cartesian dualism*. This idea that mind isn't matter and matter can't be mind was not invented by Descartes. It had seemed obvious to reflective people for thousands of years that our minds are not like the furniture of the "external" world. The doctrine that *each of us has an immaterial (and immortal) soul that resides in and controls the material body* long passed for shared knowledge, thanks to the instruction of the Church. But it was Descartes who distilled this default assumption into a positive "theory": The immaterial mind, the conscious *thinking thing* that we know intimately through introspection, is somehow in communication with the material brain, which provides all the input *but none of the understanding or experience.*

The problem with dualism, ever since Descartes, is that nobody has ever been able to offer a convincing account of how these postulated interactive transactions between mind and body could occur without violating the laws of physics. The candidates on display today offer us a choice between a revolution in science so radical that it can't be described (which is convenient, since critics are standing by, ready to pounce) or a declaration that some things are just Mysteries, beyond human understanding (which is also convenient if you don't have any ideas and want to exit swiftly). But even if, as I noted years ago, dualism tends to be regarded as a cliff over which you push your opponents, those left on the plateau have a lot of unfinished business constructing a theory that is *not* dualism in disguise. The mysterious linkage between "mind and matter" has been a battleground of scientists and philosophers since the seventeenth century.

Francis Crick, the recently deceased co-discoverer of the structure of DNA, was another of history's greatest scientists, and his last major piece of writing was *The Astonishing Hypothesis: The Sci-*

entific Search for the Soul (1994), in which he argued that dualism is false; the mind just *is* the brain, a material organ with no mysterious extra properties not found in other living organisms. He was by no means the first to put forward this denial of dualism; it has been the prevailing—but not unanimous—opinion of both scientists and philosophers for the better part of a century. In fact, many of us in the field objected to his title. There was nothing astonishing about this hypothesis; it had been our working assumption for decades! Its *denial* would be astonishing, like being told that gold was not composed of atoms or that the law of gravity didn't hold on Mars. Why should anyone expect that *consciousness* would bifurcate the universe dramatically, when even *life* and *reproduction* could be accounted for in physico-chemical terms? But Crick wasn't writing his book for scientists and philosophers, and he knew that among laypeople, the appeal of dualism was still quite overpowering. It seemed not only obvious to them that their private thoughts and experiences were somehow conducted in some medium *in addition to* the neuronal spike trains scientists had found buzzing around in their brains, but the prospect of denying dualism threatened horrible consequences as well: If "we are just machines," what happens to free will and responsibility? How could our lives have meaning at all if we are just huge collections of proteins and other molecules churning away according to the laws of chemistry and physics? If moral precepts were nothing but extrusions generated by the hordes of microbiological nano-machines between our ears, how could they make a difference worth honoring?

Crick did his best to make "the astonishing hypothesis" not just comprehensible but also palatable to the lay public. Despite his clear and energetic writing, and unparalleled gravitas, he didn't make much progress. This was largely, I think, because in spite of his book's alarm bell of a title, he underestimated the emotional turmoil this idea provokes. Crick was an excellent explainer of science to nonscientists, but the pedagogical problems in this arena are not the usual ones of attracting and holding the attention of semi-bewildered and intimidated laypeople and getting them to work through a smattering of math. When the topic of consciousness arises, the difficult task is to keep a lid on the anxieties and suspicions that seduce people—including many scientists—into distorting what we know and aiming preemptive strikes at dangerous ideas they dimly see looming. Moreover, on this topic *everybody's an expert*. People are calmly prepared to be instructed about the chemical properties of calcium or the microbiological details of cancer, but they think they have a particular personal authority about the nature of their own conscious experiences that can trump any hypothesis they find unacceptable.

Crick is not alone. Many others have tried their hand at knitting up what one of the best of them, Terrence Deacon, has called "the Cartesian wound that severed mind from body at the birth of modern science" (2011, p. 544). Their efforts are often fascinating, informative, and persuasive, but no one has yet managed to be entirely *convincing*. I have devoted half a century, my entire academic life, to the project in a dozen books and hundreds of articles tackling various pieces of the puzzle, without managing to move all that many readers from wary agnosticism to calm conviction. Undaunted, I am trying once again and going for the whole story this time.

Why do I think it is worth trying? Because, first, I think we have made tremendous scientific progress in the last twenty years; many of the impressionistic hunches of yore can now be replaced with well-researched details. I plan to rely heavily on the bounty of experimental and theoretical work that others have recently provided. And second, I think I now have a better sense of the various undercurrents of resistance that shackle our imaginations, and I plan to expose and disarm them as we go, so that, for the first time, the doubters can *take seriously* the prospect of a scientific, materialist theory of their own minds.

Cartesian gravity

Over the years, trudging back and forth over the battleground, participating in many skirmishes, I've gradually come to be able to *see* that there are powerful forces at work, distorting imagination—my own imagination included—pulling us first one way and then another. If you learn to see these forces too, you will find that suddenly things begin falling into place in a new way. You can identify the forces tugging at your thinking, and then set up alarms to alert you and buffers to protect you, so that you can resist them effectively while simultaneously exploiting them, because they are not just distorting; they can also be imagination-enhancing, launching your thinking into new orbits.

One cold, starry night over thirty years ago, I stood with some of my Tufts students looking up at the sky while my friend, the philosopher of science, Paul Churchland instructed us how to *see the plane of the ecliptic*, that is, to look at the other visible planets in the sky and picture them, and ourselves, as wheeling around the sun all on the same invisible plane. It helps to tip your head just so and remind yourself of where the sun must be, way back behind you. Suddenly, the orientation clicks into place and shazam, you *see* it![4] Of course we all knew for years that this was the situation of our planet in the solar system, but until Paul made us see it, it was a rather inert piece of knowledge. Inspired by his example, I am going to present some eye-opening (actually *mind*-opening) experiences that I hope will move your mind into some new and delightful places.

The original distorting force, which I will call *Cartesian gravity*, actually gives birth to several other forces, to which I will expose you again and again, in different guises, until you can see them clearly too. Their most readily "visible" manifestations are already familiar to most everyone—too familiar, in fact, since we tend to think we have already taken their measure. We underestimate them. We must look behind them, and beyond them, to see the way they tend to sculpt our thinking.

Let's begin by looking back at Crick's "astonishing hypothesis." Those of us who insist that we don't find it at all astonishing fuel

---

4   Churchland includes instruction and a diagram that will help you enjoy this delightful effect in his 1979 book *Scientific Realism and the Plasticity of Mind*.

our confidence by reminding ourselves of the majestic array of well-solved puzzles, well-sleuthed discoveries, well-confirmed theories of modern, materialistic science that we all take for granted these days. When you think about it, it is just amazing how much we human beings have figured out in the few centuries since Descartes. We know how atoms are structured, how chemical elements interact, how plants and animals propagate, how microscopic pathogens thrive and spread, how continents drift, how hurricanes are born, and much, much more. We know our brains are made of the same ingredients as all the other things we've explained, and we know that we belong to an evolved lineage that can be traced back to the dawn of life. If we can explain *self-repair in bacteria* and *respiration in tadpoles* and *digestion in elephants*, why shouldn't *conscious thinking in H. sapiens* eventually divulge its secret workings to the same ever-improving, self-enhancing scientific juggernaut?

That's a rhetorical question, and trying to *answer* rhetorical questions instead of being cowed by them is a good habit to cultivate. So might consciousness be more challenging than self-repair or respiration or digestion, and if so, why? Perhaps because it *seems* so different, so private, so intimately *available* to each of us in a way unlike any other phenomenon in our living bodies. It is not all that hard these days to imagine how respiration works even if you're ignorant of the details: you breathe in the air, which we know is a combination of different gases, and we breathe out what we can't use—carbon dioxide, as most people know. One way or another the lungs must filter out and grab what is needed (oxygen) and exude the waste product (carbon dioxide). Not hard to grasp in outline. The phenomenon of smelling a cookie and suddenly remembering an event in your childhood seems, in contrast, not at all mechanical. "Make me a nostalgia-machine!" "What? What could the parts possibly do?" Even the most doctrinaire materialists will admit that they have only foggy and programmatic ideas about how brain activities might amount to nostalgia or wistfulness or prurient curiosity, for example.

Not so much an astonishing hypothesis, many might admit, as a dumbfounding hypothesis, a hypothesis about which one can only wave one's hands and hope. Still, it's a comfortable position to maintain, and it's tempting to diagnose those who disagree—the self-appointed Defenders of Consciousness from Science—as suffering from one or another ignominious failing: narcissism ("I refuse to have *my* glorious mind captured in the snares of science!"); fear ("If my mind is just my brain, I won't be in charge; life will have no meaning!"); or disdain ("These simple-minded, *scientistic* reductionists! They have no idea how far short they fall in their puny attempts to appreciate the world of meaning!").

These diagnoses are often warranted. There is no shortage of pathetic bleats issuing from the mouths of the Defenders, but the concerns that motivate them are not idle fantasies. Those who find Crick's hypothesis not just astonishing but also deeply repugnant are onto something important, and there is also no shortage of anti-dualist philosophers and scientists who are not yet comfortable with materialism and are casting about for something in between, something that can actually make some progress on the science of consciousness without falling into either. The trouble is that they tend to misdescribe it, inflating it into something deep and metaphysical.[5]

What they are feeling is a way of thinking, an overlearned habit, so well entrenched in our psychology that denying it or abandoning it is literally unthinkable. One sign of this is that the confident scientific attitude expressed by the "other side" begins to tremble the closer the scientists get to a certain set of issues dealing with consciousness, and they soon find themselves, in spite of themselves, adopting the shunned perspective of the Defenders. I am going to

5 Working with Nick Humphrey some years ago on what was then called multiple personality disorder, I discovered the almost irresistible temptation, even in Nick and myself, to exaggerate anything that strikes one as both important and uncanny. I surmise that whenever we human beings encounter something truly strange and unsettling, our attempts to describe *to ourselves* what we are experiencing *tend* to err on the side of exaggeration, perhaps out of a subliminal desire to impress ourselves with the imperative that this is something we ignore at our peril and must get to the bottom of.

describe this dynamic process metaphorically at the outset to provide a simple framework for building a less metaphorical, more
explicit and factual understanding of what is happening.

Suppose the would-be mind-explainer starts with her *own* mind.
She stands at Home, on Planet Descartes, meditating on the task
ahead and looking at the external universe from the "first-person
point of view." From this vantage point, she relies on all the familiar
furniture of her mind to keep her bearings, and Cartesian gravity
is the force that locks her into this egocentric point of view "from
the inside." Her soliloquy might be, echoing Descartes: "Here I
am, a conscious thinking thing, intimately acquainted with the
ideas in my own mind, which I know better than anybody else just
because they're mine." She cannot help but be a Defender of her
own Home. Meanwhile, from faraway comes the scientific explorer
of consciousness, approaching Planet Descartes confidently, armed
with instruments, maps, models, and theories, and starts moving
in for the triumphant conquest. The closer she gets, however, the
more uncomfortable she finds herself; she is being dragged into an
orientation she knows she must avoid, but the force is too strong.
As she lands on Planet Descartes she finds herself flipped suddenly
into first-person orientation, feet on the ground but now somehow
unable to reach, or use, the tools she brought along to finish the job.
Cartesian gravity is all but irresistible when you get that close to the
surface of Planet Descartes. How did she get there, and what happened in that confusing last-minute *inversion*? (Strange inversions
will be a major theme in this book.) There seem to be two competing orientations, the first-person point of view of the Defenders and
the third-person point of view of the scientists, much like the two
ways of seeing the philosophers' favorite illusions, the duck-rabbit
and the Necker cube. You can't adopt both orientations at once.

The problem posed by Cartesian gravity is sometimes called
the Explanatory Gap (Levine 1983) but the discussions under that
name strike me as largely fruitless because the participants tend
to see it as a chasm, not a glitch in their imaginations. They may
have *discovered* the "gap," but they don't see it for what it actually is

FIGURE 1.1: Duck-rabbit.

because they haven't asked "how it got that way." By reconceiving of the gap as a dynamic imagination-distorter that has arisen for good reasons, we can learn how to traverse it safely or—what may amount to the same thing—make it vanish.

Cartesian gravity, unlike the gravity of physics, does not act on things in proportion to their mass and proximity to other massy items; it acts on ideas or representations of things in proportion to their proximity *in content* to other ideas that play privileged roles in the maintenance of a living thing. (What this means will gradually become clear, I hope—and then we can set this metaphorical way of speaking aside, as a ladder we have climbed and no longer need to rely on.) The *idea* of Cartesian gravity, as so far presented, is just a metaphor, but the phenomenon I am calling by this metaphorical name is perfectly real, a disruptive force that bedevils (and sometimes aids) our imaginations, and unlike the gravity of physics, it is itself an evolved phenomenon. In order to understand it, we need to ask how and why it arose on planet Earth.

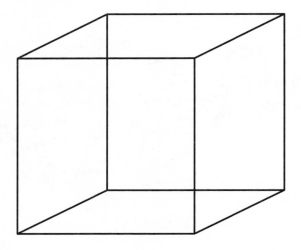

FIGURE 1.2: Necker cube.

It will take several passes over the same history, with different details highlighted each time, to answer this question. We tend to underestimate the strength of the forces that distort our imaginations, especially when confronted by irreconcilable insights that are "undeniable." It is not that we *can't* deny them; it is that we *won't* deny them, won't even *try* to deny them. Practicing on the forces that are easy to identify—species chauvinism, human exceptionalism, sexism—prepares us to recognize more subtle forces at work. In the next chapter I turn briefly to the very beginning of life on the planet and give a preliminary outline of the story to come, with scant detail, and also tackle one of the first objections that (I predict) will occur to the reader while encountering that outline. I cast evolutionary processes as *design* processes (processes of research and development, or R&D) and this *adaptationist* or *reverse-engineering* perspective has long lived under an undeserved cloud of suspicion. Contrary to widespread belief, as we shall see, adaptationism is alive and well in evolutionary biology.

2

Before Bacteria and Bach

Why Bach?

To get a sound perspective on our history, we actually have to go back to the time before bacteria, before life in any form existed, since some of the demands on getting life going at all send important echoes down the eons to explain features of our minds today. And before turning to *that* story, let me pause to draw attention to one word: "Bach." I could have chosen "From Archaea to Shakespeare" or "From *E. coli* to Einstein" or perhaps "From Prokaryotes to Picasso," but the alliteration of "Bacteria to Bach" proved irresistible.

What about the glaring fact that all the candidates I just considered in my pantheon of great minds are men? What an awkward stumble out of the starting blocks! Do I really want to alienate many readers at the outset? What was I thinking? I did it on purpose, to provide a relatively mild and simple example of *one variety* of the gravitational forces of Cartesian gravity we will deal with. If you bristled when you noticed my all-male roster of genius, then good. That means you won't forget that I've taken out a loan on your patience, to be repaid at some later time in the book. Bristling (like any acute emotional reaction, from fear to amusement) is a sort of boldface for your memory, making the offending item less likely to be forgotten. At this point I ask you to resist the urge to launch a preemptive strike. Throughout our journey, we will need to identify uncomfortable facts without indulging in premature explanations or rebuttals. While I delight in having readers who are not only paying close attention but also way ahead of me, I would much prefer that you bide your time, cutting me some slack—giving me enough rope to hang myself, if you like—instead of trying to derail my attempt at a calm, objective account with your premonitions.

So let's stand down, pause, and review a few plain facts, leaving the explanations and refutations for later. It is an obvious fact that although there have been many brilliant women of great attainment, none of them has achieved the iconic status of Aristotle, Bach, Copernicus, Dickens, Einstein. . . . I could easily list a dozen more men in the same league, but try for yourself to think of a great female thinker who could readily displace any of these men in playing the emblematic role in my title. (My favorites would be Jane Austen, Marie Curie, Ada Lovelace, and Hypatia of Alexandria. I doubt I've overlooked any obvious candidates, but time will tell.)

There have *not yet* been any female superstar geniuses. What might explain this fact? Political oppression? The self-fulfilling sexist prophecies that rob young girls of inspiring role models? Media bias over the centuries? Genes? Please don't jump to conclusions, even if you think the answer is obvious. (I don't.) We will soon see that genes, though essential players in the history of mind, are nowhere near as important as many like to think. Genes may account for basic animal competence, but *genes don't account for genius!* Moreover, the traditional view of great human societies owing their greatness to the creative brilliance of (some of) their inhabitants is, I will try to show, just about backward. Human culture *itself* is a more fecund generator of brilliant innovations than any troupe of geniuses, of either gender. This it achieves by a process of *cultural* evolution that is as much "the author" of our finest achievements as any individual thinker is.

The very idea that evolution by natural selection might have a foundational role to play in understanding human culture fills some people, even wise and thoughtful people, with loathing. They see human culture as something transcendent, the miraculous endowment that distinguishes us human beings from the beasts, the last bulwark against creeping reductionism, against genetic determinism, against the philistinism they think they see in contemporary science. And, in reaction to these "culture vultures," there are hardnosed scientists to whom any appeal to "culture" smells of mysterymongering or worse.

When I hear the word "culture" I reach for my gun.[6]

Now I must ask "both sides" to holster their guns and be patient. There is a middle ground that can do justice to the humanities and science at the same time, explaining how human culture got up and running by a process of evolution of cultural items—*memes*—that invaded human brains the way viruses invade human bodies. Yes, memes have *not* been shown to be a bad idea, and they will get their day in court in this book. Those on both sides who scoff and hoot at the idea of memes will see that there are *good* objections to memes—in addition to the flawed "refutations" that have been uncritically accepted by many who just can't stand the idea—but these good objections point to refinements that save the concept of memes after all.

So whose side am I on? Readers intent on framing the issue in these terms have missed the point. This polarization of visions, with cheering, hissing spectators egging on the combatants, is just a conveniently obvious first manifestation of the forces I am trying to render visible to all and neutralize. There is much more to come—subtler and more insidious pressures on the thinking of scientists, philosophers, and laypeople alike. Now back to my first pass at the story.

---

6 *Not* Hermann Göring (and not Heinrich Himmler either). According to Wikipedia, this oft-misattributed declaration was born as a line of dialogue in a pro-Nazi play by Hans Johst.

How investigating the prebiotic world
is like playing chess

The simplest, earliest life form capable of reproducing itself, something like a bacterium, was already a breathtakingly complex and brilliantly designed self-maintaining system. (Hang on. Did I not just give aid and comfort to the Intelligent Design crowd? No. But how can a good materialist, atheist Darwinian like me keep a straight face while declaring the earliest reproducing life forms to be *brilliantly designed?* Hold your fire.)

A well-known chicken-and-egg-problem beloved by the Intelligent Design folks challenges us to undo the "paradox" of the origin of life: evolution by natural selection couldn't get started until there were *reproducing* things, since there would be no offspring to inherit the best designs, but the simplest reproducing thing is much too complex to arise by mere chance.[7] So, it is claimed, evolution cannot get started without a helping hand from an Intelligent Designer. This is a defective argument, a combination of misdirection and failure of imagination, as we shall see. But we should concede the truly striking fact that the first concoction of molecules capable of reliably reproducing itself was—must have been—an "engineering" marvel, composed of thousands of complex parts working together.

For researchers working on the origin of life this poses a straightforward challenge: How could this *possibly* come about without a miracle? (Perhaps an intelligent designer from another galaxy was responsible, but this would just postpone the question and make it harder to address.) The way to proceed is clear: start with the minimal specification for a living, reproducing thing—a list of all the things it *has to be able to do*—and work backward, making an inventory of the available raw materials (often called the feedstock molecules of prebiotic chemistry), and asking what sequence of

---

7 I will have a lot to say about intelligent design—by human beings—in this book but almost nothing more about Intelligent Design, the latest wave of creationist propaganda. It is not worth any further rebuttal.

possible events could gradually bring together, non-miraculously, all the necessary parts in the right positions to accomplish the job. Notice that this minimal specification lists what the thing must do, a list of functions, not a list of parts and materials. A mousetrap has to trap mice, and a can-opener has to open cans, and a living thing has to capture energy and protect (or repair) itself long enough to reproduce.

How could such a living thing possibly arise? If you can answer this question, you *win*, rather like achieving checkmate in chess. This is a huge undertaking, with many gaps still to fill, but every year there are encouraging breakthroughs, so confidence runs high that the task can be done, the game can be won. There may actually be many ways life could possibly have arisen out of nonlife, but finding just one that deserves scientific allegiance (until a better alternative is discovered) would muffle the "impossible in principle" choir for good. However, finding even one way is a task so daunting that it has fueled the conviction among researchers that even though the processes that must be invoked to create the end product are utterly blind and purposeless, the product itself is not just intricate but stunningly effective at what it does—a brilliant design. It takes all the ingenuity human reverse engineers can muster to figure out how the thing got assembled. A commentary by Jack Szostak, one of the leading researchers on one of the biggest breakthroughs of recent years (by Powner, Gerland, and Sutherland 2009), illustrates the attitude perfectly. (Don't worry about the chemical details mentioned; just see how the investigation is conducted, as revealed by the phrases I have italicized.)

> For 40 years, efforts to understand the prebiotic synthesis of the ribonucleotide *building blocks* of RNA have been *based on the assumption that they must have assembled from their three molecular components*: a nucleobase (which can be adenine, guanine, cytosine or uracil), a ribose sugar and phosphate. Of the many difficulties encountered by those in the field, the most frustrating has been *the failure to find any way of properly joining the*

pyrimidine nucleobases—cytosine and uracil—to ribose. . . . But Powner et al. revive the prospects of the "RNA first" model by exploring a pathway for pyrimidine ribonucleotide synthesis in which the sugar and nucleobase emerge from a common precursor. In this pathway, the complete ribonucleotide structure forms without using free sugar and nucleobase molecules as intermediates. This central insight, combined with a series of additional innovations, *provides a remarkably efficient solution to the problem of prebiotic ribonucleotide synthesis.* (Szostak 2009)

Greg Mayer (2009), an evolutionary biologist commenting on this, makes an important point:

> John Sutherland, one of Powner's coauthors, and in whose lab the work was done, worked on the problem for twelve years before he found the solution. What if he had given up after ten? Could we have concluded that no synthesis was possible? No. This work demonstrates the futility of all the various sorts of arguments—the argument from design, the God of the gaps, the argument from personal incredulity—that rely on ignorance as their chief premise.

Throughout this book I will exploit the perspective of *reverse engineering,* taking on the premise that every living thing is a product of nonmysterious physical processes that gradually brought all the elements together, refining them along the way, and eventually arrived at the working system we observe, or at some hypothesized intermediate system, a stepping-stone that would represent clear *progress* toward the living things we know exist. The cascade of processes must make changes that we can see, in retrospect, to be *improvements* in the design of the emerging systems. (We're on the way to checkmate. Are we making progress?) Until there were systems that could be strictly called *reproducing systems*, the processes at work were only proto-evolutionary, semi-Darwinian, partial *analogues* of proper evolution by natural selection; they were processes that raised the likelihood that various combinations of ingredients would arise and persist, concentrating the feedstock molecules until this eventually led to the origin of life. A living thing must capture *enough* energy and materials, and fend off its own destruction *long enough* to construct a *good enough* replica of itself. The reverse-engineering perspective is ubiquitous in biology and is obligatory in investigations of the origin of life. It always involves some kind of optimality considerations: What is the *simplest* chemical structure that could *possibly* do *x*? Or would phenomenon *x* be *stable enough* to sustain process *y*?

In a highly influential essay, Stephen Jay Gould and Richard Lewontin (1979) coined the phrase "Panglossian paradigm" as a deliberately abusive term for the brand of biology—adaptationism—that relies on the methodological principle of assuming, until proven otherwise, that all the parts of an organism are *good for something.* That is, they have useful roles to play, such as pumping blood, improving speed of locomotion, fending off infection, digesting food, dissipating heat, attracting mates, and so forth. The assumption is built right into the reverse-engineering perspective that sees all living things as efficiently composed of parts with functions. (There are well-recognized exceptions: for instance, features that *used* to be good for something and are now vestigial, along for the ride unless they are too expensive to maintain, and features that have no actual function but just "drifted" to "fixation" by chance.)

Gould and Lewontin's joke was a recycled caricature. In *Candide*, Voltaire created Dr. Pangloss, a wickedly funny caricature of the philosopher Leibniz, who had maintained that our world was the *best* of all possible worlds. In Dr. Pangloss's overfertile imagination, there was no quirk, no deformity, no catastrophe of Nature that couldn't be seen, in retrospect, to have a function, to be a blessing, just what a benevolent God would arrange for us, the lucky inhabitants of the perfect world. Venereal disease, for instance, "is indispensable in this best of worlds. For if Columbus, when visiting the West Indies, had not caught this disease, which poisons the source of generation, which frequently even hinders generation, and is clearly opposed to the great end of Nature, we should have neither chocolate nor cochi-neal" (quoted by Gould and Lewontin 1979, p. 151). Leibniz scholars will insist, with some justice, that Voltaire's parody is hugely unfair to Leibniz, but leave that aside. Was Gould and Lewontin's reuse of the idea an unfair caricature of the use of optimality assumptions in biology? Yes, and it has had two unfortunate effects: their attack on adaptationism has been misinterpreted by some evolution-dreaders as tantamount to a refutation of the theory of natural selection, and it has convinced many biologists that they should censor not only their language but also their thinking, as if reverse engineering was some sort of illicit trick they should shun if at all possible.

Those working on the origin of life have ignored the "Pangloss" critique of their methods, knowing that their strategic assump-tions serve to direct the investigation away from fruitless wan-dering. There is no point in looking at chemical reactions that couldn't possibly generate a target structure presumed to be a necessary component. Admittedly, there are risks to this strategy; as Szostak notes, for years the researchers made the mistaken assumption that the obviously best, most efficient way of uniting the nucleobases to the ribose was directly, and they overlooked the more devious path of having the ribonucleotide emerge from a common precursor, without involving the intermediate steps that had *seemed* necessary.

In chess, a *gambit* is a strategy that gives up material—a step backward, it seems—in order to take a better, forward step from an improved position. When trying to calculate what your opponent is going to do, gambits are hard to spot, since they seem at first to be losing moves that can be safely ignored, since one's opponent is not that stupid. The same risk of ignoring devious but fruitful trails besets the reverse engineer in biology, since, as Francis Crick famously said, enunciating what he called Orgel's Second Rule: "Evolution is cleverer than you are." The uncanny way the blind, purposeless churn of evolution (including prebiotic chemical evo-lution) uncovers off-the-wall solutions to problems is not evidence for an Intelligent Designer, nor is it grounds for *abandoning* reverse engineering, which would mean giving up the inquiry altogether; it is grounds for persisting and improving your reverse-engineering game. As in chess, don't give up; learn from your mistakes and keep on exploring, as imaginatively as you can, bearing in mind that your hypotheses, however plausible, still risk disconfirmation, which should be conscientiously sought.

Here is an example of a possible *gambit* in the origin of life. It is initially tempting to assume that the very first living thing capable of reproducing must have been the *simplest possible* living thing (given the existing conditions on the planet at the time). First things first: Make the simplest replicator you can imagine and then build on that foundation. But this is by no means necessary. It is possible, and more likely, I think, that a rather inelegantly complicated, expensive, slow, Rube-Goldberg conglomeration of *objets trouvés* was the first real replicator, and after it got the replication ball rolling, this ungainly replicator was repeatedly simplified in competition with its kin. Many of the most baffling magic tricks depend on the audience not imagining the ridiculously extravagant lengths magicians will go to in order to achieve a baffling effect. If you want to reverse engineer magicians, you should always remind yourself that they have no shame, no abhorrence of bizarre expenditures for "tiny" effects that they can then exploit. Nature, similarly, has no shame—and no budget, and all the time in the world.

Talk of improvements or progress in the slow, uncertain process of biogenesis is not indulging in illicit value judgments (which have no place in science, let's agree) but is rather an acknowledgment of the ever-present requirements of stability and efficiency in anything living. If you like, you can imagine biochemists working on how something utterly *terrible* might come into existence, a doomsday device or self-replicating death ray. They would still have to discipline their search by imagining possible paths to construct this horror. And they might well marvel at the brilliance of the design they finally figured out. I will have more to say about the presuppositions and implications of reverse engineering in biology later. Here I hope to forestall premature dismissal of my project by any who has been persuaded, directly or by hearsay, that Gould and Lewontin's propaganda against adaptationism was fatal. Contrary to the opinion widely engendered by their famous essay, adaptationism is alive and well; reverse engineering, when conducted with due attention to the risks and obligations, is still the royal road to discovery in biology and the only path to discovery in the demanding world of prebiotic chemistry of the origin of life.[8]

Next I want to look at the phenomenon of the origin of life from a more philosophical perspective, as the origin of *reasons*. Is there design in Nature or only apparent design? If we consider evolutionary biology to be a species of reverse engineering, does this imply that there are reasons for the arrangements of the parts of living things? Whose reasons? Or can there be reasons without a reasoner, designs without a designer?

---

8 Nikolai Renedo has suggested to me that the take-home message of Gould and Lewontin's famous essay is "be on the lookout for gambits," which is certainly good advice for any adaptationist to follow. If that is what Gould and Lewontin intended, however, they failed to convey it to their audiences, both lay and scientific, where the opinion persists that it was an authoritative demotion of adaptationism as a central feature of evolutionary thinking.

On the Origin of Reasons

The death or rebirth of teleology?

D arwin is often credited with overthrowing Aristotle's all-too-influential doctrine that everything in the world has a purpose, an "end" (in the sense that the ends justify the means), or as the French say, a *raison d'être*, a reason for being.

Aristotle identified four questions we might want to ask about anything:

1. What is it made of, or its *material cause?*
2. What is its structure, or its *formal cause?*
3. How did it get started, or what is its *efficient cause?*
4. What is its purpose, or its *final*, or *telic, cause?*

The Greek for the fourth cause is *telos* from which we derive the term *teleology*. Science, we are often told, has banished the *telos*, and we have Darwin to thank for this. As Karl Marx (1861) once famously put it, in his appreciation of Darwin's *Origin of Species*: "Not only is a death blow dealt here for the first time to 'Teleology' in the natural sciences but their rational meaning is empirically explained."

But a closer look shows that Marx is equivocating between two views that continue to be defended:

We should banish all teleological formulations from the natural sciences

or

now that we can "empirically explain" the "rational meaning" of natural phenomena without ancient ideology (of entelechies, Intelligent Creators and the like), we can replace old-fashioned teleology with new, post-Darwinian teleology.

This equivocation is firmly knitted into the practice and declarations of many thoughtful scientists to this day. On the one hand, biologists routinely and ubiquitously refer to the *functions* of behaviors such as foraging and territory marking, organs such as eyes and swim bladders, subcellular "machinery" such as ribosomes, chemical cycles such as the Krebs cycle, and macromolecules such as motor proteins and hemoglobin. But some thoughtful biologists and philosophers of biology are uneasy about these claims and insist that all this talk of functions and purposes is really only shorthand, a handy metaphor, and that strictly speaking there are no such things as functions, no purposes, no teleology at all in the world. Here we see the effect of another of the imagination-distorting forces spawned by Cartesian gravity. So seductive is the lure of Cartesian thinking that in order to resist it, some think we should follow the abstemious principle that whenever there is any risk of *infection* by prescientific concepts—of souls and spirits, Aristotelian teleology and the like— it is best to err on the side of squeaky-clean, absolute quarantine. This is often a fine principle: the surgeon excising a tumor takes out a generous "margin" around the suspect tissue; political leaders institute buffer zones, to keep dangerous weapons—or dangerous ideologies—at arm's length.

A little propaganda can help keep people vigilant. Among the epithets hurled at unrepentant teleologists are "Darwinian paranoia" (Francis 2004; Godfrey-Smith 2009) and "conspiracy theorists" (Rosenberg 2011). It is of course open to defend an intermediate position that forbids certain teleological excesses but licenses more staid and circumscribed varieties of talk about functions, and philosophers have devised a variety of such views. My informal sense is that many scientists assume that some such sane middle position is in place and must have been adequately defended in some book or article that they probably read years ago. So far as I know, however, no such consensus classic text exists,[9] and many of the scientists who guiltlessly allude to the functions of whatever they are studying still insist that they would *never* commit the sin of teleology.

One of the further forces in operation here is the desire not to give aid and comfort to creationists and the Intelligent Design crowd. By speaking of purpose and design in Nature, we (apparently) give them half their case; it is better, some think, to maintain a stern embargo on such themes and insist that *strictly speaking* nothing in the biosphere is designed unless it is designed by human artificers. Nature's way of generating complex systems (organs, behaviors, etc.) is so unlike an artificer's way that we should not use the same language to describe them. Thus Richard Dawkins speaks (on occasion—e.g., 1996, p. 4) of *designoid* features of organisms, and in *The Ancestors' Tale* (2004) he says, "the illusion of design conjured by Darwinian natural selection is so breathtakingly powerful" (p. 457). I disagree with this overkill austerity, which can backfire badly. A few years ago I overheard some Harvard Medical School students in a bar marveling at the intricacies to be found in the protein machinery inside a cell. One of them exclaimed, "How could anybody believe in evolution in the face of all that design!" The others did not demur, whatever their private thoughts. Why would anyone say that? Evolutionists aren't embarrassed by the intricacy of Nature. They revel in it! Discovering and explaining the evolution

9   Biologists and philosophers have written often about function talk, and although there are persistent disagreements about how to license it, there is something of a consensus that evolutionary considerations do the trick one way or another for natural functions, and facts about both history and current competence anchor attributions of function to artifacts. For a good anthology of the best work by both biologists and philosophers, see Allen, Bekoff, and Lauder 1998.

of the intracellular complexities that govern the life of a cell has been one of the glories of evolutionary microbiology in recent years. But this fellow's remark suggests that one of the themes gaining ground in common understanding is that evolutionary biologists are reluctant to "admit" or "acknowledge" the manifest design in Nature. People should know better, especially medical students!

Consider in this regard Christoph Schönborn, Catholic archbishop of Vienna, who was seduced by the Intelligent Design folks into denying the theory of natural selection on the grounds that it couldn't explain all the design. He said, notoriously, in a *New York Times* op-ed piece entitled "Finding Design in Nature" (July 7, 2005):

> The Catholic Church, while leaving to science many details about the history of life on earth, proclaims that by the light of reason the human intellect can readily and clearly discern purpose and design in the natural world, including the world of living things. Evolution in the sense of common ancestry might be true, but evolution in the neo-Darwinian sense—an unguided, unplanned process of random variation and natural selection—is not. Any system of thought that denies or seeks to explain away the overwhelming evidence for design in biology is ideology, not science.

Which battle do we want to fight? Do we want to try to convince lay people that they don't really see the design that is stunningly obvious at every scale in biology, or would we rather try to persuade them that what Darwin has shown is that there can be design—real design, as real as it gets—without an Intelligent Designer? We have persuaded the world that atoms are not atomic, and that the Earth goes around the Sun. Why shrink from the pedagogical task of showing that there can be design without a designer? So I am defending here (once again, with new emphasis) the following claim:

> *The biosphere is utterly saturated with design, with purpose, with reasons.* What I call the "design stance" predicts and explains fea-

tures throughout the living world using the same assumptions that work so well when reverse-engineering artifacts made by (somewhat) intelligent human designers.

There are three different but closely related strategies or stances we can adopt when trying to understand, explain, and predict phenomena: the physical stance, the design stance, and the intentional stance (Dennett 1971, 1981, 1983, 1987, and elsewhere). The physical stance is the least risky but also the most difficult; you treat the phenomenon in question as a physical phenomenon, obeying the laws of physics, and use your hard-won understanding of physics to predict what will happen next. The design stance works only for things that are designed, either artifacts or living things or their parts, and have functions or purposes. The intentional stance works *primarily* for things that are designed to use information to accomplish their functions. It works by treating the thing as a rational agent, attributing "beliefs" and "desires" and "rationality" to the thing, and predicting that it will act rationally.

Evolution by natural selection is not itself a designed thing, an agent with purposes, but it acts as if it were (it occupies the role vacated by the Intelligent Designer): it is a set of processes that "find" and "track" reasons for things to be arranged one way rather than another. The chief difference between the reasons found by evolution and the reasons found by human designers is that the latter are typically (but not always) represented in the minds of the designers, whereas the reasons uncovered by natural selection are represented for the first time by those human investigators who succeed in reverse engineering Nature's productions. Dawkins's title, *The Blind Watchmaker* (1986), nicely evokes the apparently paradoxical nature of these processes: on the one hand they are blind, mindless, without goals, and on the other hand they produce designed entities galore, many of which become competent artificers (nest-builders, web-spinners, and so forth) and a few become intelligent designers and builders: us.

Evolutionary processes brought purposes and reasons into existence the same way they brought color vision (and hence colors) into existence: gradually. If we understand the way our human world of reasons grew out of a simpler world where there were no reasons, we will see that purposes and reasons are as real as colors, as real as life. Thinkers who insist that Darwin has banished teleology should add, for consistency's sake, that science has also demonstrated the unreality of colors and of life itself. Atoms are all there is, and atoms aren't colored, and aren't alive either. How could mere large conglomerations of uncolored, unalive things add up to colored, live things? This is a rhetorical question that should be, and can be, answered (eventually). Now I want to defend the claim that there are reasons for what proteins do, and there are reasons for what bacteria do, what trees do, what animals do, and what we do. (And there are colors as well, of course, and yes, Virginia, life really exists.)

Different senses of "why"

Perhaps the best way of seeing the reality, indeed the ubiquity in Nature, of *reasons* is to reflect on the different meanings of "why." The English word is equivocal, and the main ambiguity is marked by a familiar pair of substitute phrases: *what for?* and *how come?"*

> "Why are you handing me your camera?" asks *what* are you doing this *for?*

> "Why does ice float?" asks *how come*: what it is about the way ice forms that makes it lower density than liquid water?

The *how come* question asks for a *process narrative* that explains the phenomenon without saying it is *for* anything. "Why is the sky blue?" "Why is the sand on the beach sorted by size?" "Why did the ground just shake?" "Why does hail accompany thunderstorms?" "Why is this dry mud cracked in such a fashion?" And also, "Why did this turbine blade fail?" Some folks might wish to treat the question of why ice floats as inviting a *what for* reason—God's reason, presumably—for this feature of the inanimate world. ("I guess God wanted fish to be able to live under the ice in the winter, and if ponds froze from the bottom up, this would be hard on the fish.") But as long as we have an answer to the *how come* question, in terms of physics and chemistry, it really would be something like paranoia to ask for more.

Compare four questions:

1. Do you know the reason why planets are spherical?
2. Do you know the reason why ball bearings are spherical?
3. Do you know the reason why asteroids aren't spherical?
4. Do you know the reason why dice aren't spherical?

The word "reason" is acceptable in all four questions (at least to my ear—how about yours?), but the answers to (1) and (3) don't give *reasons* (there aren't any reasons); they give *causes*, or process narratives. In some contexts the word "reason" can mean *cause*, unfortunately. You can answer questions (2) and (4) with process narratives along the lines of "well, the ball bearings were made on a lathe of sorts, which spun the metal . . . and the dice were cast in boxlike molds . . ." but those are not *reasons*. Sometimes people confuse the different questions, as in a memorable exchange that occurred in a debate I had with an ardent champion of Skinnerian behaviorism, Lou Michaels, at Western Michigan University in 1974. I had presented my paper "Skinner Skinned" (in *Brainstorms* 1978), and Michaels, in his rebuttal, delivered a particularly bold bit of behaviorist ideology, to which I responded, "But why do you say that, Lou?" to which his instant reply was "Because I have been rewarded for saying that in the past." I was demanding a reason—a *what for*—and getting a process narrative—a *how come*—in reply. There is a difference, and the Skinnerians' failed attempt to make it go away should alert positivistically minded scientists that they pay a big price in understanding if they try to banish "what for."

The first two sentences of this book are "How come there are minds? And how is it possible for minds to ask and answer this question?" It is asking for a process narrative, and that is what I am going to provide. But it will be a process narrative that also answers the questions how come there are "what for?" questions, and what are "what for?" questions for?

The evolution of "why": from *how come* to *what for*

Evolution by natural selection starts with *how come* and arrives at *what for*. We start with a lifeless world in which there are no reasons, no purposes at all, but there are processes that happen: rotating planets, tides, freezing, thawing, volcanic eruptions, and kazillions of chemical reactions. Some of those processes happen to generate other processes that happen to generate other processes until at some "point" (but don't look for a bright line) we *find it appropriate* to describe the *reasons* why some things are arranged as they now are. (*Why* do we find it appropriate, and how did we get into that state of mind? Patience, the answer to that will come soon.)

A central feature of human interaction, and one of the features unique to our species, is the activity of asking others to explain themselves, to justify their choices and actions, and then judging, endorsing, rebutting their answers, in recursive rounds of the "why?" game. Children catch on early, and often overdo their roles, trying the patience of their parents. "Why are you sawing the board?" "I'm making a new door." "Why are you making a new door?" "So we can close the house up when we go out." "Why do you want to close the house up when we go out?" . . . "Why don't we want strangers taking our things?" . . . "Why do we have things?" The fluency with which we all engage in this mutual reason-checking testifies to its importance in conducting our lives: our capacity to *respond* appropriately in this reason-checking activity is the root of *responsibility*. (Anscombe 1957) Those who cannot explain themselves or cannot be moved by the reasons offered by others, those who are "deaf to" the persuasions of advisors, are rightly judged to be of diminished responsibility and are treated differently by the law.

This activity of demanding and evaluating each other's reasons for action does not occupy our every waking hour, but it does play a major role in coordinating our activities, initiating the young into their adult roles, and establishing the norms by which we judge one another. So central is this practice to our own way of life that it is sometimes hard to imagine how other social species—dolphins, wolves, and chimpanzees, for instance—can get along without it. How do the juveniles "learn their place," for instance, without being *told* their place? How do elephants settle disagreements about when to move on or where to go next? Subtle instinctual signals of approval and disapproval must suffice, and we should also remember that no other species engages in the level of complex cooperative behaviors that we human beings have achieved.

Wilfrid Sellars, a philosopher at the University of Pittsburgh, described this activity of reasoning with one another as creating or constituting "the logical space of reasons" (1962) and inspired a generation of Pittsburgh philosophers, led by Robert Brandom and John Haugeland, to explore this arena in detail. What are the permissible moves, and why? How do new considerations enter the space, and how are transgressions dealt with? The space of reasons is bound by *norms*, by mutual recognition of how things *ought* to go—the right way, not the wrong way, to play the reason-giving game. Wherever there are reasons, then, there is room for, and a need for, some kind of *justification* and the possibility of *correction* when something goes wrong.

This "normativity" is the foundation of ethics: the ability to appreciate how reason-giving *ought to go* is a prerequisite for appreciating how life in society ought to go. Why and how did this practice and its rules arise? It hasn't existed forever, but it exists now. How come and what for? The Pittsburgh philosophers have not addressed this question, asking how "it got that way," so we will have to supplement their analysis with some careful speculation of our own on the evolution of the reason-giving game. I will try to show that ignoring this question has led the Pittsburgh philosophers to elide the distinction between two different kinds of norms and their associated modes of correction, which I will call *social normativity* and *instrumental normativity*. The former, analyzed and celebrated at Pittsburgh, is concerned with the *social* norms that arise within the practice of communication and collaboration (hence Haugeland [1998] speaks of the "censoriousness" of members of society as the force that does the correcting). The latter is concerned with quality control or efficiency, the norms of engineering, you could say, as revealed by market forces or by natural failures. This is nicely illustrated by the distinction between a good deed and a good tool. A good deed might be clumsily executed and even fail in its purpose, while a good tool might be an efficient torture device or evil weapon. We can see the same contrast in negative cases, in the distinction between *naughty* and *stupid*. People may punish you for being naughty, by their lights, but Nature itself may mindlessly punish you for being stupid. As we shall see, we need both kinds of norms to create the perspective from which reasons are *discernible* in Nature.

Reason-appreciation did *not* coevolve with reasons the way color vision coevolved with color. Reason-appreciation is a later, more advanced product of evolution than reasons.

Wherever there are reasons, an implicit norm may be invoked: real reasons are supposed always to be good reasons, reasons that justify the feature in question. (No demand for justification is implied by any "how come" question.) When we reverse engineer a newly discovered artifact, for instance, we may ask why there is a conspicuous knob in the corner that doesn't seem to "do anything" (anything useful—it makes a shadow when light falls on it, and changes the center of gravity of the artifact, but has no apparent function). We expect, until we learn otherwise, that the designer had a reason, a good reason, for that knob. It might be that there used to be a good reason, but that reason has lapsed and the manufacturers have forgotten this fact. The knob is vestigial, functionless, and present only because of inertia in the manufacturing process. The same expectations drive the reverse-engineering explorations of living things, and biologists often permit themselves to speak, casually, about what "Nature intended" or what "evolution had in mind" when it "selected" some puzzling feature of a living thing.[10] No doubt the biologists' practice is a direct descendant of the reverse engineering of artifacts designed and made by other human beings, which is itself a direct descendant of the societal institution of asking for and giving reasons for human activities. That *might* mean that this practice is an outdated vestige of prescientific thinking—and many biologists surmise as much—or it might mean that biologists have found a brilliant extension of reverse engineering into the living realm, using the thinking tools Nature has endowed us with to discover real patterns in the world that can well be called the *reasons* for the existence of other real patterns. To defend the latter claim, we need to take a look at how evolution itself could get going.

Go forth and multiply

In *Darwin's Dangerous Idea* (1995), I argued that natural selection is an *algorithmic* process, a collection of sorting algorithms that are themselves *composed* of generate-and-test algorithms that exploit randomness (pseudo-randomness, chaos) in the generation phase, and some sort of mindless quality-control testing phase, with the winners advancing in the tournament by having more offspring. How does this cascade of generative processes get under way? As noted in the last chapter, the actual suite of processes that led to the origin of life are still unknown, but we can dissipate some of the fog

10   For instance, biologist Shirley Tilghman, in the 2003 Watson Lecture, said: "But clearly, what is immediately apparent when you look at any part of those two genomes that have been compared is that evolution has indeed been hard at work, conserving far more of the genome than we could explain by genes and their closely allied regulatory elements. . . . Scientists should have a field day trying to understand what evolution had in mind when she paid so much attention to these little segments of DNA."

by noting that, as usual, a variety of gradual processes of revision are available to get the ball rolling.

The prebiotic or abiotic world was not utter chaos, a random confetti of atoms in motion. In particular there were *cycles*, at many spatio-temporal scales: seasons, night and day, tides, the water cycle, and thousands of chemical cycles discoverable at the atomic and molecular level. Think of cycles as "do-loops" in algorithms, actions that return to a starting point after "accomplishing" something—accumulating something, or moving something, or sorting something, for instance—and then repeating (and repeating and repeating), gradually changing the conditions in the world and thus *raising the probability that something new will happen*. A striking abiotic example is illustrated by Kessler and Werner in *Science* 2003.

These stone circles would strike anyone as a highly improbable scattering of stones across the landscape; it looks "man-made"— reminiscent of the elegant outdoor sculptures by Andy Goldsworthy—but it is the natural outcome of hundreds or thousands of mindless cycles of freezing and thawing on Spitsbergen in the Arctic. New England farmers have known for centuries about frost driving a "fresh crop" of stones up to the soil surface every winter; stones that have to be removed before plowing and planting. The classic New England "stone walls" we still see today along field edges and marching through former fields now reforested, were never meant to keep anything in or out; they are really not walls but very long narrow piles of boulders and small rocks hauled to the nearest part of the edge of the cultivated field. They are clearly the result of deliberate, hard human work, which had a purpose. Ironically, if the farmers hadn't removed the stones, over many cycles of freezing and thawing the stones might have formed one of the "patterned ground" phenomena illustrated here, not always circles, but more often polygons, and sometimes labyrinths and other patterns. Kessler and Werner provide an explanation of the process with a model—an algorithm—that produces these different sorting effects by varying the parameters of stone size, soil moisture and

FIGURE 3.1: Kessler and Werner, stone circles. © *Science magazine and Mark A. Kessler.*

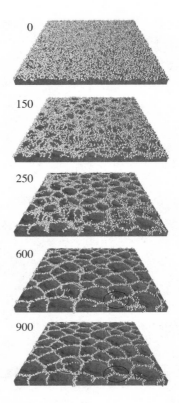

FIGURE 3.2: Kessler and Werner's stone-sorting algorithm at work. © *Science magazine and Mark A. Kessler.*

density, temperature, speed of freezing, and hillslope gradient. So we have a pretty good idea *how come* these phenomena exist where they do, and anybody who encountered these stone circles and concluded that there had to be a purposeful artificer behind them, an answer to the *what for* question, would be wrong.

In the abiotic world, many similar cycles occur concurrently but asynchronously, wheels within wheels within wheels, with different periods of repetition, "exploring" the space of chemical possibility. This would be a variety of parallel processing, a little bit like the *mass production* of industry, which makes lots of different parts in different places at different rates of production and then brings them together for assembly, except that this abiotic mass production is utterly unplanned and unmotivated.

There is no differential *re-production* in the abiotic world, but we do get varieties of differential *persistence*: some temporary combinations of parts hang around longer than others, thereby having more time to pick up revisions and adjustments. The rich can get richer, in short, even though they can't yet bequeath their riches to descendants. Differential persistence must then somehow gradually turn into differential *reproduction*. The proto-Darwinian algorithms of differential "survival" of chemical combinations can give rise to auto-catalytic reaction cycles that in turn give rise to differential *replication* as just a special case of differential *persistence*, a very special case, a particularly explosive type that multiplies its advantage by . . . multiplication! It generates many near-duplicate persisters, which can then "explore" many more slightly different corners of the world than any one or two persisters could do.

"A diamond is forever" according to the advertising slogan, but that is an exaggeration. A diamond is magnificently persistent, much more persistent than its typical competition, but its persistence is simply modeled by its linear descent through time, Tuesday's diamond being like its parent, Monday's diamond, and so forth. It never multiplies. But it can accumulate changes, wear and tear, a coating of mud that hardens, and so forth, which may make it more or less persistent. Like other durable things, it is affected by many cycles, many do-loops that involve it in one way or another. Usually these effects do not accumulate for long but rather get wiped out by later effects, but sometimes a barrier happens to get erected: a wall or membrane of some sort that provides extra shielding.

In the world of software, two well-recognized phenomena are *serendipity* and its opposite *clobbering*. The former is the chance collision of two unrelated processes with a happy result, and clobbering is such a collision with a destructive result. Walls or membranes that tend for whatever reason to prevent clobbering will be particularly persistent and will permit internal cycles (do-loops) to operate without interference. And so we see the engineering necessity of membranes to house the collection of chemical cycles—the Krebs cycle and *thousands* of others—that together permit life to emerge. (An excellent source on this algorithmic view of chemical cycles in cells is Dennis Bray, *Wetware* 2009.) Even the simplest bacterial cells have a sort of nervous system composed of chemical networks of exquisite efficiency and elegance. But how could just the right combination of membranes and do-loops ever arise in the prebiotic world? "Not in a million years!" some say. Fair enough, but then how about once in a hundred million years? It only has to happen once to ignite the fuse of reproduction.

Imagine we are back in the early days of this process where persistence turns gradually into multiplication, and we see a proliferation of some type of items where before there were none and we ask, "Why are we seeing these improbable things here?" The question is equivocal! For now there is both a process narrative answer, *how come*, and a justification, *what for*. We are confronting a situation in which some chemical structures are present while chemically possible alternatives are absent, and what we are looking at are things that are *better* at persisting in the local circumstances than their contemporary alternatives. *Before we can have competent reproducers, we have to have competent persisters, structures with enough stability to hang around long enough to pick up revisions.* This is not a very impressive competence, to be sure, but it is just what the Darwinian account needs: something that is only sorta competent, nothing too fancy. We are witnessing an "automatic" (algorithmic) paring away of the *nonfunctional*, crowded out by the functional. And by the time we get to a reproducing bacterium, there is functional virtuosity galore. In other words, there are *reasons why* the parts are shaped and ordered as they are. We can reverse engineer any reproducing entity, determining its good and its bad, and saying *why* it is good or bad. This is the birth of reasons, and it is satisfying to note that this is a case of what Glenn Adelson has aptly called Darwinism about Darwinism (Godfrey-Smith 2009): we see the gradual emergence of the species of reasons out of the species of mere causes, *what fors* out of *how comes,* with no "essential" dividing line between them. Just as there is no Prime Mammal—the first mammal that didn't have a mammal for a mother—there is no Prime Reason, the first feature of the biosphere that helped something exist because it made it better at existing than the "competition."

Natural selection is thus an automatic reason-finder, which "discovers" and "endorses" and "focuses" reasons over many generations. The scare quotes are to remind us that natural selection doesn't have a mind, doesn't itself have reasons, but is nevertheless competent to perform this "task" of design refinement. Let's be sure we know how to cash out the scare quotes. Consider a population of beetles with lots of variation in it. Some do well (at multiplying); most do not. Take the minority (typically) that do well, reproductively, and ask about each one: *why* did it do better than average. Our question is equivocal; it can be interpreted as asking *how come* or *what for.* In many cases, most cases, the answer is *no reason at all*; it's just dumb luck, good or bad. In which case we can have only a *how come* answer to our question. But if there is a subset, perhaps very small, of cases in which there is an answer to the *what for* question, a *difference that happens to make a difference,* then those cases have in common the germ of a reason, a proto-reason, if you like. The process narrative explains how it came about and also, in the process, points to why these are better than those, why they won the competition. "Let the best entity win!" is the slogan of the evolution tournament, and the winners, being better, wear the justification of their enhancements on their sleeves. In every generation, in every lineage, only some competitors manage to reproduce, and each descendant in the next generation is either just lucky or lucky-to-be-gifted in some way. The latter group was *selected* (for cause, you might say, but better would be for a *reason*). This process accounts for the accumulation of function by a process that blindly tracks reasons, creating things that have purposes but don't need to know them. The Need to Know principle made famous in spy novels also reigns in the biosphere: an organism doesn't need to know the reasons why the gifts it inherits are beneficial to it, and natural selection itself doesn't need to know what it's doing.

Darwin understood this:

> The term "natural selection" is in some respects a bad one, as it seems to imply conscious choice; but this will be disregarded after a little familiarity. No one objects to chemists speaking of "elective affinity"; and certainly an acid has no more choice in combining with a base, than the conditions of life have in determining whether or not a new form be selected or preserved. . . . For brevity sake I sometimes speak of natural selection as an intelligent power;—in the same way as astronomers speak of the attraction of gravity as ruling the movements of the planets. . . . I have, also, often personified Nature; for I have found it difficult to avoid this ambiguity; but I mean by nature only the aggregate action and product of many natural laws,—and by laws only the ascertained sequence of events. (1868, pp. 6–7)

So there were reasons long before there were reason-representers— us. The reasons tracked by evolution I have called "free-floating rationales," a term that has apparently jangled the nerves of some few thinkers, who suspect I am conjuring up ghosts of some sort. Not at all. Free-floating rationales are no more ghostly or problematic than numbers or centers of gravity. Cubes had eight corners before people invented ways of articulating arithmetic, and asteroids had centers of gravity before there were physicists to dream up the idea and calculate with it. Reasons existed long before there were reasoners. Some find this way of thinking unnerving and probably "unsound," but I am not relenting. Instead I am hoping here to calm their fears and convince them that we should all be happy to speak of the reasons uncovered by evolution before they were ever expressed or represented by human investigators or any other minds.[11] Consider the strikingly similar constructions in figures 3.3 and 3.4 of the color insert following p. 238.

The termite castle and Gaudí's La Sagrada Familia are very similar in shape but utterly different in genesis and construction. *There are reasons* for the structures and shapes of the termite castle, but they are not represented by any of the termites who constructed it. There is no Architect Termite who planned the structure, nor do any individual termites have the slightest clue about why they build the way they do. This is competence without comprehension, about which more later. There are also reasons for the structures and shapes of Gaudí's masterpiece, but they are (in the main) Gaudí's reasons. Gaudí *had* reasons for the shapes he ordered created; *there are* reasons for the shapes created by the termites, but the termites didn't *have* those reasons. There are reasons why trees spread their branches, but they are not in any strong sense the trees' reasons. Sponges do things for reasons, bacteria do things for reasons; even viruses do things for reasons. But they don't *have* the reasons; they don't need to have the reasons.

Are *we* the only reason-representers? This is a very important question, but I will postpone an answer until I have provided a wider setting for the perspective shift I am proposing here. So far, what I take to have shown is that Darwin didn't extinguish teleology; he naturalized it, but this verdict is not as widely accepted as it should be, and a vague squeamishness leads some scientists to go overboard avoiding design talk and reason talk. The space of reasons is created by the human practice of reason-giving and is bound by norms, both social/ethical and instrumental (the difference between being naughty and being stupid). Reverse engineering in biology is a descendant of reason-giving-judging.

---

11 Philosophers who are skeptical about my intransigence on this score might like to read T. M. Scanlon's recent book, *Being Realistic about Reasons* (2014), for an exhaustive and exhausting survey of the problems one confronts if one ignores engineering reasons and concentrates on *having* moral reasons for action.

The evolution of *what for* from *how come* can be seen in the way we interpret the gradual emergence of living things via a cascade of prebiotic cycles. Free-floating rationales emerge as the reasons why some features exist; they do not presuppose intelligent designers, even though the designs that emerge are extraordinarily good. For instance, there are reasons why termite colonies have the features they do, but the termites, unlike Gaudí, do not have or represent reasons, and their excellent designs are not products of an intelligent designer.

4

Two Strange Inversions of Reasoning

How Darwin and Turing broke a spell

The world before Darwin was held together not by science but by tradition. All things in the universe, from the most exalted ("man") to the most humble (the ant, the pebble, the raindrop) were the creations of a still more exalted thing, God, an omnipotent and omniscient intelligent creator—who bore a striking resemblance to the second-most exalted thing. Call this the trickle-down theory of creation. Darwin replaced it with the bubble-up theory of creation. Robert MacKenzie Beverley,[12] one of Darwin's nineteenth-century critics, put it vividly:

> In the theory with which we have to deal, Absolute Ignorance is the artificer; so that we may enunciate as the fundamental principle of the whole system, that, IN ORDER TO MAKE A PERFECT AND BEAUTIFUL MACHINE, IT IS NOT REQUI-SITE TO KNOW HOW TO MAKE IT. This proposition will be found, on careful examination, to express, in condensed form, the essential purport of the Theory, and to express in a few

---

12   I have been misidentifying this author as Robert Beverley MacKenzie for over thirty years; I thank the fact checkers at Norton for correcting my error.

words all Mr. Darwin's meaning; who, by a strange inversion of
reasoning, seems to think Absolute Ignorance fully qualified
to take the place of Absolute Wisdom in all the achievements
of creative skill. (Beverley 1868)

This was indeed a "strange inversion of reasoning" and the incre-
dulity expressed by Beverley is still echoing through a discourag-
ingly large proportion of the population in the twenty-first century.

When we turn to Darwin's bubble-up theory of creation, we can
conceive of all the creative design work metaphorically as lifting in
what I call Design Space. It has to start with the first crude replicators,
as we saw in chapter 3, and gradually ratchet up, by wave after wave
of natural selection, to multicellular life in all its forms. Is such a process really capable of having produced all the wonders we observe in
the biosphere? Skeptics ever since Darwin have tried to demonstrate
that one marvel or another is simply unapproachable by this laborious and unintelligent route. They have been searching for something
alive but *unevolvable.* My term for such a phenomenon is a *skyhook,*
named after the mythical convenience you can hang in the sky to
hold up your pulley or whatever you want to lift (Dennett 1995). A
skyhook floats high in Design Space, unsupported by ancestors, the
direct result of a special act of intelligent creation. And time and
again, these skeptics have discovered not a miraculous skyhook but
a wonderful *crane,* a nonmiraculous innovation in Design Space that
enables ever more efficient exploration of the possibilities of design,
ever more powerful lifting in Design Space. Endosymbiosis is a crane;
it lifted simple single cells into a realm of much complexity, where
multicellular life could take off. Sex is a crane; it permitted gene pools
to be stirred up, and thus much more effectively sampled by the blind
trial-and-error processes of natural selection. Language and culture
are cranes, evolved novelties that opened up vast spaces of possibility
to be explored by ever more intelligent (but not miraculously intelligent) designers. Without the addition of language and culture to the
arsenal of R&D tools available to evolution, there wouldn't be glow-
in-the-dark tobacco plants with firefly genes in them. These are not

miraculous. They are just as clearly fruits of the Tree of Life as spider webs and beaver dams, but the probability of their emerging without the helping hand of *Homo sapiens* and our cultural tools is nil.

As we learn more and more about the nano-machinery of life that makes all this possible, we can appreciate a second strange inversion of reasoning, achieved almost a century later by another brilliant Englishman: Alan Turing. Here is Turing's strange inversion, put in language borrowed from Beverley:

IN ORDER TO BE A PERFECT AND BEAUTIFUL COMPUT-
ING MACHINE, IT IS NOT REQUISITE TO KNOW WHAT
ARITHMETIC IS.

Before Turing's invention there were computers, by the hundreds or thousands, employed to work on scientific and engineering calculations. Computers were people, not machines. Many of them were women, and many had degrees in mathematics. They were human beings who knew what arithmetic was, but Turing had a great insight: they didn't need to know this! As he noted, "The behavior of the computer at any moment is determined by the symbols which he is observing, and his 'state of mind' at that moment" (Turing 1936, 5). That "state of mind" (in Turing's scare quotes) was a dead-simple set of if-then instructions about what to do and what "state of mind" to go into next (and repeat until you see the instruction to STOP). Turing showed that it was possible to design mindless machines that were Absolutely Ignorant, but that could do arithmetic perfectly, following "instructions" that could be mechanically implemented. More importantly, he showed that if their instructions included *conditional branching* (if-then instructions, such as "if you observe 0, replace it with 1 and move left, and if you observe 1 leave it as is and move right, and change to state *n*."), then these machines could pursue indefinitely complex paths determined by the instructions, which gave them a remarkable competence: they could do *anything* computational. In other words, a programmable digital computer is a Universal Turing Machine, capable of mimicking any special-purpose digital computer by following a set of instructions that implement that special-purpose computer in software.[13] (You don't have to rewire your smartphone to get it to do new tasks; just download an app and turn it into a star finder or translator or hand calculator or spell-checker or. . . .) A huge Design Space of information-processing was made accessible by Turing, and he foresaw that there was a traversable path from Absolute Ignorance to Artificial Intelligence, a long series of lifting steps in that Design Space.

Many people can't abide Darwin's strange inversion. We call them creationists. They are still looking for skyhooks—"irreducibly complex" (Behe 1996) features of the biosphere that could not have evolved by Darwinian processes. Many more people can't abide Turing's strange inversion either, and for strikingly similar reasons. They want to believe that the wonders of the mind are inaccessible by mere material processes, that minds are, if not literally miraculous, then mysterious in ways that defy natural science. They don't want the Cartesian wound to be healed.

Why not? We've already noted some of their less presentable motives: fear, pride, the misplaced love of unsolved mystery. Here is another reason (is it *how come* or *what for?*): Both Darwin and Turing claim to have discovered something truly unsettling to a human mind—*competence without comprehension*. Beverley expressed his outrage with gusto: the *very idea* of creative skill without intelligence! Consider how shockingly this goes against an idea enshrined in our educational policies and practices: *comprehension as the (best) source of competence.* We send our children to universities so that they will gain an understanding of all the ways the world works that will stand them in good stead throughout their lives, generating competences

13   The standard jargon for asserting this is known as the Church-Turing Thesis, formulated by logician Alonzo Church: "all effective procedures are Turing-computable"—though of course many of them are not feasible since they take too long to run. Since our understanding of what counts as an effective procedure (basically, a computer program or algorithm) is unavoidably intuitive, this thesis cannot be proved, but it is almost universally accepted, so much so that Turing-computability is typically taken as an acceptable operational definition of effectiveness.

as needed from the valuable store of comprehension we have inculcated in them. (I am using "comprehension" and "understanding"
as synonymous, by the way, favoring "comprehension" for its alliteration in the slogan, which will come up again and again.) Why do we
disparage rote learning these days? Because we have seen—haven't
we?—that getting children to *understand* a topic or a method is the
way (the only way or just the best way?) to make them competent with
regard to that topic or method. We disparage the witless memorizer
who just fills in the blanks on the template without knowing what the
point is. We scoff at the idea that paint-by-numbers kits are the way to
train creative artists. Our motto might well be

If you make them comprehend, their competence will follow!

Note that there is more than a smidgen of ideology at play here.
We are quite familiar with some disastrous misapplications of our
hallowed principle, such as the "new math," which tried—unsuc-
cessfully—to teach children set theory and other abstract concepts
first, instead of drilling them on addition and subtraction, the multiplication table, fractions, and simple algorithms like long division,
or counting by twos and fives and tens.

The armed forces are some of the most effective educational
institutions in the world, turning average high school students into
reliable jet-engine mechanics, radar operators, navigators, and a
host of other technical specialists thanks to heavy doses of "drill
and practice." In due course, a valuable variety of comprehension
arises out of the instilled competences in these practiced practitio-
ners, so we have good empirical evidence that competence doesn't
always depend on comprehension and sometimes is a precondition
for comprehension. What Darwin and Turing did was envisage the
most extreme version of this point: *all* the brilliance and comprehension in the world arises ultimately out of uncomprehending
competences compounded over time into ever more competent—
and *hence* comprehending—systems. This is indeed a strange inversion, overthrowing the pre-Darwinian mind-first vision of Creation

with a mind-*last* vision of the eventual evolution of us, intelligent designers at long last.

Our skepticism about competence without comprehension has causes, not reasons. It doesn't "stand to reason" that there cannot be competence without comprehension; it just feels right, and it feels right *because* our minds have been shaped to think that way. It took Darwin to break the spell cast by that way of thinking, and Turing shortly thereafter came along and broke it again, opening up the novel idea that we might invert the traditional order and build comprehension out of a cascade of competences in much the way evolution by natural selection builds ever more brilliant internal arrangements, organs, and instincts without having to comprehend what it is doing.

There is one big difference between Darwin's strange inversion and Turing's. Darwin showed how brilliant designs could be created by cascades of processes lacking all intelligence, but the system for Turing's cascades of processes was the product of a very intelligent designer, Turing. One might say that while Darwin *discovered* evolution by natural selection, Turing *invented* the computer. Many people contend that an intelligent God had to set up all the conditions for evolution by natural selection to occur, and Turing appears to be playing that role in setting up the underlying idea of a (material, non-living, non-comprehending) computer which can then become the arena in which comprehension might arise by something a little bit like evolution, a series of design improvements concocted from the basic building blocks of computation. Doesn't Turing's role as intelligent designer *oppose* rather than *extend* the reach of Darwin's strange inversion? No, and answering this important question is a major task for the rest of the book. The short explanation is that Turing himself is one of the twigs on the Tree of Life, and his artifacts, concrete and abstract, are indirectly products of the blind Darwinian processes in the same way spider webs and beaver dams are, so there is no *radical* discontinuity, no need for a skyhook, to get us from spiders and beaver dams to Turing and Turing machines. Still, there is a large gap to be filled, because Turing's way of making things was strikingly different from the spider's way and the

beaver's way, and we need a good evolutionary account of that difference. If competence *without* comprehension is so wonderfully
fecund—capable of designing nightingales, after all—why do we
need comprehension—capable of designing odes to nightingales
and computers? Why and how did human-style comprehension
arrive on the scene? First, let's make the contrast sharp and vivid.

If termites are impressive exemplars of competence without comprehension, capable of building strong, safe, air-conditioned homes
without benefit of blueprints or bosses (the Queen termite is more
like the Crown Jewels than a boss), Antoni Gaudí is a near-perfect
model of the Intelligent Designer, a Godlike boss, armed from the
outset with drawings and blueprints and manifestos full of passion-
ately articulated reasons. His great church in Barcelona is an example of top-down creation that is hard to surpass, but Turing's original
computer, the Pilot ACE (which can now be seen in the Science
Museum in London), might beat it out for first prize. One of the first
truly useful computers, it became operational in 1950 at the National
Physical Laboratory in England, and it rivaled La Sagrada Familia in
originality, intricacy—and cost. Both creators had to convince back-
ers to fund their ambitious designs and both worked out elaborate
diagrams, along with supporting explanations. So in each case, the
eventual reality depended on the existence of prior representations,
in the mind of a genius, of the purpose of the design, and hence the
raison d'être of all the parts.[14] When it came to the actual construction of the artifacts, there were workers who were *relatively* uncomprehending, who had rather minimal appreciation of the point of their
labors. Comprehension was distributed, of course: Gaudí didn't have
to understand as much about how to mix mortar or carve stone as the
masons on the job did, and Turing didn't have to be a virtuoso with
a soldering gun or an expert on the techniques for manufacturing

14   Gaudí died in 1926 but left drawings and instructions and models that are
still guiding the completion of the unfinished church; Turing left NPL before
the Pilot ACE was completed, but he also left representations of the artifact to
guide its completion.

vacuum tubes. Distribution of expertise or understanding of this sort is a hallmark of human creative projects, and it is clearly essential for today's high-tech artifacts, but not for all earlier artifacts. A lone artificer can make a spear, or even a kayak or a wooden wagon or a thatched hut, understanding every aspect of the design and construction, but not a radio or an automobile or a nuclear power plant.

A closer look at a few examples of human artifacts and the technology we have invented to make them will clarify the way-stations on the path from clueless bacteria to Bach, but first we need to introduce a term that began in philosophy and has been extended to several scientific and engineering enterprises.

Ontology and the manifest image

"Ontology" comes from the Greek word for *thing*. In philosophy, it refers to the set of "things" a person believes to exist, or the set of things defined by, or assumed by, some theory. What's in your ontology? Do you believe in ghosts? Then ghosts are in your ontology, along with tables and chairs and songs and vacations, and snow, and all the rest. It has proved more than convenient to extend the term "ontology" beyond this primary meaning and use it for the set of "things" that an animal can recognize and behave appropriately with regard to (whether or not animals can properly be said to have beliefs) and—more recently—the set of "things" a computer program has to be able to deal with to do its job (whether or not *it* can properly be said to have beliefs). Vacations are not in the ontology of a polar bear, but snow is, and so are seals. Snow is probably not in the ontology of a manatee, but outboard-motor propellers may well be, along with seaweed and fish and other manatees. The GPS system in your car handles one-way streets, left and right turns, speed limits, and the current velocity of your car (if it isn't zero, it may not let you put in a new target address), but its ontology also includes a number of satellites, as well as signals to and from those satellites, which it doesn't bother you with, but needs if it is to do its job.

The ontology of the GPS was intelligently designed by the programmers who built it, and the R&D process probably involved a lot of trial and error as different schemes were attempted and found wanting. The ontology of a polar bear or manatee was designed by some hard-to-sort-out combination of genetic evolution and individual experience. Manatees may have seaweed in their ontology the way human babies have nipples in theirs, instinctually, genetically designed over the eons. Any manatee with outboard-motor-propeller in its ontology has gained that from experience. We human beings have extremely varied ontologies. Some believe in witches and some believe in electrons and some believe in morphic resonances and abominable snowmen. But there is a huge common core of ontology that is shared by all normal human beings from quite an early age—six years old will capture almost all of it.

This common ontology was usefully named the *manifest image* by Wilfrid Sellars (1962). Consider the world we live in, full of other people, plants, and animals, furniture and houses and cars . . . and colors and rainbows and sunsets, and voices and haircuts, and home runs and dollars, and problems and opportunities and mistakes, among many other such things. These are the myriad "things" that are easy for us to recognize, point to, love or hate, and, in many cases, manipulate or even create. (We can't create sunsets, but in the right conditions we can create a rainbow with some water and a little ingenuity.) These are the things we use in our daily lives to anchor our interactions and conversations, and, to a rough approximation, for every noun in our everyday speech, there is a kind of thing it refers to. That's the sense in which the "image" is "manifest": it is obvious to all, and everybody knows that it is obvious to all, and everybody knows *that*, too. It comes along with your native language; it's the world according to *us*.[15] Sellars contrasted this

15 In fact, Sellars distinguished a "pre-scientific, uncritical, naïve conception of man-in-the-world . . . [which] might be called the 'original' image" (1962, p. 6ff) from what he called the manifest image, a "refinement or sophistication" of that original image. What he was mainly getting at in this distinction is that philosophers have been reflecting critically on the naïve conception for millennia, so the manifest image was not just folk metaphysics.

with the *scientific image*, which is populated with molecules, atoms, electrons, gravity, quarks, and who knows what else (dark energy, strings? branes?). Even scientists conduct most of their waking lives conceiving of what is going on in terms of the manifest image. ("Pass the pencil, please" is a typical bit of communication that depends on the manifest image, with its people and their needs and desires; their abilities to hear, see, understand, and act; the characteristic identifying marks of pencils, their size and weight, their use; and a host of other things. Making a robot that can understand and accede to such a request is far from trivial, unless you make a robot that can "understand" only that sentence and a few others.)

The scientific image is something you have to learn about in school, and most people (laypeople) acquire only a very cursory knowledge of it. These two versions of the world are quite distinct today, rather like two different species, but they were once merged or intertwined in a single ancestral world of "what everyone knows" that included all the local fauna and flora and weapons and tools and dwellings and social roles, but also goblins and gods and miasmas and spells that could jinx your life or guarantee your hunting success. Gradually our ancestors learned which "things" to oust from their ontologies and which new categories to introduce. Out went the witches, mermaids, and leprechauns, and in came the atoms, molecules, and germs. The early proto-scientific thinkers, such as Aristotle, Lucretius, and, much later, Galileo, conducted their inquiries without making a crisp distinction between the ontology of everyday life (the manifest image) and the ontology of science, but they were bold proposers of new types of things, and the most persuasive of these caught on. Undoing some of their most tempting mistakes, while creating the ontology of the scientific image, has been a major task of modern science.

Unlike the term "ontology," "manifest image" and "scientific image" have not yet migrated from philosophy to other fields, but I'm doing my best to export them, since they have long seemed to me to be the best way I know to clarify the relationship between "our" world and the world of science. Where did the prescientific manifest image come from? Sellars concentrated on the manifest image of human beings or societies. Should we extend the concept to other species? They have ontologies, in the extended sense. Do they also have manifest images, and how might they differ from ours? These questions are important to our inquiry because to understand what a great feat Darwin's strange inversion of reasoning was, we need to understand what Darwin was inverting and how *it* got that way.

Automating the elevator

It will help to start dead simple, with an example that has nothing to do with consciousness or even life: the electronic device that controls an automatic elevator. When I was a youngster, there were still human elevator operators, people whose job was to go up and down in an elevator all day, stopping at the right floors to take on and let off passengers. In the early days they manipulated a curious handle that could be swung clockwise or counterclockwise to make the elevator go up or down, and they needed skill to stop the elevator at just the right height. People often had to step up or down an inch or two on entering and leaving, and operators always warned people about this. They had lots of rules about what to say when, and which floors to go to first, and how to open the doors, and so forth. Their training consisted in memorizing the rules and then practicing: following the rules until it became second nature. The rules themselves had been hammered out over the years in a design process that made a host of slight revisions and improvements. Let's suppose that this process had more or less settled down, leaving an ideal rule book as its product. It worked wonderfully. Anybody who followed the rules exactly was an excellent elevator operator. (I located one of these antique rule books online, a US Army publication—not surprisingly, considering their pioneering role in drill and practice. Figure 4.1 reproduces a page.)

Now imagine what happened when a simple computer program could take over all the control tasks of the operator. (In fact, this happened gradually, with various automatic mechanical devices

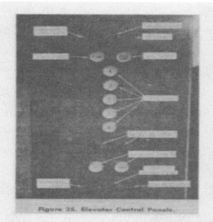

4.2.2.3 As Car Stops. As car stops operator should:

(1) Say, "Please wait until car stops," if passengers attempt to alight from or enter while it is still leveling.

(2) Say, "Step up, please," or "Step down, please," if car does not stop level with landing sill. This is important as few people watch door sill when car stops.

4.2.3   Operating Procedures.

4.2.3.1 General:

(1) Parked elevator is never placed in service except under direction of supervisor.

(2) When at main floor, operator stands at attention well within the car.

(3) Operator never steps outside the car except when relieved from duty. Relieving operator steps into car and takes over control before dismissed operator leaves. Passengers are never allowed to remain in car without operator.

(4) When more than one car in bank is at main floor terminal, operators in cars other than next car to be loaded should close gates, and extinguish car lights.

(5) Cars should never be overloaded. Certificate of inspection is authority for weight load or number of persons permitted to ride in elevator.

(6) Floor signals are not passed without instructions from supervisor, unless car is full and signal "Transfer" switch is thrown.

(7) Passengers should not be hurried. It is both dangerous and discourteous.

(8) Operators never give information or make statements, either written or verbal, in connection with accidents occurring in the building. If statements are to be made, they must be given in presence of building manager or supervisor.

(9) When the car is out of service, the control mechanism is left inoperative by pulling "Emergency Switch." Where a motor generator is installed, supervisor shuts down set.

(10) Operators should make complete trips to top floor unless instructed otherwise

Figure 25. Elevator Control Panels.

numbers given. Be sure buttons for all stops requested are pressed before doors are closed.

(3) Say, "Next car, please," if more than maximum number of passengers attempt to enter car.

(4) Say, "Step back in car, please," in order to prevent crowding at car door.

(5) Ask passengers to, "Face front, please," if car is crowded and passengers are facing back or side of car.

4.2.2.2 Approaching Floor. As elevator approaches floor, operator should:

(1) Announce, "First floor," "Second floor," etc, as car slows to stop.

(2) Announce, "Street floor," as well as floor number, as, "First, street floor." This is necessary particularly in case of buildings on grade where street floor at one end is on different level from street level at other end of building.

59

FIGURE 4.1: Elevator operator manual page.

being introduced to take the less skilled tasks away from the operator, but we'll imagine that elevators went from human operators to completely computer-controlled systems in one leap.) The elevator manufacturer, let's suppose, calls in a team of software engineers—programmers—and hands them the rule book that the human elevator operators have been following: "Here are *the specs*—this is a specification of the performance we want; make a computer program that follows all the rules in this book as well as the best human operators and we'll be satisfied." As the programmers go through the rule book, they make a list of all the actions that have to be taken, and the conditions under which they are prescribed or forbidden. In the process they can clean up some of the untidiness in the rule book. For instance, if they build in sensors to ensure that the elevator always stops at exactly the right level, they can eliminate the loop that requires the operator to say, "Please step up" or "please step down," but they might leave in its place a simple (recorded) voice saying "[n]th floor; watch your step."

The rule book has instructions about how many passengers may be allowed on the elevator at any one time, and the programmers have to confront issues, such as do we install a turnstile so that the program can count people getting on and off? Probably not a good idea. A scale that weighs the occupants together is better, easier, and less intrusive. Look what that does to elevator ontology: instead of having a "count noun" like "passenger" or "occupant," it has a "mass noun" like "freight" or "cargo." We can say, metaphorically, that the elevator keeps asking itself "how much cargo?" not "how many passengers?" Similarly, we can note that the polar bear doesn't try to count snowflakes but is cognizant of the presence or absence of snow, or that the anteater slurps up a lot of ant on its tongue, unlike the insectivorous bird that tracks individual insects. And notice that just as we don't have to speculate about elevator *consciousness* to draw this distinction, we can treat the animals as having different ontologies without *settling* issues of whether they are conscious of their ontologies or simply the beneficiaries of designs that can be interpreted (by reverse engineers or forward engineers) as having those ontologies.

Back to elevator ontology. It may rely on "cargo" for some purposes, but it still needs to keep track of individual requests to which it must respond appropriately: "up," and "down," from outside; "five" or "ground floor" and "keep door open" from inside. And for safety it needs to self-monitor, to check its various organs periodically, to see if they are working correctly and actually in the state they are supposed to be in. It needs to light up buttons when they are pushed and turn the light off when the task ordered by the button is complete (or for other reasons). How conscientious (or obsessive-compulsive) the controller is can vary, but programs that are designed to be negligent about interference or failure will not make the grade for long. And if there are other elevators lined up in a common lobby (as in a large office building or hotel), it will be important that the elevators communicate with each other, *or* that there is a master director that issues all the orders. (Designing the elevators to use "deictic" reference along the lines of "Where are you in relation to *where I am now?*" turns out to simplify and enhance the "cooperation" between individual elevators and eliminate the role of the omniscient master controller.)

It is useful to write the emerging control scheme in *pseudo-code,* a sort of mongrel language that is halfway between everyday human language and the more demanding system of source code. A line of pseudo-code might be along the lines of "if CALLFLOOR > CURRENTFLOOR, THEN ASCEND UNTIL CALLFLOOR = CURRENTFLOOR AND STOP; OPENDOOR. WAIT. . . ."

Once the plan is clear in pseudo-code and seems to be what is wanted, the pseudo-code can be translated into source code, which is a much more rigorous and structured system of operations, with definitions of terms—variables, subroutines, and so forth. Source code is still quite readily deciphered by human beings—after all, they write it—and hence the rules and terms of the rule book are still quite explicitly represented there, if you know how to look for them. This is made easier by two features: First, the names chosen for the variables and operations are usually chosen to wear their intended meaning on their sleeves (CALLFLOOR, WEIGHTSUM, TELLFLOOR . . .). Second, programmers can add *comments* to their source code, parenthetical explanations that tell other human readers of the source code what the programmer had in mind, and what the various parts are supposed to do. When you program, it is wise to add comments for yourself as you go, since you may easily forget what you thought the line of code was doing. When you go back to correct programming errors, these comments are very useful. Source code has to be carefully composed according to a strict syntax, with every element in the right place and all the punctuation in the right order since it has to be fed to a *compiler* program, which takes the source code and translates *it* into the sequences of fundamental operations that the actual machine (or virtual machine) can execute. A compiler can't guess what a programmer means by a line of source code; the source code must tell the compiler exactly what operations to perform—but the compiler program may have lots of different ways of performing those tasks and will be able to figure out an efficient way under the circumstances.

Somewhere in the pseudo-code, amongst thousands of other statements, you will find a statement along the lines of

IF WEIGHT-IN-POUNDS > n THEN STOP. OPEN DOOR.

{Prevents elevator from moving if over maximum weight.

After somebody steps out, reducing weight, normal operation resumes.}

The sentence in brackets is a *comment* that vanishes when the source code is compiled. Similarly, the capitalized terms don't survive in the code fed by the compiler to the computer chip that runs the program; they are also for the programmers, to help them remember which variable is which, and "IN-POUNDS" is in there to remind the programmers that the number they put in the program for maximum weight allowed better be in pounds. (In 1999, NASA's $125-million Mars Climate Orbiter got too close to Mars because one part of the control system was using meters and another part was using feet to represent the distance from the planet. The spacecraft got too close and destroyed itself. People make mistakes.) In short, the comments and labels help *us* understand the rationale of the design of the system but are ignored/invisible to the hardware. Once the program is finished and tested and deemed satisfactory, the compiled version can be burned into ROM (read-only-memory) where the CPU (central processing unit) can access it. The "rules" that were so explicit, so salient early in the design process, have become merely implicit in the zeroes and ones that get read by the hardware.

The point of this digression into elementary programming is that the finished working elevator has some interesting similarities to living things yet also a profound difference. First, its activities are remarkably appropriate to its circumstances. It is a *good* elevator, making all the *right* moves. We might almost call it *clever* (like the best human elevator operators of yore). Second, this excellence is due to the fact that its design has the *right ontology.* It uses variables that keep track of all the features of the world that matter to getting its job done and is oblivious to everything else (whether the passengers are young or old, dead or alive, rich or poor, etc.). Third, it has *no need to know* what its ontology is, or why—the *rationale* of the program is something only the program's designers have to understand. They need to understand the rationale because of the nature of the R&D process that produced the finished program: it is a process of (quite) intelligent design. That is the profound difference we must clarify as we turn to the ontology of simple living things, products of evolution by natural selection, not intelligent design.

Even bacteria are good at staying alive, making the right moves, and keeping track of the things that matter most to them; and trees and mushrooms are equally clever, or, more precisely, cleverly designed to make the right moves at the right time. They all have elevator-type "minds," not elevated minds like ours.[16] They don't need minds like ours. And their elevator-minds are—must be— the products of an R&D process of trial and error that gradually structured their internal machinery to move from state to state in a way highly likely—not guaranteed—to serve their limited but vital interests. Unlike the elevator, their machinery was not designed by intelligent designers, who worked out, argued about, and thought about the rationales for the designs of the component pieces, so there is nothing—nothing at all, anywhere—that plays the roles of the labels or comments in a source code program. This is the key

16   Let me acknowledge that this claim is somewhat peremptory; I see no reason to believe that trees or bacteria have control systems that are more like our minds than elevator-control systems are, but I concede that it is *possible* that they do. I am treating this possibility as negligible, a non-zero strategic risk I am prepared to take.

to the transformation that Darwin and Turing achieved with their strange inversions of reasoning.

Elevators can do remarkably clever things, optimizing their trajectories, thereby saving time and energy, automatically adjusting their velocity to minimize discomfort of their passengers, "thinking of everything" that needs to be thought about, and obeying instructions and even answering the frequently asked questions. Good elevators earn their keep. They do this without any neurons, sense organs, dopamine, glutamate, or the other organic components of brains. So it seems fair to say that what they do so "cleverly" is a perfect case of competence without the slightest smidgen of comprehension or consciousness. Unless, of course, the machinery that provides them with this limited competence counts as having a smidgen, or maybe even two smidgens, of comprehension. (And in the same spirit, its prudent self-monitoring can be seen to be an elementary step towards consciousness.)

Whether or not we want to concede a minor, negligible touch of comprehension to the elevator, we should take the same line with bacteria, and with trees and mushrooms. They exhibit impressive competence at staying-alive-in-their-limited-niches, thanks to the well-designed machinery they carry with them, thanks to their genes. That machinery was designed by the R&D process of natural selection, however, so there is nothing anywhere at any time in that R&D history that represents the *rationales* of either the larger functions of whole systems or component functions of their parts the way comments and labels represent these functions for human designers. The rationales are nevertheless there to be discovered by reverse engineering. You can more or less count on it that there will be a *reason why* the parts are shaped as they are, why the behaviors are organized as they are, and that reason will "justify" the design (or have justified an earlier design that has now become either vestigial or transformed by further evolution to serve some newer function). The justification will be straightforward, in engineering terms: if you remove the element, or reshape the element, the system won't work, or won't work as well. Claims about such free-floating rationales should be, and can be, testable, and are confirmed beyond reasonable doubt in many cases.

Back to our elevator, successfully automated. *Tada!* One actual human being—not a figurative homunculus—has been replaced by a machine. And the machine *follows the same rules* as the human operator. Does it really? OK, it doesn't. It *sorta* follows the same rules. This is a nice intermediate case between a human being who memorizes—and hence literally represents in her mind, and consults—the rules that dictate her behavior, and the planets, whose orbits are elegantly described by equations that the planets "obey." We human beings also often occupy this intermediate level, when we have internalized or routinized through practice a set of explicit rules that we may then discard and even forget. (*i* before *e* except after *c*, or when it sounds like *a* as in "neighbor" and "weigh.") And it is also possible to sorta follow rules that have still not been made explicit: the rules of English grammar, for instance, which continue to challenge linguists. Put in terms of this example, linguists today are still thrashing around trying to write a satisfactory version of the "rule book" for speaking English, while every ten-year-old native English speaker has somehow installed and debugged a pretty good version of the executable object code for the control task of speaking and understanding the language.

Before we take up the minds of animals, I want to turn to some further examples of the design of artifacts that will help isolate the problem that evolution solved when it designed competent animals.

The intelligent designers of Oak Ridge and GOFAI

After seventy years there are still secrets about World War II that have yet to emerge. The heroic achievements of Alan Turing in breaking the German Enigma code at Bletchley Park are now properly celebrated even while some of the details are still considered too sensitive to make public. Only students of the history of atomic energy engineering are likely to be well acquainted with the role that General Leslie Groves played in bringing the Manhattan Project to a successful conclusion. It took only six years from the day in August of 1939 when the Einstein-Szilard letter arrived on President Roosevelt's desk informing him of the prospect of an atomic bomb until the dropping of the first bomb on Hiroshima on August 6, 1945. The first three years went into basic research and "proof of concept," and almost everybody involved in those early years knew exactly what they were trying to accomplish. In 1942, Leslie Groves was appointed director of what came to be called the Manhattan Project, and in three incredibly pressured years intertwining further R&D with the colossal (and brand new) task of refining weapons grade uranium, thousands of workers were recruited, trained, and put to work, mostly controlling the newly invented machines for separating out the isotope uranium 235, which was a fraction of 1% of the previously refined uranium 238.

At the height of operations over 130,000 people worked full time on the project, and only a tiny percentage of them had any idea at all what they were making. Talk about competence without comprehension! The Need to Know principle was enforced to the maximum degree. In K-25, the gaseous diffusion plant in the instant city of Oak Ridge, Tennessee, tens of thousands of men and women worked around the clock attending to dials and pushing buttons and levers on a task that they came to perform with true expertise and no understanding at all. As their reactions in the aftermath of Hiroshima made clear, they didn't know if they were making airplane parts or crankcase oil for submarines or what. Think of the planning required to create a training system that could turn them into experts without letting on what kind of experts they were. The level of secrecy was higher than ever before (or since, probably). Leslie Groves and the planners all needed to know a great deal about the project, of course; they were intelligent designers, armed with detailed and precise understanding of the specs of the task; only by using that understanding could they create the sheltered environment for uncomprehending competence.

The project proved one point beyond a shadow of a doubt: it is possible to create very reliable levels of high competence with almost no comprehension for rather insulated tasks. So far as I can determine, to this day the precise distribution of understanding through the entire workforce of the Manhattan Project is a closely guarded secret. What did the engineers and architects who designed the K-25 building need to know? It was the largest building in the world when they constructed it in a matter of months. Some of them obviously needed to know what the miles of highly specialized pipes were going to be used for, but probably the designers of the roof, the foundation, and the doors had no inkling. It is pleasant to reflect that while Groves and his team were intelligently designing a system of thousands of human jobs that required minimal comprehension, Turing and his team on the other side of the Atlantic were intelligently designing a system that could replace those clueless homunculi with electronics. A few years later, scientists and engineers, most of whom had contributed to one or another of these pathbreaking wartime projects, began exploiting Turing's invention of uncomprehending competent building blocks to create the audacious field of Artificial Intelligence.

Turing himself (1950) prophesied that by the turn of the century "the use of words and general educated opinion will have altered so much that one will be able to speak of machines thinking without expecting to be contradicted." The early work in the field was brilliant, opportunistic, naïvely optimistic, and, one might well say, saturated with hubris. An artificial intelligence ought to be able to see, as well as think, certainly, so let's first design a seeing machine. The notorious "summer vision project" of the mid-1960s at MIT was an attempt to "solve vision" over one long vacation, leaving harder problems for later! By today's standards the "giant electronic brains" on which the early work was conducted were tiny and achingly slow, and one of the side effects of these limitations was that efficiency was a high-priority goal. Nobody would bother creating a computer model that would take days to respond to a realistic set of inputs, especially if the goal was to harness the computer's unprecedented speed to handle real-world, real-time problems.

Early AI, or GOFAI (Good Old-Fashioned AI [Haugeland 1985]), was a "top-down," "intellectualist" approach to Artificial Intelligence: write down what human experts know, in a language the computer can manipulate with *inference engines* that could patrol the "huge" memory banks stocked with this carefully handcrafted *world knowledge*, deducing the theorems that would be needed to make informed decisions and control appropriately whatever limbs or other effectors the intelligence had. GOFAI can be seen in retrospect to have been an exercise in creating something rather Cartesian, a rationalistic expert with myriads of *propositions* stored in its memory, and all the *understanding* incorporated in its ability to draw conclusions from the relevant axioms and detect contradictions in its world knowledge—as efficiently as possible. What is an intelligent agent after all, but a well-informed rational being, which can think fast enough, using the propositions it knows, to plan actions to meet whatever contingencies arise? It seemed like a good idea at the time, and to some researchers in the field, it still does.[17]

The premium on speed and efficiency dictated working first on "toy problems." Many of these ingeniously scaled-down problems were more or less solved, and the solutions have found applications in the not particularly demanding world of controllers in restricted environments (from elevators and dishwashers to oil refineries and airplanes), and in medical diagnosis, game playing, and other carefully circumscribed areas of investigation or interaction: making airline reservations, spell-checking and even grammar checking, and the like. We can think of these designs as rather indirect descendants of the heavily insulated systems created by Groves and his elite team of intelligent designers, adhering to the Need to Know principle, and relying on the comprehension of the *designers* to contrive systems composed of subsystems that were foresightedly equipped with exactly the competences they would need in order to handle

17   Douglas Lenat's CYC project is the ongoing attempt to create such an artificial intelligence, and after thirty years of work by hundreds of coders (known as CYClists), it has over a million hand-defined concepts in its memory.

the problems they might face. Since for all their brilliance the early AI designers weren't omniscient (and time was of the essence), they restricted the range and variety of inputs each subsystem had to accept, and deal with, creating programs harboring thousands of sheltered workshops to protect the idiot savants (subroutines) that worked there.

Much was learned, and a lot of good practices and techniques were invented and refined, but they mainly helped dramatize how truly difficult the task of designing a free-wheeling, imaginative, open-ended human mind is. The dream of a hand-coded, top-down-organized, bureaucratically efficient know-it-all, a walking (or at least talking) encyclopedia, is not yet entirely extinguished, but as the size of the project became clearer, there has been a salutary shift of attention to a different strategy: using colossal amounts of Big Data and the new statistical pattern-finding techniques of data-mining and "deep learning" to eke out the necessary information in a more bottom-up way.

I will have much more to say about these developments later; for the time being, the point that we need to recognize is that the vast increase in speed and size of computers over the years has opened up the prospect of exploring "wasteful," "mindless," less "bureau-cratic," more evolution-like processes of information extraction, and these are achieving impressive results. Thanks to these new perspectives, we can now think in some detail about the question of how the *relatively* simple systems that control bacteria, worms, and termites, for example, might have evolved by the bottom-up, fore-sightless, brute force processes of natural selection. In other words, we want to see how evolution might play the Leslie Groves role in organizing clueless operatives into effective teams *without* the lux-ury of Groves's understanding and foresight.

Top-down intelligent designing works. The policy of planning ahead, articulating the problems, refining the tasks, and clearly representing the reasons for each step is a strategy that has not just seemed obvious to inventors and problem-solvers for millennia; it has proven itself in countless triumphs of foresight and ingenuity in every field of human endeavor: from science and engineering to political campaigns and cooking, farming, and navigation. Before Darwin, it was seen as the only way design could be accomplished; design without an intelligent designer was deemed impossible. But top-down design is in fact responsible for much less of the design in our world than is commonly appreciated, and for some of the "achievements of creative skill," to echo Beverley once again, victory has so far eluded it. Darwin's "strange inversion of reasoning" and Turing's equally revolutionary inversion were aspects of a single discovery: competence without comprehension. Comprehension, far from being a Godlike talent from which all design must flow, is an emergent effect of systems of uncomprehending competence: natural selection on the one hand, and mindless computation on the other. These twin ideas have been proven beyond a reasonable doubt, but they still provoke dismay and disbelief in some quarters, which I have tried to dispel in this chapter. Creationists are not going to find commented code in the inner workings of organisms, and Cartesians are not going to find an immaterial *res cogitans* "where all the understanding happens."

5

The Evolution of Understanding

Animals designed to deal with affordances

Animals are designed by natural selection, of course, but such a declaration of confidence in evolution is not informative. How, more particularly, might evolution turn this trick? One of the fruits of our interlude on the designing of an elevator controller and its artifactual kin is a sharper sense of how different that R&D process is from evolution by natural selection. The computer on which the designers—the programmers—test and run their solutions is itself a product of intelligent design, as we have noted, and its initial set of building-block competences—arithmetic and conditional branching—invite all would-be programmers to conceive of their tasks in the top-down way as well, as *problem-solving* in which they try to embody *their* understanding of the problem in the solutions they build.

"How else?" one might well ask. Intelligent design of this sort starts with a goal (which may well be refined or even abandoned along the way) and works top-down, with the designers using everything they know to guide their search for solutions to the design problems (and sub-problems, and sub-sub-problems . . .) they set for themselves. Evolution, in contrast, has no goals, no predefined problems, and no comprehension to bring to the task; it myopically and undirectedly muddles along with what it has already created, mindlessly trying out tweaks and variations, and keeping those that prove useful, or at least not significantly harmful.

Could something as intellectually sophisticated as a digital computer, for instance, ever evolve by bottom-up natural selection? This is very hard to imagine or even to take seriously, and this has inspired some thinkers to conclude that since evolution couldn't create a computer (or a computer program to run on it), human minds must not be products of natural selection alone, and the aspirations of Artificial Intelligence must be forlorn. The mathematician and physicist Roger Penrose (1989) is the most illustrious example. For the sake of argument let's concede that evolution by natural selection could not *directly* evolve a living digital computer (a Turing machine *tree* or a Turing machine *turtle*, for example). But there is an indirect way: let natural selection first evolve human minds, and then *they* can intelligently design *Hamlet*, La Sagrada Familia, and the computer, among many other wonders. This bootstrapping process seems almost magical at first, even self-contradictory. Isn't Shakespeare, or Gaudí, or Turing a more magnificent, brilliant "creation" than any of their brainchildren? In some regards, yes, of course, but it is also true that their brainchildren have features that couldn't come into existence without them.

If you landed on a distant planet and were hunting along its seashore for signs of life, which would excite you more, a clam or a clam rake? The clam has billions of intricate moving parts, while the clam rake has just two crude, fixed parts, but it must be an artifact of some living thing, something much, much more impressive than a clam. *How could a slow, mindless process build a thing that could build a thing that a slow mindless process couldn't build on its own?* If this question seems to you to be unanswerable, a rhetorical question only, you are still in thrall to the spell Darwin broke, still unable to adopt Darwin's "strange inversion of reasoning." Now we can see how strange and radical it is: a process with no Intelligent Designer can create intelligent designers who can then design things that permit us to understand how a process with no Intelligent Designer can create intelligent designers who can then design things.

The intermediate steps are instructive. What about the clam rake gives away its artifactual status? Its very simplicity, which indicates its dependence on something else for its ability to defy the Second Law of Thermodynamics, persisting as uniform and symmetrical collections of atoms of elements in improbable juxtapositions. Something gathered and refined these collections. Something complicated.

Let's return once more to simple organisms. The idea that every organism has its ontology (in the elevator sense) was prefigured in Jakob von Uexküll's (1934) concept of the organism's *Umwelt*,

FIGURE 5.1: Clam rake. © *Daniel C. Dennett.*

the behavioral environment that consists of all the things that matter to its well-being. A close kin to this idea is the psychologist J. J.
Gibson's (1979) concept of *affordances*: "What the environment offers
the animal for good or ill." Affordances are the relevant opportunities in the environment of any organism: things to eat or mate with,
openings to walk through or look out of, holes to hide in, things to
stand on, and so forth. Both von Uexküll and Gibson were silent
about the issue of whether consciousness (in some still-to-be-defined
sense) was involved in having an *Umwelt* populated by affordances,
but since von Uexküll's case studies included amoebas, jellyfish, and
ticks, it is clear that he, like Gibson, was more interested in char-
acterizing the *problems faced and solved* by organisms than on how,
internally, these solutions were carried out. The sun is in the ontology of a honey bee; its nervous system is designed to exploit the
position of the sun in its activities. Amoebas and sunflowers also
include the sun in their *Umwelten*; lacking nervous systems, they use
alternative machinery to respond appropriately to its position. So
the engineer's concept of elevator ontology is just what we need at
the outset. We can leave until later the questions of whether, when,
and why the ontology of an organism, or a lineage of organisms,
becomes *manifest* in consciousness of some sort and not just *implicit*
in the designed responses of its inner machinery. In other words,
organisms can be the beneficiaries of design features that imply
ontologies without themselves *representing* those ontologies (consciously, semiconsciously, or unconsciously) in any stronger sense.
The shape of a bird's beak, together with a few other ancillary features of anatomy, *imply* a diet of hard seeds, or insects or fish, so we
can stock the *Umwelten* of different species of birds with hard seeds,
insects, and fish, as species-specific *affordances* on the basis of these
anatomical features alone, though of course it is wise to corroborate
the implication by studying behavior if it is available. The shape of
the beak does not in any interesting sense *represent* its favored food-
stuff or way of obtaining it.

Paleontologists draw conclusions about the predatory preferences and other behaviors of extinct species using this form of inference, and it is seldom noted that it depends, ineliminably, on making adaptationist assumptions about the designs of the fossilized creatures. Consider Niles Eldredge's (1983) example of Fisher's (1975) research on horseshoe crab swimming speed. He cites it to demonstrate that asking the historical question "what has happened?" ("how come") is a better tactic than asking the adaptationist question ("what for"), with its optimality assumptions. But Fisher's conclusion about how fast the ancient horseshoe crabs swam

> depends on a very safe adaptationist assumption about what is good: *Faster is better—within limits.* The conclusion that Jurassic horseshoe crabs swam faster depends on the premise that they would achieve maximal speed, given their shape, by swimming at a certain angle, *and* that they would swim so as to achieve maximal speed. So . . . [Fisher needs an] entirely uncontroversial, indeed tacit, use of optimality considerations to get *any purchase at all* on "what happened" 150 million years ago. (Dennett 1983)

Remember, biology is reverse engineering, and reverse engineering is methodologically committed to optimality considerations. "What is—or was—this feature *good for*?" is always on the tip of the tongue; without it, reverse engineering dissolves into bafflement.

As I said in the opening paragraph of the book, bacteria don't know they are bacteria, but of course they respond to other bacteria in bacteria-appropriate ways and are capable of avoiding or tracking or trailing things they distinguish in their *Umwelt*, without needing to have any idea about what they are doing. Bacteria are in the ontology of bacteria the same way floors and doors are in the ontology of elevators, only bacteria are much more complicated. Just as there are reasons why the elevator's control circuits are designed the way they are, there are reasons why the bacteria's internal protein control networks are designed the way they are: in both cases the designs have been optimized to handle the problems encountered efficiently and effectively.[18] The chief difference is that the design of the elevator circuits was done by intelligent designers who had worked out descriptions of the problems, and representations of *reasoned* solutions, complete with justifications. In the R&D history of the bacteria, there was no source code, and no comments were ever composed, to provide hints of what Mother Nature intended. This does not stop evolutionary biologists from assigning functions to some evolved features (webbed feet are for propulsion in water), and interpreting other features as mistakes of Nature (a two-headed calf). Similarly, literary editors of texts of long-dead authors don't have to rely on autobiographical divulgences left behind in the author's papers to interpret some unlikely passages as deliberately misleading and others as typographical errors or memory lapses.

Software development is a relatively new domain of human endeavor. While still in its infancy many foibles and glitches have been identified and corrected, and a Babel Tower of new programming languages has been created, along with a host of software-writing tools to make the job easier. Still, programming is an "art," and even commercially released software from the best purveyors always turns out to have "bugs" in it that require correction in post-purchase updates. Why hasn't debugging been automated, eliminating these costly errors from the outset? The most intelligent human designers, deeply informed about the purposes of the software, still find debugging code a daunting task, even when they can examine carefully commented source code produced under strictly regimented best practices (Smith 1985, 2014). There is a reason why debugging cannot be completely automated: what counts as a bug depends on all the purposes (and sub-purposes, and sub-sub-purposes) of the

18   There is much controversy about using the term "optimize" when referring to the "good enough" products of natural selection. The process of natural selection cannot "consider all things" and is always in the midst of redesign, so it is not guaranteed to find the *optimal* solution to any specific design problem posed, but it does amazingly well, typically better than intelligent human designers who are striving for optimal design.

software, and specifying in sufficient detail what those purposes are (in order to feed them to one's imagined automated debugger program) is, at least for practical purposes, the very same task as writing debugged code in the first place![19] Writing and debugging computer code for ambitious systems is one of the most severe tests of human imagination yet devised, and no sooner does a brilliant programmer devise a new tool that relieves the coder of some of the drudgery than the bar is raised for what we expect the coder to create (and test). This is not an unprecedented phenomenon in human activity; music, poetry, and the other arts have always confronted the would-be creator with open-ended spaces of possible "moves" that did not diminish once musical notation, writing, and ready-made paints were made available, nor does artistic creation become routinized by the addition of synthesizers and MIDI files, word-processing and spell-checking, and million-color, high-resolution computer graphics.

How does Nature debug its designs? Since there is no source code or comments to read, there can be no debugging by brilliant intellectual explanation; design revision in Nature must follow the profligate method of releasing and test-driving many variants and letting the losers die, *unexamined*. This won't necessarily find the globally optimal design but the best locally accessible versions will thrive, and further test-driving will winnow the winners further, raising the bar slightly for the next generation.[20] Evolution is, as Richard Dawkins's (1986) memorable title emphasizes, the Blind Watchmaker, and given the R&D method used, it is no wonder that evolution's products are full of opportunistic, short-sighted, but deviously effective twists and turns—effective except when they

19   Legendary software designer Charles Simonyi, the principal creator of Microsoft Word, has devoted more than twenty years to the task of creating what he calls "Intentional Software," which would ideally solve this problem or a valuable subset of these problems. The fact that several decades of high-quality work by a team of software engineers has not yet yielded a product says a lot about the difficulty of the problem.

20   Evolution explores the "adjacent possible," see Kauffman (2003).

aren't! One of the hallmarks of design by natural selection is that it is full of bugs, in the computer programmer's sense: design flaws that show up only under highly improbable conditions, conditions never encountered in the finite course of R&D that led to the design to date, and hence not yet patched or worked around by generations of tinkering. Biologists are very good at subjecting the systems they are studying to highly improbable conditions, imposing extreme challenges to see where and when the systems fail, *and why.*

What they typically discover, when reverse engineering an organism, is like the all-but-undecipherable "spaghetti code" of undisciplined programmers. If we make the effort to decipher spaghetti code, we can usually note which unlikely possibilities *never occurred to* the designers in their myopic search for the best solution to the problems posed for them. *What were they thinking?* When we ask the same question about Mother Nature, the answer is always the same: nothing. No thinking was involved, but nevertheless she muddled through, cobbling together a design so effective that it has survived to this day, beating out the competition in a demanding world until some clever biologist comes along and exposes the foibles.

Consider *supernormal stimuli,* a design glitch found in many organisms. Niko Tinbergen's (1948, 1951, 1953, 1959) experiments with seagulls revealed a curious bias in their perceptual/behavioral machinery. The adult female has an orange spot on her beak, at which her chicks instinctually peck, to stimulate their mother to regurgitate and feed them. What if the orange spot were bigger or smaller, brighter or less distinct? Tinbergen showed that chicks would peck even more readily at exaggerated cardboard models of the orange spot, that supernormal stimuli evoked supernormal behaviors. Tinbergen also showed that birds that laid light blue, gray-dappled eggs preferred to sit on a bright blue black polka-dotted fake egg so large that they slid off it repeatedly.

"This isn't a bug, it's a feature!" is the famous programmers' retort, and the case can be made for supernormal stimuli. As long as their *Umwelt* doesn't have sneaky biologists with vivid imaginations challenging the birds with artificial devices, the system works very well, focusing the organism's behavior on what (almost always) matters. The free-floating rationale of the whole system is clearly good enough for practical purposes, so Mother Nature was wise not to splurge on something more foolproof that would detect the ruse. This "design philosophy" is everywhere in Nature, providing the opportunities for arms races in which one species exploits a short-cut in another species' design, provoking a counter-ploy in Design Space that ratchets both species to develop ever better defenses and offenses. Female fireflies sit on the ground watching male fireflies emit patterns of flashes, showing off and hoping for an answer from the female. When the female makes her choice and flashes back, the male rushes down for a mating. But this ingenious speed-dating system has been invaded by another species of firefly, Photuris, that pretends to be a female, luring the males to their death. The Photuris prefers males with longer, stronger signals, so the males are evolving shorter love letters (Lewis and Cratsley 2008).

Higher animals as intentional systems: the emergence of comprehension

Competence without comprehension is Nature's way, both in its methods of R&D and in its smallest, simplest products, the brilliantly designed motor proteins, proofreading enzymes, antibodies, and the cells they animate. What about multicellular organisms? When does comprehension emerge? Plants, from tiny weeds to giant redwood trees, exhibit many apparently clever competences, tricking insects, birds, and other animals into helping them reproduce, forming useful alliances with symbionts, detecting precious water sources, tracking the sun, and protecting themselves from various predators (herbivores and parasites). It has even been argued (see, e.g., Kobayashi and Yamamura 2003; Halitschke et al. 2008) that some species of plants can warn nearby kin of impending predation by wafting distress signals downwind when attacked, permitting those that receive the signals to heighten their defense mechanisms in anticipation, raising their toxicity or generating odors that either repel the predators or lure symbionts that repel the predators. These responses unfold so slowly that they are hard to see as proper behaviors without the benefit of time-lapse photography, but, like the microscopic behaviors of single cells, they have clear rationales that need not be understood by the actors.

Here we see emerging something like a double standard of attribution. It is well-nigh impossible to describe and explain these organized-processes-in-time without calling them behaviors and *explaining* them the way we explain our own behaviors, by citing reasons and assuming that they are guided by something like perceptual monitoring, the intake of information that triggers, modulates, and terminates the responses. And when we do this, we *seem* to be attributing not just competence but also the comprehension that— in us—"normally goes with" such behavioral competence. We are anthropomorphizing the plants and the bacteria in order to understand them. This is not an intellectual sin. We are *right* to call their actions behaviors, to attribute these competences to the organisms, to explain their existence by citing the rationales that account for the benefits derived from these competences by the organisms in their "struggle" for survival. We are right, I am saying, to adopt what I call the intentional stance. The only mistake lies in attributing *comprehension* to the organism or to its parts. In the case of plants and microbes, fortunately, common sense intervenes to block that attribution. It is easy enough to understand how their competence can be provided by the machinery without any *mentality* intruding at all.

Let's say that organisms that have spectacular competences without any need for comprehension of their rationales are *gifted*. They are the beneficiaries of talents bestowed on them, and these talents are not products of their own individual investigation and practice. You might even say they are *blessed* with these *gifts*, not from God, of course, but from evolution by natural selection. If our imaginations need a crutch, we can rely on the obsolescing stereotype of the robot as a mindless mechanism: plants don't have understanding; they're living *robots*. (Here's a prediction: in a hundred years, this will be seen as an amusing fossil of *biocentrism*, a bit of prejudice against comprehending robots that survived well into the twenty-first century.)

While we're on this topic, it's interesting to recall that in the twentieth century one of the most popular objections to GOFAI was this:

> The so-called intelligence in these programs is really just the intelligence—the understanding—of the programmers. The programs don't understand anything!

I am adopting and adapting that theme but not granting understanding (yet) to anyone or anything:

> The so-called intelligence in trees and sponges and insects is not theirs; they are just brilliantly designed to make smart moves at the right time, and while the design is brilliant, the designer is as uncomprehending as they are.

The opponents of GOFAI thought they were stating the obvious when they issued their critique of so-called intelligent machines, but see how the emotional tug reverses when the same observation is ventured about animals. Whereas—I surmise—most readers will be quite comfortable with my observation that plants and microbes are merely gifted, blessed with well-designed competences, but otherwise clueless, when I then venture the same opinion about "higher" animals, I'm being an awful meanie, a killjoy.

When we turn to animals—especially "higher" animals such as mammals and birds—the temptation to attribute comprehension in the course of describing and explaining the competences is much greater, and—many will insist—entirely appropriate. Animals really do understand what they're doing. See how amazingly clever they are! Well, now that we have the concept of competence without comprehension firmly in hand, we need to reconsider this gracious opinion. The total weight of all life on the planet now—the biomass—is currently estimated as more than half made up of bacteria and other unicellular "robots," with "robotic" plants making up more than half the rest. Then there are the insects, including all the clueless termites and ants that outweigh the huge human population celebrated by MacCready. We and our domesticated animals may compose 98% of the *terrestrial vertebrate* biomass, but that is a small portion of life on the planet. Competence *without* comprehension is the way of life of the vast majority of living things on the planet and should be the default presumption until we can demonstrate that some individual organisms really do, in one sense or another, *understand* what they are doing. Then the question becomes when, and why, does the design of organisms start *representing* (or otherwise intelligently incorporating) the free-floating rationales of their survival machinery? We need to reform our imaginations on this issue, since the common practice is to *assume* that there is some kind of understanding in "higher animals" wherever there is a rationale.

Consider a particularly striking example. Elizabeth Marshall Thomas is a knowledgeable and insightful observer of animals (including human animals), and in one of her books, *The Hidden Life of Dogs* (1993), she permits herself to imagine that dogs enjoy a wise understanding of their ways: "For reasons known to dogs but not to us, many dog mothers won't mate with their sons" (p. 76). There is no doubt about their instinctive resistance to such inbreeding; they probably rely mainly on scent as their cue, but who knows what else contributes—a topic for future research. But the suggestion that dogs have any more insight into the reasons for their instinctual behaviors and dispositions than we have into ours is romanticism run wild. I'm sure she knows better; my point is that this lapse came naturally to her, an extension of the prevailing assumption, not a bold proposal about the particular self-knowledge of dogs. This is like a Martian anthropologist writing, "For reasons known to human beings but not to us, many human beings yawn when sleepy and raise their eyebrows when they see an acquaintance." There *are* reasons for these behaviors, *what for* reasons, but they are not *our* reasons. You *may* fake a yawn or raise your eyebrows for a reason—to give a deliberate signal, or to feign acquaintance with an attractive but unfamiliar person you encounter—but in the normal case you don't even realize you do it, and hence have no occasion to know why you do it. (We still don't know why we yawn—and certainly dogs aren't ahead of us on this point of inquiry, though they yawn just as we do.)

What about more obviously deliberate behaviors in animals? Cuckoos are *brood parasites* that don't make their own nests. Instead, the female cuckoo surreptitiously lays her egg in the nest of a host pair of some other species of birds, where it awaits the attentions of its unwittingly adoptive parents. Often, the female cuckoo will roll one of the host eggs out of the nest—in case the host parents can count. And as soon as the cuckoo chick is hatched (and it tends to incubate more quickly than the host eggs), the little bird goes to great efforts to roll any remaining eggs out of the nest. Why? To maximize the nurture it will get from its adoptive parents. The video clips of this behavior by the hatchling cuckoo are chilling demonstrations of efficient, competent killing, but there is no reason to suppose that mens rea (guilty intention, in the law) is in place. The baby bird knows not what it is doing but is nevertheless the beneficiary of its behavior. What about nest building in less larcenous species? Watching a bird build a nest is a fascinating experience, and there is no doubt that highly skilled weaving and even sewing actions are involved (Hansell 2000). There is quality control, and a modicum of learning. Birds hatched in captivity, never having seen a nest being built, will build a serviceable species-typical nest out of the available materials when it is time to build a nest, so the behavior is instinctual, but it will build a better one the next season.

How much understanding does the nest-building bird have? This can be, and is being, probed by researchers (Hansell 2000, 2005, 2007; Walsh et al. 2011; Bailey et al. 2015), who vary the available materials and otherwise interfere with the conditions to see how versatile and even foresighted the birds can be. Bearing in mind that evolution can only provide for challenges encountered during R&D, we can predict that the more novel the artificial intrusions in the bird's *Umwelt* are, the less likely it is that the bird will interpret them appropriately, *unless* the bird's lineage evolved in a highly varied selective environment that obliged natural selection to settle on designs that are not entirely *hard-wired* but have a high degree of plasticity and the learning mechanisms to go with it. Interestingly, when there isn't enough stability over time in the selective environment to permit natural selection to "predict" the future accurately (when "selecting" the best designs for the next generation), natural selection does better by leaving the next generation's design partially unfixed, like a laptop that can be configured in many different ways, depending on the purchaser's preferences and habits.[21] Learning can take over where natural selection left off, optimizing the individuals in their own lifetimes by extracting information from the world encountered and using it to make local improvements. We will soon turn to a closer examination of this path to understanding, but, first, I want to explore a few more examples of behaviors with free-floating rationales and their implications.

You may have seen video of antelopes being chased across the plains by a predator and noticed that some of the antelopes leap high in the air during their attempts to escape their pursuer. This is called stotting. Why do antelopes stot? It is clearly beneficial, because antelopes that stot seldom get caught and eaten. This is a causal regularity that has been carefully observed, and it demands a *what for* explanation. No account of the actions of all the proteins and the like in the cells of all the antelopes and predators chasing them could reveal why this regularity exists. For an answer we need the branch of evolutionary theory known as costly signaling theory (Zahavi 1975; Fitzgibbon and Fanshawe 1988). The strongest and fastest of the antelopes stot in order to advertise their fitness to the pursuer, signaling, in effect, "Don't bother chasing me; I'm too hard to catch; concentrate on one of my cousins who isn't able

21 How can I speak of evolution, which famously has no foresight, being able or unable to *predict* anything? We can cash out this handy use of the intentional stance applied to evolution itself by saying, less memorably and instructively, that highly variable environments *have no information* about future environments for natural selection to (mindlessly) exploit (see chapter 6).

to stot—a much easier meal!" and the pursuer takes this to be an honest, hard-to-fake signal and ignores the stotter. This is both an act of *communication* and an act with only a free-floating rationale, which need not be appreciated by either antelope or lion. That is, the antelope may be entirely oblivious of why it is a good idea to stot if you can, and the lion may not understand why it finds stotting antelopes relatively unattractive prey, but if the signaling wasn't honest, costly signaling, it couldn't persist in the evolutionary arms race between predator and prey. (If evolution tried a "cheap" signal, like tail flicking, which every antelope, no matter how frail or lame, could send, it wouldn't pay lions to attend to it, so they wouldn't.) This may seem an overly skeptical *killjoy* demotion of the intelligence of both antelope and lion, but it is the strict application of the same principles of reverse engineering that can account for the cuckoo and the termite and the bacterium. The rule of attribution must be then, if the competence observed can be explained without appeal to comprehension, don't indulge in extravagant anthropomorphism. Attributing comprehension must be supported by demonstrations of much more intelligent behavior. Since stotting is not (apparently) an element in a more elaborate system of interspecies or intraspecies communication on many topics, the chances of finding a need for anything that looks like comprehension here are minimal. If you find this verdict too skeptical, try to imagine some experiments that could prove you right.

How could experiments support the verdict of comprehension? By showing that the animals can do what we comprehenders can do with variations on the behavior. Stotting is a variety of showing off, or bragging, and we can do that, but we also can bluff or refrain from bragging or showing off if conditions arise that render such behavior counterproductive or worse. We can modulate our bragging, tuning it to different audiences, or do some transparently exaggerated bragging to telegraph that we don't really mean it and are making a joke. And so forth, indefinitely. Can the antelope do any of this? Can it refrain from stotting in circumstances that, in novel ways, make stotting inappropriate? If so, this is some evidence

that it has—and uses—some minimal understanding of the rationale of its actions.

A rather different free-floating rationale governs the injury-
feigning, ground-nesting bird, such as a piping plover, that lures
a predator away from her nest by seeming to have a broken wing,
keeping just out of the predator's reach until she has drawn it far
from her nest. Such a "distraction display" is found in many very
widely separated species of ground-nesting birds (Simmons 1952;
Skutch 1976). This seems to be *deception* on the bird's part, and it is
commonly called that. Its purpose is to *fool* the predator. Adopting
Dawkins's (1976) useful expository tactic of inventing "soliloquies,"
we can devise a soliloquy for the piping plover:

> I'm a low-nesting bird, whose chicks are not protectable against
> a predator that discovers them. This approaching predator can
> be *expected* soon to discover them unless I distract it; it could be
> distracted by its *desire* to catch and eat me, but only if it *thought*
> there was a *reasonable* chance of its actually catching me (it's no
> dummy); it would contract just that *belief if I gave it evidence that*
> I couldn't fly anymore; I could do that by feigning a broken
> wing, and so on.

Talk about sophistication! Not just a goal, but also a *belief* about an
expectation, and a *hypothesis* about the *rationality* of the predator and
a *plan* based on that hypothesis. It is unlikely in the extreme that
any feathered "deceiver" is capable of such mental representation.
A more realistic soliloquy to represent what is "in the mind" of the
bird would be something like: "Here comes a predator; all of a sudden I feel this tremendous urge to do that silly broken-wing dance.
I wonder why?" But even this imputes more *reflective* capacity to the
bird than we have any warrant for. Like the elevator, the bird has
been designed to make some important discriminations and do the
right thing at the right time. Early investigators, rightly deeming
the sophisticated soliloquy too good to be true as an account of
the bird's *thinking*, were tempted to hypothesize that the behavior

was not deliberate at all, but a sort of panic attack, composed of unguided spasms that had the beneficial side effect of attracting the attention of the predator. But that drastically *under*estimates the bird's grasp of the situation. Clever experiments on piping plovers by Ristau (1983, 1991) using a remote-controlled toy dune buggy with a stuffed raccoon mounted on it, demonstrated that the plover closely monitors the predator's attention (gaze direction), and modulates its injury feigning, raising the intensity and then letting the predator get closer if the predator shows signs of abandoning the hunt. And, of course, she flies away at the opportune moment, once the predator is some distance from her nest. The bird doesn't need to know the whole rationale, but it does recognize and respond appropriately to some of the conditions alluded to in the rationale. The behavior is neither a simple "knee-jerk" reflex inherited from her ancestors nor a wily scheme figured out in her rational mind; it is an evolution-designed routine with variables that respond to details in the circumstances, details that the sophisticated soliloquy captures—without excess—in the rationale of that design.

The free-floating rationale answers the reverse-engineering question: *Why* is this routine organized like this? If we are squeamish about anthropomorphism, we can pretend to put the answer in somewhat less "mentalistic" terms by liberal use of scare quotes: The routine is an "attention-grabbing" behavior that depends for its success on the likely "goals" and "perceptions" of a predator, designed to provoke the predator into "approaching" the plover and thus distancing itself from the nest; by "monitoring" the predator's "attention" and modulating the behavior to maintain the predator's "interest," the plover typically succeeds in preventing the predation of its young. (This long-winded answer is only superficially more "scientific" than the intentional-stance version expressed in the soliloquy; the two explanations depend on the same distinctions, the same optimality assumptions, and the same informational demands.) Further empirical research may reveal further appropriate sensitivities, or it may reveal the foibles in this cobbled-together device. There is some evidence that piping plovers "know enough" not to engage in injury feigning when a cow approaches, but instead fly *at* the cow, pushing, not luring, it away from the nest. Would a plover resist the urge to put on an injury-feigning display if it could see that an actually injured bird, or other vulnerable prey item, had already captured the attention of the predator? Or even more wonderful, as suggested by David Haig (2014, personal correspondence):

> One could imagine a bird with an actual broken wing unconvincingly attempting to escape with the intention that the predator interpret its actions as "this is a broken wing display therefore the bird is not easy prey but a nest is near." If the predator started to search for a nest, then the predator would have recognized that the bird's actions were a text but misunderstood the bird's motives. The interpretation of the text is "wrong" for the predator but "right" for the bird. The text has achieved the bird's intention but foiled that of the predator who has been deliberately misled.

Haig speaks unguardedly of the bird's motives and intentions and of the predator's "interpretation" of the "text," recognizing that the task of thinking up these varied opportunities for further experiments and observations *depends on* our adoption of the intentional stance, but also appreciating that there is a graceful and gradual trade-off between interpreting animals (or for that matter, plants or robots or computers) as *themselves* harboring the reasons and the reasoning, and relegating the rationale to Mother Nature, as a free-floating rationale exposed by the mindless design-mining of natural selection.

The *meaning* of the injury-feigning *signal* will have its *intended* effect only if the predator does not *recognize* it to be a *signal* but *interprets* it as an *unintentional* behavior—and this is true whether or not the bird or the predator understands the situation the way we do. It is the *risk* that the predator will *catch on* that creates the selection pressure for better *acting* by the deceptive bird. Similarly, the strikingly realistic "eye-spots" on the wings of butterflies owe their verisimilitude to the visual acuity of their predators, but of course the butterflies are the clueless beneficiaries of their deceptive gear. The deceptive rationale of the eye-spots is there all the same, and to say it is *there* is to say that there is a domain within which it is *predictive* and, hence, explanatory. (For a related discussion, see Bennett 1976, §§ 52, 53, 62.) We may fail to notice this just because of the obviousness of what we can predict: For example, in an environmental niche with bats but not birds for predators, we don't expect moths with eye-spots (for as any rational deceiver knows, visual sleight of hand is wasted on the blind and myopic).

Comprehension comes in degrees

The time has come to reconsider the slogan *competence without comprehension*. Since cognitive competence is often assumed to be an effect of comprehension, I went out of my way to establish that this familiar assumption is pretty much backward: competence comes first. Comprehension is not the *source* of competence or the *active ingredient* in competence; comprehension is *composed* of competences. We have already considered the possibility of granting a smidgen or two of comprehension to systems that are particularly clever in the ways that they marshal their competences but that may play into the misleading image of comprehension as a separable element or phenomenon kindled somehow by mounting competence.

The idea of comprehension or understanding as a separate, stand-alone, mental marvel is ancient but obsolete. (Think of Descartes's *res cogitans*, or Kant's *Critique of Pure Reason*, or Dilthey's *Verstehen*—which is just the German word for understanding, but since, like all German nouns, it is capitalized, when it is said with furrowed brow, it conjures up in many minds a Bulwark against Reductionism and Positivism, a Humanistic alternative to Science.) The illusion that understanding is some additional, separable mental phenomenon (over and above the set of relevant competences, including the meta-competence to exercise the other competences at appropriate times) is fostered by the *aha!* phenomenon, or eureka effect—that delightful moment when you suddenly recognize that you *do* understand something that has heretofore baffled you. This psychological phenomenon is perfectly real and has been studied by psychologists for decades. Such an experience of an abrupt onset of understanding can easily be misinterpreted as a demonstration that understanding is a *kind of experience* (as if suddenly learning you were allergic to peanuts would show that allergies are a kind of feeling), and it has led some thinkers to insist that there can be no genuine comprehension without consciousness (Searle [1992] is the most influential). Then, if you were to think that it is obvious that consciousness, whatever it is, sunders the universe in two—everything is either conscious or not conscious; consciousness does not admit of degrees—it would stand to reason that comprehension, *real* comprehension, is enjoyed only by conscious beings. Robots understand nothing, carrots understand nothing, bacteria understand nothing, oysters, well, we don't know *yet*—it all depends on whether oysters are conscious; if not, then their competences, admirable though they are, are competences utterly without comprehension.

I recommend we discard this way of thinking. This well-nigh magical concept of comprehension has no utility, no application in the real world. But the distinction between comprehension and incomprehension is still important, and we can salvage it by the well-tested Darwinian perspective of gradualism: comprehension comes in degrees. At one extreme we have the bacterium's sorta comprehension of the quorum-sensing signals it responds to (Miller and Bassler 2001) and the computer's sorta comprehension of the "ADD" instruction. At the other extreme we have Jane Austen's comprehension of the interplay of personal and social forces in the emotional states of people and Einstein's comprehension of relativity. But even at the highest levels of competence, comprehension is never absolute. There are always ungrasped implications and unrecognized presuppositions in any mind's mastery of a concept or topic. All comprehension is sorta comprehension from some perspective. I once gave a talk at Fermi Lab in Illinois, to a few hundred of the world's best physicists and confessed that I only sorta understood Einstein's famous formula:

$$E = mc^2$$

I can do the simple algebraic reformulations, and say what each term refers to, and explain (roughly) what is important about this discovery, but I'm sure any wily physicist could easily expose my incomprehension of some aspects of it. (We professors are good at uncovering the mere sorta understanding of our students via examinations.) I then asked how many in the audience understood it. All hands went up, of course, but one person jumped up and shouted "No, no! We theoretical physicists are the ones who understand it; the experimentalists only think they do!" He had a point. Where understanding is concerned, we all depend on something like a division of labor: we count on experts to have deep, "complete" understanding of difficult concepts we rely on every day, only half-comprehendingly. This is, in fact, as we shall see, one of the key contributions of language to our species' intelligence: the capacity to transmit, faithfully, information we only sorta understand!

We human beings are the champion comprehenders on the planet, and when we try to understand other species, we tend to model their comprehension on our experience, imaginatively filling animals' heads with wise reflections as if the animals were strangely shaped people in fur coats. The Beatrix Potter syndrome, as I have called it, is not restricted to children's literature, though I think every culture on earth has folk tales and nursery stories about talking, thinking animals. We do it because, to a first approximation, *it works*. The intentional stance works whether the rationales it adduces are free floating or explicitly represented in the minds of the agents we are predicting. When a son learns from his father how to figure out what their quarry is attending to and how to foil its vigilance, both are treating the animal as a wise fellow thinker in a battle of wits. But the success of the intentional stance does not depend on this being a faithful representation of what is going on in the animal's mind except to the extent that whatever is going on in the animal's brain has the competence to detect and respond appropriately to the information in the environment.

The intentional stance gives "the specs" for a mind and leaves the implementation for later. This is particularly clear in the case of a chess-playing computer. "Make me a chess program that not only *knows the rules* and *keeps track of all the pieces* but also *notices* opportunities, *recognizes* gambits, *expects* its opponent to make intelligent moves, *values* the pieces soundly, and *looks out for* traps. How you accomplish that is your problem." We adopt the same noncommittal strategy when dealing with a human chess player. In the midst of a chess match we rarely have hunches about—or bother trying to guess—the detailed thinking of our opponent; we expect her to see what's there to be seen, to notice the important implications of whatever changes, and to have good ways of formulating responses to the moves we choose. We idealize everybody's thinking, and even our own access to reasons, blithely attributing phantom bouts of clever reasoning to ourselves after the fact. We tend to see what we chose to do (a chess move, a purchase, parrying a blow) to have been just the right move at the right time, and we have no difficulty explaining to ourselves and others how we figured it out in advance, but when we do this we may often be snatching a free-floating rationale out of thin air and pasting it, retrospectively, into our subjective experience. Asked, "Why did you do that?," the most honest thing to say is often "I don't know; it just came to me," but we often succumb to the temptation to engage in *whig history*, not settling for *how come* but going for a *what for*.[22]

When we turn to the task of modeling the competences out of which comprehension is composed, we can distinguish four grades, schematically characterized by successive applications of the tactic

22   The useful term *Whig history* refers to interpreting history as a story of progress, typically justifying the chain of events leading to the interpreter's privileged vantage point. For applications of the term to adaptationism in evolutionary biology, both favorably and unfavorably, see Cronin (1992) and Griffiths (1995).

known in computer science as "generate and test." In the first, lowest level we find *Darwinian creatures*, with their competences predesigned and fixed, created by the R&D of evolution by natural selection. They are born "knowing" all they will ever "know"; they are gifted but not learners. Each generation generates variations which are then tested against Nature, with the winners copied more often in the next round. Next come the *Skinnerian creatures*, who have, in addition to their hard-wired dispositions, the key disposition to adjust their behavior in reaction to "reinforcement"; they more or less randomly generate new behaviors to test in the world; those that get reinforced (with positive reward or by the removal of an aversive stimulus—pain or hunger, for instance) are more likely to recur in similar circumstances in the future. Those variants born with the unfortunate disposition to mislabel positive and negative stimuli, fleeing the good stuff and going for the bad stuff, soon eliminate themselves, leaving no progeny. This is "operant conditioning" and B. F. Skinner, the arch-behaviorist, noted its echo of Darwinian evolution, with the generation and testing occurring in the individual during its lifetime but requiring no more *comprehension* (mentalism-*fie!*) than natural selection itself. The capacity to improve one's design by operant conditioning is clearly a fitness-enhancing trait under many circumstances, but also risky, since the organism must blindly try out its options in the cruel world (as blindly as evolution does) and may succumb before it learns anything.

Better still is the next grade, the *Popperian creatures*, who extract information about the cruel world and keep it handy, so they can use it to pretest *hypothetical* behaviors offline, letting "their hypotheses die in their stead" as the philosopher of science Karl Popper once put it. Eventually they must act in the real world, but their first choice is not random, having won the generate-and-test competition trial runs in the internal environment model. Finally, there are the *Gregorian creatures*, named in honor of Richard Gregory, the psychologist who emphasized the role of thinking tools in providing thinkers with what he called "potential intelligence." The Gregorian creature's *Umwelt* is well stocked with thinking tools, both abstract and concrete: arithmetic and democracy and double-blind studies, and microscopes, maps, and computers. A bird in a cage may see as many words every day (on newspaper lining the cage floor) as a human being does, but the words are not thinking tools in the bird's *Umwelt*.

The merely Darwinian creature is "hard-wired," the beneficiary of clever designs it has no need to understand. We can expose its cluelessness by confronting it with novel variations on the conditions it has been designed by evolution to handle: it learns nothing and flounders helplessly. The Skinnerian creature starts out with some "plasticity," some optionality in a repertoire of behaviors that is incompletely designed at birth; it learns by trial-and-error forays in the world and is hard-wired to favor the forays that have "reinforcing" outcomes. It doesn't have to understand why it now prefers these tried-and-true behaviors when it does; it is the beneficiary of this simple design-improvement ratchet, its own portable Darwinian selection process. The Popperian creature looks before it leaps, testing candidates for action against information about the world it has stored in its brain somehow. This looks more like comprehension because the selective process is both information-sensitive and forward-looking, but the Popperian creature need not understand how or why it engages in this pretesting. The "habit" of "creating forward models" of the world and using them to make decisions and modulate behavior is a fine habit to have, whether or not you understand it. Unless you were a remarkably self-reflective child, you "automatically" engaged in Popperian lookahead and reaped some of its benefits long before you noticed you were doing it. Only with the Gregorian creature do we find the deliberate introduction and use of thinking tools, systematic exploration of possible solutions to problems, and attempts at higher-order control of mental searches. Only we human beings are Gregorian creatures, apparently.

Here is where the hot button of human exceptionalism gets pushed, with fierce disagreements between romantics and killjoys (see chapter 1) about how much comprehension is exhibited by which species or which individual animals. The prevailing but still tentative conclusion these days among researchers in animal intelligence is that the smartest animals are not "just" Skinnerian creatures but Popperian creatures, capable of *figuring out* some of the clever things they have been observed to do. Corvids (crows, ravens, and their close kin), dolphins and other cetaceans, and primates (apes and monkeys) are the most impressive wild animals so far investigated, with dogs, cats, and parrots leading the pet parade. They engage in exploratory behavior, for instance, getting the lay of the land, and often making landmarks to ease the burden on their memories, stocking their heads with handy local information. They need not know that this is the rationale for their behavior, but they benefit from it by reducing uncertainty, extending their powers of anticipation ("look before you leap" is the free-floating maxim of their design), and thereby improving their competences. The fact that they don't understand the grounds of their own understanding is no barrier to calling it understanding, since we humans are often in the same ignorant state about how we manage to figure out novel things, and that is the very hallmark of understanding: the capacity to apply our lessons to new materials, new topics.

Some animals, like us, have something like an inner workshop in which they can engage in do-it-yourself understanding of the pre-fabricated designs with which they were born. This idea, that the individual organism has a portable design-improvement facility that is more powerful than brute trial-and-error-and-take-your-lumps, is, I submit, the core of our folk understanding of understanding. It doesn't depend on any assumptions about *conscious experience*, although that is a familiar decoration, an ideological amplification, of the basic concept. We are slowly shedding the habit of thinking that way, thanks in part to Freud's championing of unconscious motivations and other psychological states, and thanks also to cognitive science's detailed modeling of unconscious processes of perceptual inference, memory search, language comprehension, and much else. An *unconscious mind* is no longer seen as a "contradiction in terms"; it's the *conscious* minds that apparently raise all the problems. The puzzle today is "what is consciousness *for* (if anything)?" if unconscious processes are fully competent to perform all the cognitive operations of perception and control.

To summarize, animals, plants, and even microorganisms are equipped with competences that permit them to deal appropriately with the affordances of their environments. There are free-floating rationales for all these competences, but the organisms need not appreciate or comprehend them to benefit from them, nor do they need to be conscious of them. In animals with more complex behaviors, the degree of versatility and variability exhibited can justify attributing a sort of behavioral comprehension to them so long as we don't make the mistake of thinking of comprehension as some sort of stand-alone talent, a source of competence rather than a manifestation of competence.

In part II, we zero in on the evolution of us Gregorian creatures, the reflective users of thinking tools. This development is a giant leap of cognitive competence, putting the human species in a unique niche, but like all evolutionary processes it must be composed of a series of unforeseen and unintended steps, with "full" comprehension a latecomer, not leading the way until very recently.

FROM EVOLUTION TO INTELLIGENT DESIGN

6

What Is Information?

As the Chinese say, 1001 words is worth more than a picture.
—John McCarthy

Welcome to the Information Age

If we were told we were living in the Age of Analysis, we might wonder just which kind of analysis was being celebrated. Psychoanalysis? Chemical analysis? Statistical analysis? Conceptual analysis? That term has many uses. We live in the Information Age, say the pundits, and people seem to agree, but few notice that this term also has several different meanings. *Which* Information Age: The age of megabytes and bandwidth? *Or* the age of intellectual and scientific discovery, propaganda and *dis*information, universal education, and challenges to privacy? The two concepts of information are closely related and are often inextricably intertwined in discussion, but they can be made distinct. Consider a few examples:

1. The brain is an information-processing organ.
2. We are *informavores*. (George Miller, psychologist)
3. "The information is in the light." (J. J. Gibson, ecological psychologist, 1966)
4. The task of a nervous system is to extract information from the environment to use in modulating or guiding successful behavior.
5. We are drowning in information.
6. We no longer can control our personal information.

7. The task of the Central Intelligence Agency is to gather information about our enemies.

8. *Humint*, intelligence or information gathered by human agents in clandestine interaction with other human beings, is much more important than the information obtainable by satellite surveillance and other high-tech methods.

Claude Shannon's mathematical theory of information (Shannon 1948; Shannon and Weaver 1949) is duly celebrated as the scientific backbone that grounds and legitimizes all the talk about information that engulfs us, but some of that talk involves a different conception of information that is only indirectly addressed by Shannon's theory. Shannon's theory is, at its most fundamental, about the statistical relationship between different states of affairs in the world: What can be gleaned (in principle) about state A from the observation of state B? State A has to be causally related somehow to state B, and the causal relationship can vary in something like richness. Shannon devised a way of *measuring* information, independently of what the information was *about*, rather like measuring *volume* of liquid, independently of which liquid was being measured. (Imagine someone bragging about owning lots of quarts and gallons and not having an answer when asked, "Quarts of what—paint, wine, milk, gasoline?") Shannon's theory provides a way of breaking down information into uniform amounts—bits, bytes, and megabytes—that has revolutionized all systems of storing and transmitting information. This demonstrates the power of *digitization* of information by computers, but just as wine can be valuable whether or not it is bottled in uniform liters, information (in the brain, for instance) doesn't have to be digitized to be stored, transmitted, and processed.

We are living in the *Digital* Age now that CDs, DVDs, and cell phones have replaced LP phonograph records, and analog radio, telephone and television transmission, but the *Information* Age was born much earlier, when people began writing things down, drawing maps, and otherwise recording and transmitting valuable information they couldn't keep in their heads with high fidelity. Or we could place the beginning of the Information Age earlier, when people began speaking and passing on accumulated lore, history, and mythology. Or with some justice we could say the Information Age began over 530 million years ago, when eyesight evolved during the Cambrian Era, triggering an arms race of innovation in behavior and organs that could respond swiftly to the information gathered from the light. Or we could insist that the Information Age began when life began; even the simplest reproducing cells survived thanks to parts that functioned by discriminating differences in themselves and their immediate surroundings.

To distance these phenomena from our contemporary preoccupation with systems of information encoding, I will call them instances of *semantic* information, since we identify the information of interest to us on a particular occasion by specifying what it is *about* (events, conditions, objects, people, spies, products . . .). Other terms have been used, but "semantic information" is the term of choice. The information that Tom is tall is about Tom and his height, and the information that snow is white is about snow and its color. These are different items of semantic information (don't say "bits of information," because "bits," while perfectly good units, belong to the other, Shannon, sense of the term). Even before writing was widespread, people invented ways of improving their control of semantic information, using rhyme, rhythm, and musical tone to anchor valued formulations in their memories, mnemonic crutches we still encounter today, for example, Every Good Boy Deserves Fudge (the notes on the lines of the treble clef in musical notation), HOMES (Great Lakes Huron, Ontario, Michigan, Erie, and Superior), and My Very Educated Mother Just Served Us Nachos (it used to be: Nine Pies) for the order of the planets, starting at the Sun.

Shannon idealized and simplified the task of moving semantic information from point A to point B by breaking the task down into a sender and a receiver (two rational agents, note) with a channel between them and a preestablished or agreed-upon code, the alphabet or ensemble of permissible signals. The channel was susceptible to *noise* (which was anything that interfered with transmission, degrading the signal), and the task was to achieve reliable transmission that could overcome the noise. Some of the designs that accomplish this were already well understood when Shannon devised his theory, such as the Able Baker Charlie Dog Easy Fox . . . system of alphabet letters, used by the US Navy (superseded by the Alpha Bravo Charlie Delta Echo Foxtrot . . . system, the NATO Phonetic Alphabet in 1955) in voice radio transmission to minimize the confusion between the rhyming letters (in English) Bee, Cee, Dee, Eee, Gee, Pee, Tee, Vee, Zee, and so forth.

By converting all codes, including words in ordinary language, to a binary code (with an alphabet containing only two symbols, 0 and 1), Shannon showed how noise reduction could be improved indefinitely, and the costs (in terms of coding and decoding and slowing down the transmission speed) could be measured precisely, in *bits*, which is short for *binary digit*. Like the parlor game Twenty Questions, where only yes/no questions are permitted, all information transmission could be broken down into binary decisions, yes or no, 1 or 0, and the number of such decisions required to recover the message could be given a measure, in bits, of the amount of (Shannon) information in the message. "I'm thinking of a number between 0 and 8. What is it?" In the game of Twenty Questions, how many questions do you have to ask to be sure of answering this question? Not eight questions (Is it 0, is it 1, is it 2 . . . ?), but only three: Is it 4 or more? Is it 6 or more (or 2 or more, depending on the first answer)? Is it 7? Yes, yes, yes = 111 = 7 in binary notation; it takes three bits to specify a number between 0 and 8. A byte is eight bits, and a megabyte is eight million bits, so you can send a 2.5 megabyte monochrome bitmap picture file by playing Twenty Million Questions. (Is the first pixel white? . . .)

Shannon's information theory is a great advance for civilization because *semantic* information is so important to us that we want to be able to use it effectively, store it without loss, move it, transform it, share it, hide it. Informational artifacts abound—telephones, books, maps, recipes—and information theory itself began as an artifact for studying important features of those artifacts. What began as an engineering discipline has subsequently proven useful to physicists, biologists, and others not concerned with the properties of informational artifacts. We will touch lightly on some of these further applications of Shannon information, but our main quarry is semantic information.[23]

Shannon's mnemonic tricks, and their descendants, are not just good for sending information from one agent to another agent, but for "sending" information from an agent *now* to the same agent in the future. Memory can be conceived as an information channel, just as subject to noise as any telephone line. In principle, *digitization* could be done with an alphabet of three or four or seventeen or a million distinct signals, or in other ways (as we shall see in a later chapter), but binary coding proves for many reasons to be superior for most purposes and is always available. *Anything* can be coded (not perfectly, but to whatever degree of precision you like) in a scheme of 0s and 1s, or a scheme of 0s, 1s, and 2s, or . . . but the binary code is physically simpler to implement (on/off; high-voltage/low-voltage; left/right) so it has pretty much gone to fixation in human technology, although there is still competition among secondary codes composed of binary code. (ASCII code for printable characters has been surpassed by UTF 8, which includes ASCII as a subset, and HTML, the code used for websites, has two different color codes in use: HEX code and RGB triplets, for example.)

The task of translating or transducing the light and sound and other physical events that carry semantic information into the binary bit-string format (the format directly measurable as Shannon information) is a mature technology now, and there are many varieties of ADCs (analog-to-digital converters) that take continuous or analog variation in some physical event (an acoustic wave hitting a microphone, light striking a pixel in a digital camera, tem-

23 Colgate and Ziock (2010) provide a brief, useful summary of some of the history of definitions of information growing out of the work of Shannon and Weaver.

perature change, acceleration, humidity, pH, blood pressure, etc.) and transduce it into one string of bits or another. These devices are analogous to the sensitive cells that accomplish transduction on the outer input edges of the nervous systems: rods and cones in eyes, hair cells in ears, heat sensors, nociceptors for damage (pain), stretch sensors on muscles, and all manner of internal monitoring cells that feed the nervous system, including the autonomic nervous system.

The conversion in brains is not into bit strings but neuronal spike trains, voltage differences passing rather slowly—millions of times slower than bit strings moving in computers—from neuron to neuron. In 1943 (before there were any working digital computers!) the neuroscientist Warren McCulloch and the logician Walter Pitts proposed a way these neuronal signals might be operating. It seemed that when a spike train from one neuron arrived at another neuron, the effect on it was either excitatory (Yes!) or inhibitory (No!). If the receiving neuron had a *threshold* mechanism that could sum up the Yes votes and subtract the No votes and then trigger its own signal depending on the net result, it could compute a simple logical function (it could be an "AND-gate" or an "OR-gate" or a "NOT-gate," to take the simplest cases). If a cell's threshold could be raised or lowered by something in its history of inputs and outputs, then a neuron could "learn" something that changed its local behavior. McCulloch and Pitts proved that a network of these units could be wired up or "trained" to represent any proposition whatever, based on logical operations on its input.

This was an inspiring idealization, one of the great oversimplifications of all time, since while it has turned out that real neurons in interaction are *much* more complicated than the "logical neurons" defined by McCulloch and Pitts, they had demonstrated the logical possibility of a general purpose representing-and-learning-and-controlling network made out of units that performed simple, nonmiraculous, clueless tasks—a comprehender of sorts made of merely competent parts. Ever since then, the goal of computational neuroscience has been to determine just which, of the infinite variety of more complicated networks, are operating in nervous systems. The wiring diagram of the nematode worm, *C. elegans*, with its 302 neurons of 118 varieties, is now just about complete, and its operation is becoming understood at the level of individual neuron-to-neuron actions. The Human Connectome Project aspires to make an equally detailed map of the tens of billions of neurons in our brains, and the Human Brain Project in Europe aspires to "simulate the complete human brain on supercomputers," but these mega-projects are in their early days. The brain is certainly not a digital computer running binary code, but it is still a kind of computer, and I will have more on this in later chapters.

Fortunately, a lot of progress has been made on understanding the computational architecture of our brains at a less microscopic level, but it depends on postponing answers to almost all the questions about the incredibly convoluted details of individual neuron connectivity and activity (which probably differs dramatically from person to person, in any case, unlike the strict uniformity found in *C. elegans*). Learning that a specific small brain region (containing millions of neurons) is particularly active when looking at faces (Kanwisher 1997, 2013) is a valuable breakthrough, for instance. We know that *information about faces* is somehow being processed by the activity of the neurons in the *fusiform face area* even if we as yet have scant idea about just what they are doing and how they are doing it. This use of the term "information," which is ubiquitous in cognitive science (and elsewhere), does *not* refer to Shannon information. Until an encoding scheme—not necessarily a binary (0 and 1) scheme—is proposed that digitizes the ensemble of possibilities, there is no basis for distinguishing signal from noise, and no way of measuring amounts of information. Some day in the future we may find that there is a natural interpretation of transmission in the nervous system that yields a measure of bandwidth or storage capacity, in bits, of an *encoding* of whatever is being transmitted, processed, and stored, but until then the concept of information we use in cognitive science is semantic information, that is, information *identified as* being *about* something specific: faces, or places, or glucose, for instance.

In other words, cognitive scientists today are roughly in the position that evolutionary theorists and geneticists were in before the analysis of the structure of DNA: they knew that information *about* phenotypic features—shapes of bodily parts, behaviors, and the like—was somehow being transmitted down through the generations (via "genes," whatever they were), but they didn't have the ACGT code of the double helix to provide them with a Shannon measure of *how much* information could be passed from parent to offspring as genetic inheritance. Some thinkers, perhaps inspired by DNA, think that there *must be* an encoding in the nervous system like the DNA code, but I have never seen a persuasive argument for this, and as we will soon see, there are reasons for skepticism. Semantic information, the concept of information that we must start with, is remarkably independent of encodings, in the following sense: two or more observers can acquire the *same* semantic information from encounters that share no channel.[24] Here is a somewhat contrived example:

> Jacques shoots his uncle dead in Trafalgar Square and is apprehended on the spot by Sherlock. Tom reads about it in the *Guardian* and Boris learns of it in *Pravda.* Now Jacques, Sherlock, Tom, and Boris have had remarkably different experiences, but there is one thing they share: semantic information to the effect that a Frenchman has committed a murder in Trafalgar Square. They did not all *say* this, not even "to them-

---

24   Giulio Tononi (2008) has proposed a mathematical theory of consciousness as "integrated information" that utilizes Shannon information theory in a novel way and has a very limited role for *aboutness*: it measures the amount of Shannon information a system or mechanism has about its own previous state—that is, the states of all its parts. As I understand it, Tononi's theory presupposes a digital, but not necessarily binary, system of encoding, since it has a countable repertoire of output states.

selves"; *that proposition* did not, we can suppose, "occur to" any of them, and even if it had, it would have had very different import for Jacques, Sherlock, Tom, and Boris. They share no encoding, but they do share semantic information.

How can we characterize semantic information?

The ubiquity of ADCs in our lives, now taken for granted in almost all informational transmissions, probably plays a big role in entangling Shannon's mathematical concept of information with our everyday concept of semantic information in spite of numerous alerts issued. A high-resolution color photograph of confetti on a sidewalk, broken down into eight million pixels, might fill a file ten times larger than a text file of, say, Adam Smith's *The Wealth of Nations*, which can be squeezed into two megabytes. Depending on the coding systems you use (GIF or JPEG or . . . Word or PDF or . . .) a picture can be "worth a thousand words," measured in bits, but there is a better sense in which a picture can be worth a thousand words or more. Can that sense be formalized? Can semantic information be quantified, defined, and theorized about? Robert Anton Wilson, an author of science fiction and writer on science, proposed the Jesus unit, defined as the amount of (scientific) information known during the lifetime of Jesus. (Scientific information is a subset of semantic information, leaving out all the semantic information then available about who lived where, what color somebody's robe was, what Pilate had for breakfast, etc.) There was exactly one Jesus of scientific information in AD 30, by definition, an amount that didn't double (according to Wilson) until the Renaissance 1,500 years later. By 1750 it doubled again to 4 Jesus, and doubled to 8 Jesus in 1900. By 1964 there were 64 Jesus, and Lord knows how many Jesus (Jesuses?) have accumulated in the meantime. The unit hasn't caught on, thank goodness, and while Wilson is undoubtedly right to dramatize the information-explosion theme, it is far from clear that any scale or measure of the amount of (scientific) information will be an improvement on such precise-but-tangential measures as number of peer-reviewed journal pages or megabytes of text and data in online journals.

Luciano Floridi, in his useful primer (2010), distinguishes *economic* information as whatever is *worth some work*. A farmer is wise to take the time and trouble to count his cows, check the level of the water in the well, and keep track of how efficiently his farm-hands labor. If he doesn't want to do this supervisory work himself, he ought to hire somebody to do it. It "pays you" to find out something about your products, your raw materials, your competition, your capital, your location, . . . to take the obvious categories only. You can canvass public information about markets and trends, and buy examples of your competition's products to reverse engineer and otherwise test against your own products, or you can try to engage in industrial espionage. Trade secrets are a well-established legal category of information that can be stolen (or carelessly leaked or given away), and the laws of patent and copyright provide restraints on the uses others can make of information developed by the R&D of individuals and larger systems. Economic information is valuable, sometimes very valuable, and the methods of preserving information and protecting it from the prying eyes of the competition mirror the methods that have evolved in Nature for the same purpose.

The mathematical theory of games (von Neumann and Morgenstern 1944) was another brilliant wartime innovation. It highlighted, for the first time, the cardinal value of keeping one's intentions and plans secret from the opposition. Having a poker face isn't just for playing poker, and too much transparency is quite literally death to any individual or organization that has to compete in the cruel world. Survival, in short, depends on information and, moreover, depends on differential or asymmetric information: I know some things that you don't know, and you know some things that I don't know, and our well-being depends on keeping it that way. Even bacteria—even nonliving viruses—engage in devious ruses to conceal or camouflage themselves in an arms race with their intrusive, inquisitive competitors.[25]

So let's consider, as a tentative proposal, defining semantic information as *design worth getting*, and let's leave the term "design" as noncommittal as possible for the time being, allowing only for the point I stressed in chapter 3, that design without a designer (in the sense of an intelligent designer) is a real and important category. Design always involves R&D work of one sort or another, and now we can say what kind of work it is: using available *semantic information* to improve the prospects of something *by adjusting its parts in some appropriate way*. (An organism can improve its prospects by acquiring energy or materials—food, medicine, a new shell—but these are not *design* improvements. Rather, they are just instances of refueling or rebuilding one's existing design.)[26] One can actually improve one's design as an agent in the world by just *learning useful facts* (about where the fishing is good, who your friends are). Learning how to make a fishhook, or avoid your enemies, or whistle is another kind of design improvement. All learning—both learning what and learning how—can be a valuable supplement or revision to the design you were born with.

Sometimes information can be a burden, an acquisition that interferes with the optimal exercise of the design one already has, and in these cases we have often learned to shield ourselves from such unwanted knowledge. In double-blind experiments, for example, we go to great lengths to keep both subjects and investigators ignorant of

25 For instance, the Ebola virus mimics an apoptotic cell fragment in order to get itself "eaten" by a phagocytic (trash-collecting) cell, thereby hitching a safe ride in a cell that will carry it around in the body (Misasi and Sullivan 2014). There are many well-studied examples of viral and bacterial camouflage and mimicry, and biotechnologists are now copying the strategy, masking nano-artifacts that would otherwise be attacked by the immune system.

26 Suppose *bigger teeth* would be a design improvement in some organism; the raw materials required, and the energy to move the raw materials into position, do not count as semantic information, but the developmental controls to accomplish this redesign do count.

which subjects are in which conditions so that biased behavior by sub-jects or interpretation by observers is made almost impossible. This is a fine example of the power of our hard-won reflective knowledge being harnessed to improve our future knowledge-gathering abilities: we have discovered some of the limits of our own rationality—such as our inability in some circumstances to avoid being unconsciously swayed by too much information—and used that knowledge to create systems that correct for that flaw. A more dramatic, if rarer, variety of unwanted knowledge is blocked by the policy of randomly issuing a blank cartridge (or a cartridge loaded with a wax "dummy") to some of the shooters in a firing squad—and letting them know that this is the policy—so that no shooter has to live with the knowledge that their action caused the death. (Less often noted is the fact that issu-ing blanks to some shooters removes an option that might otherwise be tempting on some occasions: to rebel, turning one's rifle on the officer in command. If you knew you had a live round, that oppor-tunity would be available for informed rational choice. You can't use information that you don't have.)

What about misinformation, which can also accrue, and disinfor-mation, which is deliberately implanted in you? These phenomena seem at first to be simple counterexamples to our proposed defini-tion, but, as we will soon see, they really aren't. The definitions of "semantic information" and "design" are linked in circular defini-tions, but it's a virtuous, not vicious, circle: some processes can be seen to be R&D processes in which some aspect of the environment is singled out somehow and then exploited to improve the design of some salient collection or system of things in the sense of better equipping it for thriving/persisting/reproducing in the future.

Semantic information, then, is "a distinction that makes a dif-ference." Floridi (2010) reports that it was D. M. MacKay who first coined this phrase, which was later enunciated by Gregory Bateson (1973, 1980) and others as *a difference that makes a difference*. MacKay was yet another of the brilliant theoreticians who emerged from the intense research efforts of World War II alongside Turing and Shan-non (and von Neumann and John McCarthy, among others). He was a pioneer information theorist, physicist, neuroscientist, and even a philosopher (and one of my personal heroes, in spite of his deep religious convictions).[27] My proposed definition of semantic information is close to his 1950 definition of "information in general as that which *justifies representational activity*" (MacKay 1968, p. 158). MacKay's focus at the time was on what we now call Shannon information, but he ventured wise observations on information in the more fundamental, semantic sense, also defining it as *that which determines form* (1950, diagram, p. 159 of MacKay 1968), which points away from representation (in any narrow sense) while keeping the theme of justification or value in place.

If information is a distinction that makes a difference, we are invited to ask: A difference to whom? *Cui bono?* Who benefits?—the question that should always be on the lips of the adaptationist since the answer is often surprising. It is this that ties together economic information in our everyday human lives with biological information and unites them under the umbrella of semantic information. And it is this that permits us to characterize misinformation and disinformation as not just kinds of information but *dependent* or even *parasitic* kinds of information as well. Something emerges as *mis*information only in the context of a system that is designed to deliver—and rely on— *useful* information. An organism that simply ignores distinctions that might mislead (damage the design of) another organism has not been misinformed, even if the distinction registers somehow (ineffectually) on the organism's nervous system. In Stevie Smith's poem, "Not Waving but Drowning" (1972), the onlookers on the beach who waved back were misinformed but not the seagulls wheeling overhead. We can't be misinformed by distinctions we are not equipped to make. Understanding what disinformation is benefits doubly from our asking *cui bono?* Disinformation is the designed

27 The Ratio Club, founded in 1949 by the neurologist John Bates at Cambridge University, included Donald MacKay, Alan Turing, Grey Walter, I. J. Good, William Ross Ashby, and Horace Barlow, among others. Imagine what their meetings must have been like!

exploitation (for the benefit of one agent) of another agent's systems of discrimination, which themselves are designed to pick up useful information and *use it*. This is what makes the Ebola virus's design an instance of camouflage.

Kim Sterelny (2015, personal correspondence) has raised an important objection:

> Humans are representational magpies—think of how incredibly rich forager fauna and floras are—much of their natural history information has no practical value. Once information storage (in the head and out of the head) became cheap, the magpie habit is adaptive, because it is so hard to tell in advance which of the informational odds and ends will turn out to be valuable. But that does not alter the fact that most of it will not.

He goes on to claim most of what anybody knows is "adaptively inert. But that does not matter, since it is cheap to store, and the bits that do matter, *really* matter." Setting aside the misleading use of the term "bits" here, I largely agree but want to issue one proviso: even the dross that sticks in our heads from the flood that bombards us every day has its utility profile. Much of it sticks because it is *designed to stick*, by advertisers and propagandists and other agents whose interests are served by building outposts of recognition in other agents' minds, and, as Sterelny notes, much of the rest of it sticks because it has some non-zero probability (according to our unconscious evaluations) of being adaptive someday. People—real or mythical—with truly "photographic memories" are suffering from a debilitating pathology, burdening their heads with worse than useless information.[28]

Evolution by natural selection churns away, automatically extract-

---

28 Science has hugely expanded our capacity to discover differences that make a difference. A child can learn the age of a tree by counting its rings, and an evolutionary biologist can learn roughly how many million years ago two birds shared a common ancestor by counting the differences in their DNA, but these pieces of information about duration are not playing any role in the design of the tree or the bird; the information is not *for* them but has now become information for us.

ing tiny amounts (not "bits") of information from the interactions between phenotypes (whole, equipped organisms) and their surrounding environments. It does this by automatically letting the better phenotypes reproduce their genes more frequently than the less favored.[29] Over time, designs are "discovered" and refined, thanks to these encounters with information. R&D happens, designs are improved because they all have to "pay for themselves" in differential reproduction, and Darwinian lineages "learn" new tricks by adjusting their *form*. They are, then, in-*formed*, a valuable step up in local Design Space. In the same way, Skinnerian, Popperian, and Gregorian creatures *inform* themselves during their own lifetimes by their encounters with their environments, becoming ever more effective agents thanks to the information they can now *use* to do all manner of new things, including developing new ways of further informing themselves. The rich get richer. And richer and richer, using their information to refine the information they use to refine the information they obtain by the systems they design to improve the information available to them when they set out to design something.

This concept of useful information is a descendant of J. J. Gibson's concept of affordances, introduced in chapter 5. I want to expand his account to include not only the affordances of plants and other nonanimal evolvers but also the artifacts of human culture. Gibson says "the information is in the light" and it is by "picking up" the information that animals perceive the world.[30] Consider the reflected sunlight falling on the trunk of a tree, and on the squirrel clinging to the trunk. The same *potential* information is in the light for both tree and squirrel, but the tree is not equipped (by earlier

29   Colgate and Ziock (2010) defend a definition of information as "that which is selected," which is certainly congenial to my definition, but in order to make it fit the cases they consider, the term "selected" has to be allowed to wander somewhat.
30   Notoriously, Gibson doesn't just ignore the question of what internal machinery manages to do this pick-up; he often seems to deny that there are any difficult questions to answer here. The slogan of radical Gibsonians says it all: "It's not what's in your head; it's what your head is in." I am not endorsing this view.

R&D in its lineage) to make as much from *the information in the light* as the squirrel does. The tree does use the *energy* in the light to make sugars, by photosynthesis, and it has recently been shown that trees (and other plants) are equipped to respond appropriately to light-borne information as well: for instance, to determine whether to germinate, break dormancy, lose their leaves, and when to flower.[31]

We can think about potential utility if we like: Suppose a man with a chain saw is approaching, visible to any organism with eyes but not to the tree. Eyes are of no use to a tree unless it also has some way of using the information (if not to run and hide, perhaps to drop a heavy limb on the lumberjack, or secrete some sticky sap that will gum up the saw). The presence of the information in the light might someday "motivate" a trend in the direction of eyesight to trees, if a behavioral payoff were nearby! Unlikely, but such unlikely convergences are the heart of evolution. There has to be a difference that could make a difference if only something were present to hinge on it somehow. As noted earlier, evolution by natural selection is astonishingly good at finding needles in haystacks, almost invisible patterns that, when adventitiously responded to, yield a benefit to the responder. Just as the origin of life depends on getting the right "feedstock" molecules in the right place at the right time, there has to be raw material in the variation in the population that includes, by coincidence, some heretofore functionless (or underutilized or redundant or vestigial) feature that happens to be heritable and that covaries with the potentially useful information in the world.[32]

31   Thanks to Kim Sterelny and David Haig for drawing these facts to my attention.

32   Useful information emerges even at the molecular level. David Haig, in his fascinating essay "The Social Gene" (1997) exploits the agential perspective all the way down to what he calls the *strategic gene* concept. As he notes, "The origin of molecules that were able to discriminate between themselves and closely similar molecules greatly expanded the strategies available to genes and made possible the evolution of large multicellular bodies" (p. 294). When there is no information available to exploit, genes are unable to defect, or form coalitions, with better than chance success—an anticipation by Mother Nature of the firing squad principle.

Not everything "possible in principle" is automatically available, but given lots of time, and lots of cycles, there are *likely* to be paths of happenstance that lead to the Good Tricks in the neighborhood, but not always. That is why plausible "just-so stories" (Gould and Lewontin 1979) are only *candidate* explanations, in need of confirmation. Every well-confirmed evolutionary hypothesis (of which there are thousands) began as a just-so story in need of supporting evidence. And just as most organisms born die childless, the majority of just-so stories that get *conceived* never earn the right to reproduce. The sin of adaptationism is not conceiving of just-so stories—you can't do evolutionary biology without this—but uncritically reproducing just-so stories that haven't been properly tested.

Consider a less fantastical possibility than trees with eyes: brilliant autumn foliage. Is it an *adaptation* in trees? If so, what is it good for? It is commonly understood to be not an adaptation but merely a functionless byproduct of the chemical changes that occur in deciduous leaves when they die. The leaves stop making chlorophyll when the sunlight diminishes, and as the chlorophyll decomposes, other chemicals present in the leaves—carotenoids, flavonoids, anthocyanins—emerge to reflect the remaining light. Today, however, human beings, especially in New England, value the brilliant colors of autumn foliage, and—usually unconsciously—foster the health and reproduction of the most impressive trees by cutting down other trees first, saving the nice colors for another season, and giving them another season of reproduction. Having bright foliage is already an adaptation in the trees of northern New England, if not yet measurable directly. This is how adaptations start, imperceptible to all except the unblinking process of natural selection. There is considerable difference in how long deciduous trees hold their leaves in the autumn; in New England, the dull brown leaves of oaks are the last to fall, long after the brilliant maples have become bare. Any chemical changes that permitted some strains of maple to hold their foliage longer would become an adaptation in any environment where people exert a nonnegligible selective effect (knowingly or not). And now let's stretch our imaginations a few steps further: suppose that this foliage-prolongation capacity is itself quite energetically costly and only pays when there are people around to appreciate the colors. The evolution of a people-present-detector (more likely a pheromone sniffer than a rudimentary people-detecting eye) could roll onto the scene. This would be a step toward *self-domestication* by the tree lineage. Once offspring are favored, once reproduction is fostered or prevented by us, we're on the way to a domesticated species, alongside date palms and avocado trees. The opening moves need not be the result of conscious, deliberate, *intelligent* choice by us (and certainly not by the trees). We can, in fact, be entirely oblivious to the lineages that become *synanthropic*, evolving to thrive in human company without belonging to us or being favored by us. Bedbugs and mice are synanthropic, and so is crabgrass, not to mention the trillions of tiny things that inhabit our bodies, flying beneath the radar if possible, adapted to living in or on or next to the human body niche.

In all these cases, *semantic information about* how best to fit in has been mindlessly gleaned from the cycle of generations, and notice that it is not *encoded* directly in the organisms' nervous systems (if they have any) or even in their DNA, except by something akin to *pragmatic implication*. Linguists and philosophers of language use the term *pragmatics* to refer to those aspects of meaning that are not carried by the syntax and "lexical" meanings of the words but conveyed by circumstances of particular utterances, by the *Umwelt*, in effect, of an utterance.

If I burst into a house and yell to all assembled, "Put on the kettle!" I have uttered an imperative English sentence, but some will probably infer that I would like to have a cup of tea or other hot beverage, while another may further surmise that I feel myself at home here, and may in fact be the occupant of this house. Yet another person present, a monoglot Hungarian, may infer only that I speak English, and so does whomever I am addressing (well, it sounds like English to her), while somebody *really* in the know will be instantly informed that I have decided after all to steam open that sealed envelope and surreptitiously read the letter inside in spite of the fact that it isn't addressed to me; a crime is about to be committed. What semantic information can be gleaned from the event depends on what information the gleaner already has accumulated. Learning that somebody speaks English can be a valuable update to your world knowledge, a design improvement that may someday pay big dividends. Learning that somebody is about to commit a crime is also a valuable enhancement to anyone in a position to put it to good use. The local design improvements each can make on the basis of this interaction are wildly different, and it would be a big mistake to think the semantic information could all be extracted by looking closely at the structure of the signal as an acoustic wave *or* as an English sentence ("p-u-t-space-o-n-space . . ."). *There is no code* for all these different lessons learned.

Similarly, the DNA of a bird in a lineage that has "learned" to make a hanging nest when the time comes for breeding will not have codon sequences that describe the nest or how to build it step by step but will consist of sequences of imperatives along the lines of "then attach a lysine, then a threonine, then tryptophan, . . ." (the recipe for making one protein or another out of a string of amino acids) or "fiddle-de-dee-count-to-three" (a bit of "junk" DNA used as a timer), or just "blahblahblahblah" (a genomic parasite or other bit of junk). In any event, don't hope to "translate" any sequence of codons as "nest" or "twig" or "find" or "insert." Still, thanks to the particular string of codons inherited by the bird from its parents, and thanks to the developmental systems that had already "learned" in earlier evolutionary steps to interpret those codons, the know-how to build the hanging nest will have been transmitted from parent to offspring. The offspring inherit a manifest image with an ontology of affordances from their parents and are born ready to distinguish the things that are most important to them. Reflecting on how much one would have to know in the kettle case to understand the message "fully" is a good way of appreciating how impenetrable to analysis the transmission of know-how via DNA from one generation to the next is apt to be.

Linguists and philosophers of language have developed distinctions that might seem to tame the difficulties somewhat: there is the proposition *expressed* (e.g., "put on the kettle"), the proposition *implicated* (e.g., "I'm going to steam open the envelope"), and the proposition *justified* (e.g., "he speaks English") by the speech act.[33] I think, however, that we should resist the temptation to impose these categories from linguistics on DNA information transmission because they apply, to the extent that they do, only fitfully and retrospectively. Evolution is all about turning "bugs" into "features," turning "noise" into "signal," and the fuzzy boundaries between these categories are not optional; the opportunistic open-endedness of natural selection depends on them. This is in fact the key to Darwin's strange inversion of reasoning: creationists ask, rhetorically, "where does all the information in the DNA come from?" and Darwin's answer is simple: it comes from the gradual, purposeless, nonmiraculous transformation of noise into signal, over billions of years. Innovations must (happen to) have fitness-enhancing effects from the outset if they are to establish new "encodings," so the ability of something to convey semantic information cannot depend on its prior qualification as a code element.

There will be (so far as I can see) no privileged metric for saying *how much* semantic information is "carried" in any particular signal—either a genetic signal from one's ancestors or an environmental signal from one's sensory experience. As Shannon recognized, information is always relative to what the receiver already knows, and although in models we can "clamp" the boundaries of the signal and the receiver, in real life these boundaries with the surrounding context are porous. We will have to make do, I think, with disciplined hand-waving, relying on our everyday familiarity with how *we* can communicate so much to our fellow human beings by saying things and doing things. (Yes, *disciplined* hand-waving. There may not be any algorithms for measuring semantic information, but we can erect any number of temporary structures for *approximating* the information content of the topics of interest to

---

33   Thanks to Ron Planer for this suggestion.

us, relying on posited ontologies—the ontologies that furnish the *Umwelten* of organisms.) We already do this every day, of course, using our human categories and ontologies in a rough and ready way to say quite rigorously controlled things about what categories animals distinguish, what tasks they perform, what they fear and like and avoid and seek. For instance, if we set out to build a trap to catch a wily raccoon, we will want to pay close attention to the differences that are likely to make a difference to the raccoon. Scent is an obvious category to worry about but so is arranging the trap in such a way that the animal, on approaching it, can see (what appears to it to be) an independent escape route in addition to its point of entry. We may hope to identify chemically the scents that need to be masked or dissipated if we are to lure the raccoon into our trap, but the hallmarks of the affordance *independent escape route* are not going to be readily reducible to a simple formula.

We need to restrain ourselves from *assuming* what many theorists would like to assume: that if an organism is competent with the categories of *predator, edible, dangerous, home, mother, mate* . . . they must have a "language of thought" with terms in it for each of these categories. If DNA can convey information about how to build a nest without any terms for "build" and "nest," why couldn't a nervous system do something equally inscrutable?[34]

All evolution by natural selection is design revision, and for the most part it is design improvement (or at least design maintenance). Even the loss of organs and their functions counts as improvement when the cost of maintaining them is factored in. The famous cave

34 The definition of Colgate and Ziock (2010) has a further condition that I am explicitly denying: "To usefully select information, information must be stored (written); otherwise there is no way to decide what has been selected" (p. 58). It all depends on what "stored (written)" means. I would say that information about nest building is stored and transmitted in the bird lineage but not written down (as information about nest building). As Paul Oppenheim (personal correspondence) reminded me, F. C. Bartlett's classic *Remembering* (1932) warned against thinking of remembering as *retrieving* some thing that has been *stored* in some place ("the memory") in the brain.

fish that have abandoned vision are engaged in cost cutting, which any company executive will tell you is design improvement. Don't acquire and maintain what doesn't pay for itself. It is no accident that biologists often speak of a lineage as "learning" its instinctual behaviors over the generations, because all learning can be similarly seen to be processes of self-redesign, and in the default case, improvement of design. We see the acquisition of both know-how and factual information as learning, and it is always a matter of using the base of competence/knowledge you already have to exercise quality control over what you acquire. Forgetting is not usually considered learning any more than discarding is usually seen as design improvement, but sometimes (as in the case of the cave fish) it is. More is not always better. The legal distinction between flotsam and jetsam is apropos here: flotsam is cargo that has inadvertently or accidentally been swept off the deck or out of the hold of a ship, while jetsam is cargo that has been deliberately thrown overboard—jettisoned. Intentional mind-clearing, jettisoning information or habits that endanger one's welfare, is not an unusual phenomenon, sometimes called unlearning.[35]

Semantic information is not always valuable to one who carries it. Not only can a person be burdened with useless facts, but often particular items of information are an emotional burden as well—not that evolution cares about your emotional burdens, so long as you make more offspring than the competition. This doesn't cancel the link to utility in the definition of semantic information; it complicates it. (The value of gold coins is not put in doubt by the undeniable fact that pockets full of gold coins may drown a strong swimmer.) Still, defining semantic information as *design worth getting* seems to fly in the face of the fact that so much of the semantic information that streams into our heads every day is *not* worth get-

---

35 Robert Mathai notes (personal correspondence) that evolutionary unlearning is never really jetsam; it is flotsam that, over time, comes to resemble jetsam. It wasn't jettisoned with foresight, but its washing overboard proves to have been a good thing, saving the ship.

ting and is in fact a detestable nuisance, clogging up our control systems and distracting us from the tasks we ought to be engaged in. But we can turn this "bug" in our definition into a "feature" by noting that the very existence of information-handling systems depends on the design value of the information that justifies the expense of building them in the first place. Once in place, an information-handling system (a pair of eyes or ears, a radio, the Internet) can be exploited—parasitized—by noise of several species: sheer meaningless "random" *white noise* (the raspy "static" that interferes with your transistor radio when the signal is weak), and semantic information that is useless or harmful to the receiver. Spam and phishing e-mails on the Internet are obvious examples, both dust clouds and (deliberately released) squid ink are others. The malicious items depend for their effect on the trust the receiver invests in the medium. Since Aesop we've known that the boy who cries wolf stops commanding attention and credence after a while. Batesian mimicry (such as a nonpoisonous snake with markings that mimic a poisonous variety) is a similar kind of parasitism, getting a benefit without going to the cost of manufacturing poison, and when the mimics outnumber the genuinely poisonous snakes Aesop's moral takes hold and the deceitful signal loses its potency.

Any information-transmitting medium or channel can set off an arms race of deception and detection, but *within an organism,* the channels *tend* to be highly reliable. Since all "parties" have a common fate, sinking or swimming together, trust reigns (Sterelny 2003). (For some fascinating exceptions, see Haig 2008 on genomic imprinting.) Error is always possible, the result of simple breakdown—wear and tear—of the system, or misapplication of the system to environments it is ill equipped to handle. This is why delusions and illusions are such a rich source of evidence in cognitive neuroscience, providing hints about what is being *relied upon* by the organism in the normal case. It is often noted that the brain's job in perception is to filter out, discard, and ignore all but the noteworthy features of the flux of energy striking one's sensory organs. Keep and refine the ore of (useful) information, and leave all the noise out. Any nonrandomness in the flux is a real pattern that is *potentially useful* information for some possible creature or agent to exploit in anticipating the future. A tiny subset of the real patterns in the world of any agent comprise the agent's *Umwelt*, the set of its affordances. These patterns are the *things* that agent should have in its ontology, the things that should be attended to, tracked, distinguished, studied. The rest of the real patterns in the flux are just noise *as far as that agent is concerned*. From our Olympian standpoint (we are not gods, but we are cognitively head and shoulders above the rest of the creatures), we can often see that there is semantic information in the world that is intensely relevant to the welfare of creatures who are just unequipped to detect it. The information is indeed in the light but not for them.

Trade secrets, patents, copyright, and Bird's influence on bebop

My claims, so far, are these:

1. Semantic information is valuable—misinformation and disinformation are either pathologies or parasitic perversions of the default cases.
2. The value of semantic information is receiver-relative and not *measurable* in any nonarbitrary way but can be *confirmed* by empirical testing.
3. The *amount* of semantic information carried or contained in any delimited episode or item is also not usefully *measurable in units* but roughly comparable in local circumstances.
4. Semantic information need not be *encoded* to be transmitted or saved.

All these claims are clarified and supported when we turn to human "economic" information and consider the way human societies have enshrined these claims in their laws and practices. Consider a case of stealing a trade secret. Your competitor,

United Gadgets, has developed a new widget, an internal component of its very powerful new strimpulizer, but you can't get a
good look at it, since it is "potted" in an x-ray-opaque casing that
can only be cracked open or dissolved by something that destroys
the widget in the process. A well-kept secret, indeed. You go to
great lengths to embed a spy in United Gadgets, and she eventually encounters a bare widget. Almost home. Now how to get the
information out? A plaster piece-mold cast, a negative, would be
wonderful but too bulky to sneak out. Drawings, photographs,
or blueprints would be good, but also hard to carry out, since
security is fierce and radio signals closely monitored. A very precise recipe, in English words, for making the device, might be
encrypted and then concealed in an otherwise innocuous message, a rambling memo about health insurance options and their
many stipulations, for instance.

Another recipe system would be a CAD-CAM file; put the widget in a CAT-scanner (a computer-aided tomography machine) and
obtain a suitably high-resolution slice-by-slice tomographic representation to use as a recipe for your 3D printer. Depending on
the resolution, this could be ideal, providing, in the limit, a recipe
for making an atom-for-atom duplicate (a fantasy much fancied
by philosophers and other fans of teleportation). One virtue of
this extreme variation is that it arguably produces the maximally
detailed specification of the widget in a file whose size *can* be measured in bits. You can send "perfect" information about the widget,
specifying its every atom, in a file of only umpteen million zetabytes.
Today a widget, tomorrow the world. The idea, generalized, of the
whole universe being exhaustively (?) describable in one cosmic bitmap lies at the heart of various largely speculative but fascinating
proposals in physics. And of course it doesn't stop at big old atoms, a
relatively "low-res" recipe for reality these days. Such an application
of Shannon information theory does permit, "in principle" but not
remotely in practice, saying exactly how much (Shannon) information there is in the cubic meter of ocean and ocean floor surrounding a particular clam, for instance, but it says nothing about how

much of this information—a Vanishingly small portion—is semantic information for the clam.[36]

Back to the theft of the widget design. The CAD-CAM file, of suitably high resolution, could be stored on a tiny digital memory device that might be safely swallowed. But if she can't arrange to do the tomography, your spy may simply study the widget intensely, turning it every which way, hefting and twisting it, sniffing and tasting it, and so forth, and then memorize the shape somehow and carry the information out in her brain. (Note that this is the way most secrets get moved around: attentive observation followed by remembering.) Probably the best way to steal the information, if you can get away with it, is to borrow a widget, take it home, examine and record it in any way you like so you can make a copy that meets your requirements, and then return it to United Gadgets. *All you took was the information you needed.*

The more your spy already knows about good widgets, the less Shannon information she has to transmit or transport from United Gadget to your company's R&D department. Perhaps she can tell at a glance that the size and shape of the output orifice is the only novelty that matters in these circumstances; it is the only opportunity for design improvement that is worth the espionage. This example gives us a clearer picture of the relationship between semantic information and Shannon information. Shannon's idealized restriction to a sender and receiver presupposes, in effect, that the sender has already found the needle in the haystack that would be valued by the receiver. Finding the needle, detecting the pattern that can be exploited, is backstage, not part of the model, and so is the receiver's task of finding an appropriate use for what is

---

36   I coined the term *Vast* (with a capital *V*) as an abbreviation of *Very much more than ASTronomically* to stand for *finite* but almost unimaginably larger numbers than such merely astronomical quantities as the number of microseconds since the Big Bang times the number of electrons in the visible universe (Dennett 1995, p. 109). The Library of Babel is finite but Vast. *Vanishing* is its reciprocal (like *infinitesimal* in relation to *infinite*).

received, even though this research and development is what "pays for" all information transmission.

Suppose your spy succeeds. The information you acquired enables you to improve the design of your own strimpulizer, and thereby improve your market share and make a living. United Gadgets finds out (you didn't bother potting yours), and sues you—or worse, has you arrested for industrial espionage. It can be beyond reasonable doubt that you stole the design of the widget, even if the prosecution cannot definitely establish how you did it. If the widget is that unusual, the case will be just like plagiarism, which can be proven on the basis of design replication (with improvements and variations) alone. In fact, if United Gadgets anticipates theft from the outset, they would be wise to incorporate a distinctive but non-functional knob or pit or slot in its design, which would be a dead giveaway of theft if it appears in your version. (This is the tactic long used by encyclopedias to catch competitors copying their entries illicitly; a fictitious animal or poet or mountain that appeared with roughly the same description in both books would be hard for the culprit to explain. Look up Virginia Mountweazel in Google for details.) Notice that this telltale feature carries information of great utility to the original "sender" only so long as the original receiver does not recognize it as an intended signal (like the signal of the injury-feigning bird), while the copier unwittingly "sends" a self-damaging message back, conveying the information that it has illicitly copied the original.

These regularities, these strategic patterns in the interactions between agents, depend on *which* information is copied, not just on *how much* information is copied, so while Shannon's measure can be applied as a limiting condition, it cannot explain the free-floating rationales of the ploys and counterploys. Finding telltale information in biological systems, information that we can conclude *must be there* even when we haven't yet any detailed knowledge of how the information is "encoded" or embodied, has applications in many areas of biology. For instance, my colleague Michael Levin (2014; Friston, Levin, et al. 2015) has been developing models of morphogenesis that treat "patterning systems as primitive cognitive agents" (simple intentional systems). Neurons aren't the only cells with "knowledge" and "agendas" (see chapter 8).

We can learn another lesson, I think, from the law of patent and copyright. First, these laws were enacted to protect designs, in this case designs created by intelligent designers, people. People are often designers, and designing takes time and energy (and a modicum of intelligence unless you are an utter trial-and-error plodder, an R&D method which almost never bears interesting fruit except over evolutionary time). The designs that result are typically valuable (to somebody), so laws protecting the owners/creators of these designs are reasonable.

A few features of these laws stand out. First, you have to demonstrate the *utility* of your brainchild to get a patent. And you have to demonstrate that nobody else has already invented it. How useful and original does it have to be? Here is where there is ineliminable hand-waving in the law. For instance, the Canadian law of patent excludes as not sufficiently novel (to warrant a patent) any invention for which there has been *anticipation*, which can be demonstrated if any of eight conditions obtains (Wikipedia, s.v. "novelty [patent]"). Two of the conditions are officially described thus: anticipation may

> convey information so that a person grappling with the same problem must be able to say "that gives me what I wish."

> give information to a person of ordinary knowledge so that he must at once perceive the invention.

It should not surprise us that patent law has this awkward issue of definition. Novelty, like semantic information in general, depends very directly and heavily on the competence of the parties involved. In the Land of the Dunces you could patent a brick as a doorstop; in Engineers' Heaven, a flying house powered by sunlight might be regarded as a trivial extension of existing knowledge and practices. What can be seen "at once" by a "person of ordinary knowledge" will vary with what counts as ordinary knowledge at any time and place.

You can patent an idea (for a process, a gadget, a tool, a method) without actually building a working prototype, though it certainly is advisable to build one, to support your sketches and descriptions. You can't copyright an idea, but only a particular expression of an idea. A song can be copyrighted, but probably not a four-note sequence. If Beethoven were alive today, could he copyright the first four notes of his Fifth Symphony (Ta-ta-ta-DAH)? Has NBC copyrighted its three-note chime or is that just a trademark—a different legally protected informational item? A book title by itself cannot be copyrighted. A short poem can. How about this one:

> This verse
>
> is terse.

It is no doubt covered by the copyright on this entire book. As a stand-alone "literary work," I wonder. Copyright law has been amended many times, and vexed issues continue to be problematic. Books and articles, music and painting, drawing, sculpture, choreography, and architecture, can be copyrighted *as long as there has been some fixed expression.* An improvised jazz solo line in an unrecorded performance cannot be copyrighted, nor can an unchoreographed and unfilmed dance routine. This is in line with Colgate and Ziock's definition (see fn. 29, p. 119) of information but seems driven by legal requirements of evidence, not by any natural requirement for information to be "stored (written)." Charlie ("Bird") Parker couldn't copyright his solos, but many saxophone players and other jazz musicians who heard them were heavily influenced by him, which is to say: valuable semantic information flowed from his performances to theirs (and not to the tin ears in his audiences who couldn't pick up the affordances made available). You can't copyright an idea or discovery, but "where do we draw the line?" As Judge Learned Hand once said: "Obviously, no principle can be stated as to when an imitator has gone beyond copying the 'idea,' and has borrowed its 'expression.' Decisions must therefore inevitably be ad hoc" (*Peter Pan Fabrics, Inc. v. Martin Weiner Corp.*, 274 F.2d 487 [2d Cir. 1960], cited in Wikipedia, s.v. "copyright").

Interestingly, when considering whether an item is copyrightable, utility or function works *against* a creation, since copyright is intended to protect "artistic" creation, which must be "conceptually separable" from functional considerations, where the law of patent applies (under more stringent conditions). This is an interestingly truncated conception of "function," since *aesthetic* effects are obviously functional in most if not all contexts. This reminds me of the myopic rejection by some early evolutionists of Darwin's ideas about sexual selection, since it implied—mistakenly in their eyes—that the perception of beauty could have a functional role to play. But of course it does; females have evolved heightened competence in discernment of good properties in prospective mates while the males have evolved ever more impressive (aesthetically impressive, good for nothing but show) displays. Mating successfully is not an optional adventure in life; it's the finish line, the goal, and whatever it takes to hit the tape is functional whatever cost or burden it places on its bearer.[37] The law of copyright tries to separate "utilitarian" function from aesthetic function, and while there are good legal reasons for trying to make this a bright-line distinction, there are good theoretical reasons for seeing this as an ad hoc undertaking, as Learned Hand observes about the "idea"/"expression" distinction.

---

37 I find that some people resist the quite obvious free-floating rationale for why it is females that do the evaluating and males that do the strutting and costly advertising: asymmetrical parental investment. In species where females normally invest more time and energy per offspring than males do (making eggs versus making sperm, lactating, incubating, rearing, and so forth) the females should be more choosy when accepting a mate. A female can make only so many eggs or bear only so many litters of young, and if she chooses to mate with a second-rate male she may waste most or all her bearing capacity, while a male who chooses to mate with a second-rate female has lost only a little of his precious time and can always make more sperm. In species where parental investment is roughly equal, the males and females look and act pretty much the same.

When we turn to cultural evolution we will encounter this sort of codeless transmission on a grand scale. As I have often noted, a wagon with spoked wheels doesn't just carry grain or freight from place to place; it carries the brilliant idea of a wagon with spoked wheels. The wagon no more carries this idea—this information— to the dog in the road than the sunlight carries the information about the approaching lumberjack to the tree. You have to be informed to begin with, you have to have many competences installed, before you can avail yourself of this information, but it is there, embodied in the vehicle that rolls by. One of the tasks remaining for us is to understand how come we human beings are so much better at extracting information from the environment than any other species.

Are we really better? Beyond a doubt, if we measure our prowess in numbers of affordances. In addition to those we share with our mammalian cousins (water to drink, food to eat, holes to hide in, paths to follow, . . .) there are all our recognizable, familiar artifacts. A hardware store is a museum of affordances, with hundreds of different fasteners, openers, closers, spreaders, diggers, smoothers, reachers, grippers, cutters, writers, storers, and so forth, all recognizable and usable in appropriate circumstances, including novel circumstances in which we invent and construct new affordances ad lib, using predesigned, premanufactured parts. Richard Gregory, whose reflections on intelligence and equipment inspired me to name Gregorian creatures after him, emphasized that it not only took intelligence to use a pair of scissors, the pair of scissors *enhanced* the "intelligence" of users by giving them a large increase in available competence. These tools, like the hermit crab's shell, the bird's nest, the beaver's dam, are acquired design improvements, but they are not part of our *extended phenotypes* (Dawkins 1982, 2004b); the *general* talent to recognize such things and put them to good use is the phenotypic feature that is transmitted by our genes.

I sometimes suggest to my students that evolution by natural selection is nothing but "universal plagiarism": if it's of use to you, copy it, and use it. All the R&D that went into configuring whatever it is that you copy is now part of your legacy; you've added to your wealth without having to "reinvent the wheel." That's what Nature has been doing for billions of years, refining and articulating and spreading billions of good design features into every corner of the planet. This tremendous creativity would not be possible without almost unimaginable amounts of copying. Nature's Good Trick is against the law, for us, and there is a good reason for this: semantic information is costly to create and valuable, so unauthorized copying is theft. It is worth noting that this is not a free-floating rationale. The first laws of patent and copyright (and trade secret and trademark) were devised and enacted following extended, explicit, rational debate and discussion about the need for them. They themselves are products of intelligent design, designed to protect other intelligent designs.

Shannon information provides us with the mathematical framework for distinguishing signal from noise and for measuring capacity and reliability. This clarifies the physical environment in which all R&D must take place, but the R&D itself, the development of pattern-detection "devices" that can refine the ore, find the needles, is a process that we are only now beginning to understand in a bottom-up way. Up until now we have been able to reason about the semantic-level information needed for various purposes (to inform rational choices, to steer, to build a better mousetrap, to control an elevator) independently of considerations of how this semantic information was physically embodied. As Norbert Wiener, the father of cybernetics, once put it (1961, p. 132): "Information is information, not matter or energy. No materialism that does not admit this can survive at the present day."

7

Darwinian Spaces: An Interlude

A new tool for thinking about evolution

The basic framework for understanding evolution by natural selection was provided by Darwin in *Origin of Species* in his summary to chapter 4. It is worth repeating.

If during the long course of ages and under varying conditions of life, organic beings vary at all in the several parts of their organisation, and I think this cannot be disputed; if there be, owing to the high geometrical powers of increase of each species, at some age, season, or year, a severe struggle for life, and this certainly cannot be disputed; then, considering the infinite complexity of the relations of all organic beings to each other and to their conditions of existence, causing an infinite diversity in structure, constitution, and habits, to be advantageous to them, I think it would be a most extraordinary fact if no variation ever had occurred useful to each being's own welfare, in the same way as so many variations have occurred useful to man. But if variations useful to any organic being do occur, assuredly individuals thus characterised will have the best chance of being preserved in the struggle for life; and from the strong principle of inheritance they will tend to produce

offspring similarly characterised. This principle of preservation, I have called, for the sake of brevity, Natural Selection.

Over the years, this great insight has been refined and generalized,
and various compact formulations have been composed. The best,
for simplicity, generality, and clarity is probably philosopher of biology Peter Godfrey-Smith's (2007) trio:

Evolution by natural selection is change in a population due to

(i)   variation in the characteristics of members of the population,
(ii) which causes different rates of reproduction, and
(iii) which is heritable.

Whenever all three factors are present, evolution by natural selection is the inevitable result, whether the population is organisms,
viruses, computer programs, words, or some other variety of things
that generate copies of themselves one way or another. We can put it
anachronistically by saying that Darwin discovered the fundamental
algorithm of evolution by natural selection, an abstract structure that
can be implemented or "realized" in different materials or media.

We will be looking at the evolution by natural selection of culture—cultural items, such as words, methods, rituals, styles—and
to aid our imaginations in thinking about this treacherous topic, I
want to introduce a thinking tool that I find particularly valuable
for orienting ourselves in this tricky terrain: the Darwinian Spaces
invented by Godfrey-Smith in his book *Darwinian Populations and
Natural Selection* (2009).

It is tempting to say that the three conditions above define the
essence of Darwinian natural selection, and the classic examples of
natural selection used to illustrate the theory in action are perfect
fits. But one of Darwin's most important contributions to thought
was his denial of *essentialism*, the ancient philosophical doctrine
that claimed that for each type of thing, each natural kind, there is
an *essence*, a set of necessary and sufficient properties for being that

kind of thing. Darwin showed that different species are historically connected by a chain of variations that differed so gradually that there was simply no principled way of drawing a line and saying (for instance) dinosaurs to the left, birds to the right. What then about Darwinism itself? Does it have an essence, or is it kin to variations that blend imperceptibly into non-Darwinian explanations? What about phenomena that are not quite paradigm cases of Darwinian evolution? There are many, and Godfrey-Smith shows how to organize them so that their differences and similarities stand out clearly and even point to their own explanations.

For instance, all evolving entities need to "inherit" something—a copy of something—from their parents, but some varieties of copying are low fidelity, with lots of distortion or loss, and others produce near-perfect replicas. We can imagine lining up the different cases in order, from low-fidelity copying to high-fidelity copying, on the x-axis of a simple graph:

$(0,0)$_____x

. . . hearsay . . . stenography . . . vinyl record . . . DNA . . . digital file . . .

Evolution depends on the existence of high-fidelity copying but not *perfect* copying, since mutations (copying *errors*) are the ultimate source of all novelty. Digital copying technology is perfect, for all practical purposes: if you copy a copy of a copy of a copy . . . of a Word file, it will be letter-for-letter identical to the original file. Don't expect mutations to accumulate, for better or for worse. DNA copies itself almost perfectly, but without its very occasional errors (not one in a billion nucleotides), evolution would grind to a halt.

Here's another example of variation: the differences in fitness between members of a population may depend on "luck" or "talent" or any combination in between. In a selective environment where most of those who don't reproduce are struck by lightning before they get an opportunity to mate, evolution cannot act on whatever differences there are in prowess. We could line up varying mixtures of luck and talent on another axis and note that just as copying "noise" (low-fidelity replication) hinders evolutionary R&D, so does environmental "noise" (lightning strikes and other accidents that remove otherwise worthy candidates from the reproduction tournament). While we're at it, we could also add a third dimension and draw a cube with x, y, and z axes, on which we can locate various phenomena. (Unfortunately, most of us can't visualize four or more dimensions gracefully, so we have to stop with three, and look at only three kinds of variation at once, swapping them in and out of the diagram to suit our purposes.)

Using these three-dimensional arrays, we can display pure Darwinian phenomena, quasi-Darwinian phenomena, proto-Darwinian phenomena, and (beyond the fuzzy boundaries) phenomena that aren't at all Darwinian. Being able to see at a glance how these phenomena resemble each other and differ from each other is a fine aid to thinking about evolution. Instead of trying to draw bright lines that separate mere pseudo-Darwinian phenomena from phenomena that exhibit all three of the "essential" features of natural selection, we contrive gradients on which we can place things that are sorta Darwinian in some regard or another. Then we can see whether Darwinian theory applies to the intermediate cases, looking for trade-offs and interactions that can explain how and why these phenomena occur as they do. Godfrey-Smith reminds us that "evolutionary processes are themselves evolutionary products" and as a result emerge gradually and transform gradually. In short, his Darwinian Spaces give us a valuable tool to help us maintain and elaborate our "Darwinism about Darwinism," to use Glenn Adelson's excellent phrase.

In all these diagrams, the dimensions are treated as taking values between 0 (not Darwinian at all) and 1 (maximally Darwinian), so the paradigmatic, "most Darwinian" phenomena appear in the upper right corner, maxing out the three dimensions (1,1,1), and all phenomena that utterly lack the features represented belong at (0,0,0). We can choose any three variable features we like and assign them to the three dimensions, x, y, and z. In figure 7.1, H, fidelity

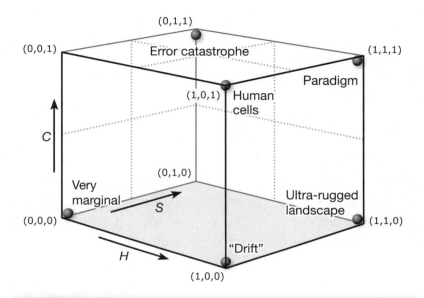

H: Fidelity of heredity
S: Dependence of realized fitness diferences on intrinsic properties
C: Continuity (smoothness of fitness landscape)

FIGURE 7.1: Darwinian space. © *Godfrey-Smith.*

of heredity, is mapped from left to right on the *x* dimension, and if fidelity is too low to maintain signal in the presence of noise, evolution can't happen at all; useful mutations may arise but disappear in the noise before selection can drive them into succeeding generations. Even if the other dimensions shown are at good Darwinian values (close to 1), we get the "error catastrophes" (see figure 7.1) that line up near the left wall. (And as we just noted, if fidelity is perfect, evolution stalls as well, for lack of new variation, so the paradigm cases should be very near, but not *at*, the right wall.) The vertical *y* dimension represents continuity or "smoothness of fitness landscape." Natural selection is a *gradual* process; it depends on blindly taking "small steps," and if the selective environment is organized so that small steps occur on a steady or smooth slope (so that a small step sideways in any direction is also a small step up or down or neutral in fitness), then a sequence of small steps can do "hill-climbing" and arrive, in spite of its myopia (the blind watchmaker), at a peak. It will arrive at the only peak, the global optimum, in a particularly smooth fitness landscape, and at a local optimum in a landscape with several peaks. When the landscape is "rugged," evolution is next to impossible since small steps are uncorrelated with progress or even maintaining one's current fitness.

On the z dimension, Godfrey-Smith has placed S, standing (don't ask why "S") for dependence on "intrinsic properties," and this feature captures the luck-versus-talent dimension. In a tennis tournament, the better athletes tend to advance, thanks to their competence and strength; in a coin-tossing tournament mere luck, not any intrinsic property of the contestants, determines the outcome, and the winner is no more likely to beat any of the losers in a rematch than any other contestant. For example, in genetic "drift" (especially in small populations where the "sampling error" will be relatively large) the "winning" feature in a population is no better than the alternatives; luck has just happened to accumulate in one direction, boosting one feature into fixation in that population. For instance, in a small population of rabbits a few of which are darker gray than the others, it just happens that some of the darker ones get wiped out as road kill, another drowns, and the last carrier of the gene for dark gray falls off a cliff on its way to mating, eliminating the gene from the local gene pool, but not for cause.

One of the most valuable features of these Darwinian Spaces is that they help us see the phenomenon of "de-Darwinizing," in which a lineage that evolved for generations under paradigmatic Darwinian conditions moves into a new environment where its future comes to be determined by a less Darwinian process. Human cells, shown in the diagram, are a fine example. These cells are all direct but very distant descendants of single-celled eukaryotes ("microbes" or "protists") that fended for themselves as independent individuals. These distant ancestors were paradigmatic Darwinian evolvers (up in the 1,1,1 corner) driven by all three features, full strength. The cells that currently make up your body are the direct descendants (the "daughter cells") of cells that are themselves descendants of earlier cells, going back to the zygote that formed when ovum and sperm were united at your conception. During development, in your mother's womb and in infancy, there is a proliferation of cells, many more than will be needed ultimately to compose your organs, and a selection process ruthlessly culls the extras, leaving in place the winners who get the various jobs, while the losers are recycled to make raw materials for the next "generation." This is particularly clear in the brain, where many brand new neurons are given the opportunity to wire themselves up usefully (e.g., as part of a path from some point on one of your retinas to some *corresponding* point in your visual cortex). It's rather like a race, with many neurons trying to grow from A to B, following molecular clues like a trail of breadcrumbs. Those that get there first win, and survive, while the rest die and become feedstock for the next wave.

How do the winners "know" how to grow from A to B? They don't; it's luck (like winning a coin toss). Many neurons attempt the growth journey and most fail; those that make the right connection are saved. The cells that survive may be "intrinsically" just like their competitors (no stronger, no faster); they just *happen* to be in the right place at the right time. So the developmental process that wires up your brain is a de-Darwinized version of the process that evolved the eukaryotes that joined forces a billion years ago to form multicellular creatures. Notice that genetic drift, in which a feature goes to fixation as a result of sheer luck (like the human cells), is also shown in the diagram as low on hill climbing; human cells are selected for location, location, location, as the real estate agents say, but the winners in genetic drift are *just* lucky. Genetic drift has been around as long as evolution has, so it is not a case of de-Darwinizing.

Depicted in figure 7.2 is another of Godfrey-Smith's Darwinian Spaces. This time the *x* dimension is B for bottleneck: Does reproduction funnel through a narrowing of one kind or another—in the extreme case, a single cell? A single sperm cell and a single ovum, out of all the several trillion cells that composed your father

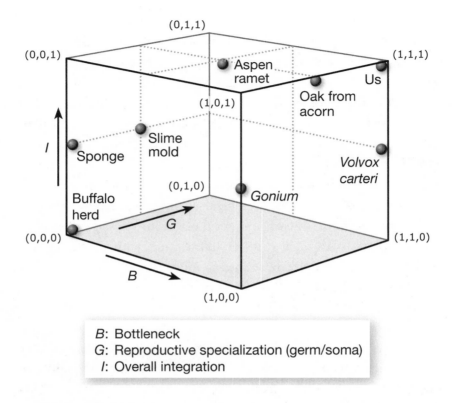

FIGURE 7.2: Darwinian space with other dimensions. © *Godfrey-Smith.*

and mother at the time of your conception, united to inaugurate your generation. In contrast, a buffalo herd can become two herds (which can later become four herds) by simply dividing in two, *or* a pair of buffalo may get isolated somehow and become the founders of a new herd, much less genetically diverse. Any part of a sponge can also become a descendant sponge, but a sponge is somewhat more integrated—more an independent *thing*—than a herd is. An oak tree is highly integrated and has a reproductive bottleneck—the acorn—but like plants in general can also be propagated from a "scion" twig planted in the ground. There is no way (yet) to make a descendant human being—a clone—from a sliver of ear or toe. We are high on all three dimensions shown. Aspen groves are interesting since they are all connected underground by root systems so that they are not genetically individuals but more like conjoined identical mega-twins, a single "plant" covering a whole hilltop.

Figure 7.3 is an application of Darwinian Spaces to the origin of life, as discussed in the first three chapters. This must be a set of processes that takes us from pre-Darwinian to proto-Darwinian to Darwinian.

When we get to bacteria, in the upper right corner, we have full-fledged reproduction, energy capture, and lots of complexity. The unanswered question concerns what route might have been taken to get there. We could plot the trajectories of the components (membrane, metabolism, replicator mechanisms, . . .) and see which joined up first and which were late additions or refinements. Other dimensions could be plotted, of course; two important dimensions might turn out to be size and efficiency. As noted in chapter 2, perhaps the first things/collections that (sorta) reproduced were relatively huge Rube-Goldberg contraptions that were then stable enough for natural selection to prune or streamline into highly efficient, compact bacteria. Populating the space with examples would be premature at this point, and even my placement

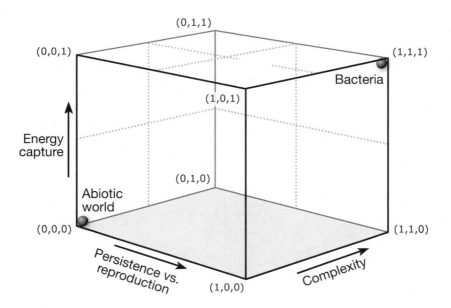

FIGURE 7.3: Darwinian space for origin of life.

of bacteria in the upper right corner might turn out to be misleading if *archaea* (or something else) turn out to have been the first clearly Darwinian reproducers.

Cultural evolution: inverting a Darwinian Space

Now let's consider an application of Darwinian Spaces to the phenomena of cultural evolution, leaving the details of explanation and defense for later.

On the x axis in figure 7.4 I put growth versus reproduction. Aspen groves grow larger "bodies" instead of having "offspring" (usually). It's sorta "just growth" and sorta reproduction. Mushrooms are another example, but I don't know of any animals that can double themselves, Siamese twin style, instead of reproducing the usual way. On the z axis I put degree of internal complexity, and on the vertical y axis is cultural versus genetic evolution. (Are there phenomena midway between? Yes, as we shall see.) Slime molds, like aspen groves, mix differential reproduction with differential

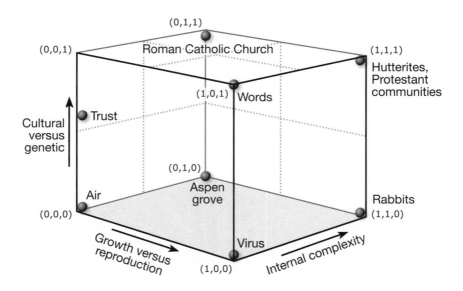

FIGURE 7.4: Darwinian space with religions.

growth—getting larger and larger instead of breaking up into individuated descendants. In culture the Roman Catholic Church grows
and grows (until recently) but seldom spawns descendants these days
(though it had some spectacular offspring in the sixteenth century).
In contrast, the Hutterites are designed to send off daughter communities whenever a community gets big enough (for the evolutionary
details, see Wilson and Sober 1995). Religions (or religious communities) are large, complex social entities. Words are more like viruses,
simpler, unliving, and dependent on their hosts for reproduction.
(Don't make the common mistake of thinking that all viruses are
bad; for every toxic virus there are millions in us that are apparently
harmless, and some may be helpful or even essential.)

Trust is (mainly) a cultural phenomenon—it no doubt gets
a boost from our genes—and, like the air we breathe, is not very
thinglike until we run out of it. It is (largely) a product of cultural
evolution, just the way the air we breathe is a product of genetic
evolution over billions of years. When life began, it was *anaerobic*
(it didn't require oxygen), and the atmosphere was almost oxygen-
free, but once photosynthesis evolved, living things began pumping
oxygen (in the form of CO_2 and O_2) into the atmosphere. This took
several billion years, and some of the O_2 in the upper atmosphere
was turned into O_3, or ozone, and without it, deadly radiation would
reach the Earth's surface and make our kind of life impossible.
The oxygen level 600 million years ago was only 10% of its current
level, so although the change is imperceptibly slow, it is dramatic
over time. We can consider the atmosphere as a *fixed* feature of the
selective environment in which evolution takes place, but it is actually also evolving and a product of earlier evolution on the planet.
These phenomena—we might call them the biological and cultural
"atmosphere" in which evolution can occur—don't themselves
reproduce, but they wax and wane locally and evolve over time in a
non-Darwinian way.

Figure 7.5 is another preview.

In this diagram, I have *inverted* the poles of the earlier Darwinian diagrams; the paradigmatic Darwinian phenomena are in the

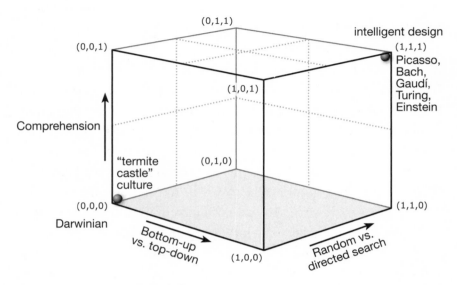

FIGURE 7.5: Inverted Darwinian space with Darwinian phenomena at (0,0,0) and intelligent design at (1,1,1).

(0,0,0) corner, and the paradigmatically *non*-Darwinian phenomena, the instances of intelligent design (not Intelligent Design—this has nothing to do with religion) are in the upper right corner.

The claim that I defend is that human culture started out profoundly Darwinian, with uncomprehending competences yielding various valuable structures in roughly the way termites build their castles, and then gradually de-Darwinized, becoming ever more comprehending, ever more capable of top-down organization, ever more efficient in its ways of searching Design Space. In short, as human culture evolved, it fed on the fruits of its own evolution, increasing its design powers by utilizing information in ever more powerful ways.

In the upper right corner, the intelligent design extreme, we have a well-known ideal that is in fact *never* reached: the Godlike genius, the Intelligent Designer who bestowed gifts on the rest of us mortals from the top down. All the *real* cultural phenomena occupy the middle ground, involving imperfect comprehension, imperfect search, and much middling collaboration.

I put Picasso at the pinnacle not because I think he was smarter than the other geniuses, but because he once said "*Je ne cherche pas. Je trouve.*" ("I don't search. I find!") This is a perfect boast of genius, compressing into a few syllables the following message: "*I* don't have to grub around with trial and error! *I* don't trudge my way up the slope to the global summit of designs. I just leap to the top of Mount Everest, every time! I comprehend all; I even comprehend my comprehension!" Baloney, of course, but inspired baloney. In the case of Picasso, a barefaced lie. He often made hundreds of sketches with variations on a theme, nibbling away in Design Space until he found something he thought was a good stopping point for his journey. (He was a very great artist but part of his genius lay in signing—and selling—so many of his trials instead of discarding them.)

There is much more to be explored in this evolution of cultural evolution and its role in creating our minds, but first we should look more closely at how it got started. Like the origin of life, this is an unsolved problem, and a very difficult one. Our minds are *in some regards* as different from other minds as living things are from non-living things, and finding even one defensible path from our common ancestor with the chimpanzees is a challenging task. There is no shortage of competing hypotheses, and we will consider the best (by my lights). Which came first: language or cooperation or tool making or fire tending or stone throwing or confrontational scavenging or trade or . . . ? We shouldn't be surprised if there proves to be no single "magic bullet" answer, but rather a coevolutionary process with lots of contributing factors feeding on each other. The fact remains that we are the only species so far that has developed explosively cumulative culture—yet another replication bomb—and any speculative story that explains why (both *how come* and *what for*) we have culture must also explain why we alone have culture. Culture has obviously been a Good Trick for us, but eyesight and flight are also obvious Good Tricks, and they have each evolved several times in different species. What barriers have stood in the way of other evolutionary lineages developing the same Good Trick?

Brains Made of Brains

Top-down computers and bottom-up brains

A bacterium is equipped to sense gradients that matter to it, to discriminate a few vital differences, to make itself at home in the familiar surroundings of its tiny *Umwelt*, capturing the energy and materials it needs to thrive and reproduce. Plants and other sessile organisms organize the cells that compose them into larger armies of slaves with particular tasks to perform, feeding and protecting them in return for their services as monitors, expediters, growth controllers, and so forth, without needing to invest in the sort of distal perception systems that permit an organism to be mobile without bumping into all the wrong things. Swift control is the key competence of mobile organisms, so nervous systems, with a headquarters, are obligatory. (Note that while plants have elaborate information-transmission systems controlling their responses to variations in the world, these are distributed systems, not centralized around a hub of any kind.) Brains are control centers for dealing swiftly and appropriately with the opportunities and risks—the affordances—of a mobile life.

As we have noted, brains are designed by natural selection to have, or reliably develop, equipment that can extract the semantic information needed for this control task. An insect is typically born with a mature set of competences, its "expectations" already installed by the history of encounters of its ancestors. An oviparous fish has no time for swimming lessons in its youth and has no parents around to teach it; it has to have a swimming "instinct" built in. A newborn wildebeest must almost literally hit the ground running, with scant opportunity to explore the world and learn of its marvels; if she can't keep up with the herd, she's dead meat. Other mammals, and birds, can afford to be *altricial* (in contrast to *precocial*); they are designed to be fed and protected by parents during a prolonged infancy, picking up semantic information that doesn't have to come through their genes and doesn't have to be learned by *unsheltered* trial and error in the dangerous world. Even if they are purely Skinnerian creatures, they get to try out a relatively unrisky selection of learning environments, chosen for safety (competently but without much comprehension) by their parents. One way or another, brains develop competences, including the meta-competences needed to acquire and hone further competences.

Before turning to some of the specialized ways brains are designed to extract semantic information, it is time to address the issue of how dramatically brains differ from the computers that have flooded our world. Control tasks formerly performed by human brains have recently been usurped by computers, which have taken charge of many things, from elevators to airplanes to oil refineries. Turing's theoretical idea, made operational by John von Neumann's implementation, the serial stored program computer, has multiplied exponentially in the last sixty years and now occupies every environment on Earth and has sent thousands, perhaps millions, of descendants into space; the most traveled brainchildren in history. The brilliant idealizations of Shannon, Turing, von Neumann, McCulloch, and Pitts have led to such an explosion in information-handling competence that today it is commonly supposed not only that brains are just organic digital computers of one sort or another but that also silicon-based computers will soon embody Artificial Intelligence that will surpass human brains in "all the achievements of creative skill" (to echo Beverley's outraged charge that Darwin thought that "Absolute Ignorance" could do the same trick). I take it that Beverley stands refuted; the Absolute Ignorance of evolution by natural selection is indeed capable of creating not just daisies and fish but also human beings who in turn have the competence to build cities and theories and poems and airplanes, and computers, which in turn could *in principle* achieve Artificial Intelligences with even higher levels of creative skill than their human creators.

But the kind of Artificial Intelligence that was originally developed to be executed by Turing's and von Neumann's wonderful machines—GOFAI—is not likely to be the right species of software for this momentous task, and the underlying hardware—the von Neumann machine and its billions of descendants[38]—may not prove to be the best platform. Turing, as we noted in chapter 4, was the epitome of a top-down intelligent designer, and the computer that he invented was an ideal tool for implementing top-down designs of all kinds. The elevator control system project was a top-down exercise in problem-solving: the programmers were *given* the "specs" in advance. They were thus able to use their intelligence to *anticipate* difficulties that might arise, going through the work cycles of the elevator in their imaginations, looking out for affordances positive and negative. When should the elevator stop and pick up passengers in the middle of a route up or down? What should it do when it receives simultaneous requests? Under what conditions might it reverse direction without disgorging all its passengers? The program designers were Popperian creatures trying out hypotheses off-line, but also Gregorian creatures using a suite of thinking tools to enhance their powers, no tool more potent than the programming language they chose to work in. The glory of a programming language is that once you get your design clearly written in the language of choice—Java or C++ or Python, say—you can count on the compiler program to take it from there, creating a file in machine

---

38   The genealogy of artifacts is not, of course, just like the genealogy of living things, but the resemblances between the R&D of both kinds of lineages are striking, a topic for a later chapter.

language that can be executed.[39] Computer programming is thus not so much top-down design as top-halfway-down design; the grubby details of the "bottom" of the design (the engine room, if you like) is something you can ignore, unless the program you are writing is a new compiler.

There is nothing *quite* like this convenience-to-the-designer in evolution by natural selection, but as Herbert Simon pointed out many years ago, in his brilliant little book *The Sciences of the Artificial* (1969), complex evolvable systems (basically *all* living, evolvable systems) depend on being organized "hierarchically": composed of parts that have some stability independently of the larger system of which they are parts, and that are themselves composed of similarly stable parts composed of parts. A structure—or a process—need be designed only once, and then used again and again, copied and copied not just *between* an organism and its offspring, but *within* an organism as it develops. As Richard Dawkins has observed, a gene is like a toolbox subroutine in a computer:

> The Mac has a toolbox of routines stored in ROM (Read Only Memory) or in System files permanently loaded at startup time. There are thousands of these toolbox routines, each one doing a particular operation, which is likely to be needed, over and over again, in slightly different ways, in different programs. For example, the toolbox routine called Obscure-Cursor hides the cursor from the screen until the next time the mouse is moved. Unseen by you, the Obscure-Cursor "gene" is called every time

39 The distinction between compiled and interpreted computer languages used to loom large, but doesn't matter here, and most languages today are hybrid. A compiled program takes a whole source code program as input and churns out an executable machine language program as output. An interpreted language in effect compiles each instruction as soon as it is composed (sometimes called "just-in-time compiling"), allowing the programmer to test the execution of each statement as it is written, making development very friendly (sometimes) at a (small, sometimes) cost in ultimate program efficiency. Lisp was originally an interpreted language.

you start typing and the mouse cursor vanishes. (Dawkins 2004, pp. 155–156)

This hierarchical systematicity is found throughout Nature; it is in the genome and in the developmental processes that the genome directs. The vertebra-making subroutine can be called a few more times, and we're on the way to making a snake. People with a sixth finger—a quite common mutation—have a hand-making subroutine that executed the finger-making subroutine one time more than usual. Once eyelid making is a debugged program, it can be used with variations all over the animal world. So evolution by natural selection does have something a bit like source code in having relatively simple instructions that trigger a cascade of further instructions to accomplish some "modular" task, but it is all "in machine language," executed whenever it is called on; it doesn't need to be compiled and there is no reader who needs a mnemonic crutch (CALLFLOOR, WEIGHT-IN-POUNDS . . . see chapter 4) in order to understand it. The developing organism sorta understands the commands of its genes the way a von Neumann machine sorta understands its machine language instructions—it (sorta) obeys them.

Competition and coalition in the brain

Evolution by natural selection is not top-down R&D, like computer programming, in spite of its invention and heavy use of modules. It is bottom-up R&D, Darwin's strange inversion. Brains, moreover, are not like digital computers in several ways. Three differences are most frequently cited, but are not, in my estimation, very important:

1. Brains are analog; computers are digital. This is true, maybe, depending on what you mean by "analog"; if you mean *not binary* (using just 0 and 1, on and off), then probably brains are analog, but they may well prove to be digital in others ways. Any finite alphabet of signals with equivalence classes is a kind of digitizing (A,a,*a*,**A**,A,**a** . . . all count as A). Th**i**s iS *a d*IgiTIzeD sIGnaL

and, as we will see in the next chapter, this kind of digitizing is a very important feature of language.

2. Brains are parallel (they execute many millions of "computations" simultaneously, spread over the whole fabric of the brain); computers are serial (they execute one single simple command after another, a serial single-file stream of computations that makes up for its narrowness by its blinding speed). There are exceptions; some special-purpose parallel-architecture computers have been created, but the computers that are now embedded in everything from alarm clocks and toaster ovens to automobiles are all serial architecture "von Neumann machines." You may well own a hundred von Neumann machines, hidden away inside your smart appliances and consigned to menial tasks that use a tiny fraction of their power. (It is much cheaper to throw in a whole computer on a mass-produced chip than to design special purpose hardware.) It is true that the brain's architecture is massively parallel, with a vision system about a million channels wide, for instance, but many of the brain's most spectacular activities are (roughly) serial, in the so-called stream of consciousness, in which ideas, or concepts or thoughts float by not quite in single file, but through a von Neumann bottleneck of sorts. You can simulate a virtual serial machine on a parallel architecture—that's what the brain does, as I showed in *Consciousness Explained*—and virtual parallel computers can be implemented on serial machines, to any breadth you like, at a cost of operating speed. It is no accident that brains are massively parallel in their architecture, given the need for speed in controlling risky lives, but the "moving parts" of today's von Neumann machines have a basic cycle time billions of times faster than the response time of a neuron, so they have speed to burn. In recent years, artificial intelligence researchers have developed *connectionist* networks that loosely mimic neural networks in the brain, and they have proven to be very good at learning to discriminate things in images, and other sorts of pattern recognition that our brains excel at, and they undoubt-

edly demonstrate the power of parallel processing (see chapter 15), but in fact, the parallel processing done by these networks is almost all simulated on conventional serial von Neumann machines. Von Neumann machines are universal mimics that can simulate any other architecture; today's speedy von Neumann machines, in spite of their severe bottlenecks, can pretend they are parallel-processing neural networks so swiftly that they can keep up with, or surpass, the parallel-processing brains that designed them.

3. Brains are carbon based (protein, etc.); computers are silicon. This is true for the time being, though striking progress has been made by nanotechnologists intent on making computers out of proteins, and the protein networks inside cells do seem to be computing (see Bray 2009 for a vivid account). And nobody has yet demonstrated that the difference in underlying chemistry gives an edge to carbon.

How about this?

4. Brains are alive; computers are not.

Some would want to reply—and I used to reply—an artificial heart isn't alive, and it functions fine. A replacement hip or knee or shoulder doesn't have to be alive. You can replace or splice the auditory nerve with nonliving wire, suitably attached at both ends. Why not the rest of the brain? Is there some super-special piece of brain that must be alive for a brain to function? The standard working assumption of Artificial Intelligence has always been that any living organ is really just a very sophisticated bit of carbon-based machinery that can be replaced, piece by piece or all at once, by a nonliving substitute that has the same input-output profile—does all and only the same things with the same inputs in the same time frame—without loss. If the brain were an organ for secreting bile or purifying the blood, the chemistry and physics of the working parts would matter greatly, and it might be impossible to find suitable substitutes for the materials used, but the brain is an information-processor, and information is medium-neutral. (A warning or a declaration of love or a promise can be "made out of anything," provided the recipient has the know-how to pick it up.)

But there is one feature of living things that might matter greatly in our quest to understand how minds work, and this has recently been illuminated by Terrence Deacon, in his difficult but important book *Incomplete Nature: How Mind Emerged from Matter* (2012). According to Deacon, the very moves I've been describing as brilliant strokes of simplification have pushed theory into the wrong regions of Design Space for half a century and more. Shannon's enabling stroke was to abstract the concept of information away from thermodynamics, away from the concept of energy (and matter, as noted above); information is information whether electrons or photons or smoke signals or magnetic regions or microscopic pits in plastic disks are involved. It takes energy to transmit and transform information (it isn't magic, after all) but we can isolate our understanding of the information processing from whatever energetic considerations are in play. Norbert Wiener created the field, and the term, *cybernetics*. He took the Greek verb for steer or govern—κυβερνάω—kybernao—from which the English word *govern* is also derived—and noted that while the controller (of a ship, a city, a body, an oil refinery) needs to use energy at some interface (to push the rudder, utter the decree, turn up the temperature), the energy required to run the control system itself is ad lib—whatever you like, and not very much. It is this isolation of the computation from the dynamics of physical control that makes it possible to control almost anything today with your battery-powered smartphone—plus transducers and effectors of sufficient strength to do the hard work. There is a tiny drain on your smartphone battery when you use it remotely to send a radio signal to open your heavy garage door; the electric motor that is controlled by that weak signal draws considerable current in order to do its work. Deacon grants that this has permitted the blossoming of all the high-tech intelligent design we have been exploring, but he insists that by divorcing information processing from thermodynamics, we restrict our theories to basically *parasitical* systems, artifacts that depend on a user for their energy, for their structural maintenance, for their interpretation, and for their *raison d'être*. Living things, in contrast, are autonomous, and moreover they are composed of living things (cells) that are themselves somewhat autonomous.

Researchers in artificial intelligence could plausibly reply that they are postponing consideration of phenomena such as energy capture, reproduction, self-repair, and open-ended self-revision, expecting to reap the benefits of this simplification by first getting clear about the purely informational phenomena of learning and self-guidance (a *kind* of autonomy, however incomplete). Nobody would dream of complicating the design of a chess-playing computer by requiring it to obtain its energy from sandwiches and soft drinks, obliging it to monitor not only its time but also its depleting reserves of usable energy. Human chess players have to control their hunger pangs, and emotions such as humiliation, fear, and boredom, but computers can finesse all that with impunity, can they not? Yes, but at a huge cost, says Deacon: by taking these concerns off their hands, system designers create architectures that are brittle (they can't repair themselves, for instance), vulnerable (locked into whatever set of contingencies their designers have anticipated), and utterly dependent on their handlers.[40]

This makes a big difference, Deacon insists. Does it? In some ways, I think, it does. At the height of GOFAI back in the 1970s, I observed that AI programs were typically disembodied, "bedridden" aspirants to genius that could only communicate through reading and writing typed messages. (Even work on computer vision was often accomplished with a single immovable video camera eye or by simply loading still images into the system the way you load pictures into your laptop, a vision system without any kind of eyes.) An

---

40 It might be more than amusing to note that one thing chess-playing computers can't do well is resign from hopeless games! Their human "operators" are typically permitted to intervene when all is lost since the program has no shame.

embodied mobile robot using sense "organs" to situate itself in its world would find some problems harder and other problems easier. In 1978 I wrote a short commentary entitled "Why not the whole iguana?" in which I argued for turning from truncated human microcompetences (answering questions about baseball, playing checkers) to attempting to model the computational architecture of a whole, self-protecting, energy-capturing robotic animal, as rudimentary as you like. (You have to get your simplicity somewhere; AI is hard.) The animal could be imaginary, if that made the task easier—a three-wheeled Martian iguana, for instance.

Some roboticists took up the challenge. For instance, Owen Holland started the SlugBot project, attempting to make a robot that would harvest slugs from grain fields and digest them to produce the electricity needed to power its computer chips, a project that has spawned other ingenious designs. That's energy capture, but what about self-repair and self-maintenance? Work on these and other requirements of life has proceeded for several decades, creating "animats" and "nanobots" that instantiate various hugely simplified versions of basic life processes, but nobody took the "whole iguana" idea as far as Deacon claims they should have: in their headquarters these artificial creatures still relied on CPU chips—von Neumann hardware—even when they simulated parallel architectures. It is important, Deacon claims, that a brain be made of cells that are themselves autonomous little agents with agendas, chief of which is staying alive, which spawns further goals, such as finding work and finding allies. Deacon's insistence on making brains (or brain substitutes) out of living neurons might look at first like some sort of romanticism—protein chauvinism, in effect—a fatal step or two away from vitalism, but his reasons are practical and compelling.

We can illuminate Deacon's point with one phenomenon: the remarkable plasticity of the brain. If an area is damaged, a neighboring region can often (not always) take on the duties of the missing tissue, swiftly and gracefully. If a region is underutilized, its neighbors soon recruit the cells there to help with their own projects. Neurons that are destroyed are not, in general, replaced the way skin cells and bone cells and blood cells (among others) are. Nerve "regeneration" is still mainly a dream of bioengineers, not a normal feature of nervous systems, so the plasticity observed in many experiments must be due to neurons being *reassigned* to new tasks or required to take on additional workloads. Is there a Master Personnel Manager figuring out these new job descriptions and issuing commands from on high to the subservient neurons down on the shop floor? The computer scientist Eric Baum, in *What Is Thought?* (2004), calls such arrangements "politburo control" and notes that this top-down way to accomplish complex tasks is not the brain's way. Economists have shown why centrally planned economies don't work as well as market economies, and centrally planned (top-down) architectures are ineffective brain-organizers for much the same reasons.

As roboticist Rodney Brooks has observed (personal communication) the hardware of existing digital computers depends critically on millions (or billions) of identical elements, perfect clones of each other almost down to the atomic level, so that they will always respond as they are supposed to respond: robotically! Engineers have managed to create the technology to print microscopic computer circuits with millions of identical flip-flops each reliably storing a 0 or 1 until ordered (from on high) to "flip the bit." These are the ultimate moving parts of a computer and they have no individuality, no idiosyncrasies at all. Neurons, in contrast, are all different; they come in a variety of quite clearly defined structural types—pyramidal, basket, spindle, and so on—but even within types, no two neurons are exactly alike. How can such a diverse population get organized to accomplish anything? Not by bureaucratic hierarchies, but by bottom-up coalition-formation, with lots of competition.

Neurons, mules, and termites

Neurons—like all the rest of the cells that compose our bodies—are domesticated descendants of the free-living, single-celled eukaryotes that thrived on their own, fending for themselves as wildlife in a cruel world of unicellular organisms. François Jacob famously said that the dream of every cell is to become two cells, but neurons can't normally have offspring. Like mules they have parents (well, a mother and a grandmother and so on) but are themselves sterile so their *summum bonum* is just staying alive in their de-Darwinized niche. They have to work to capture the energy they need, and if they are not already fully employed, they will take on whatever odd jobs are available.

Versions of this idea have recently blossomed in several different corners of cognitive science, and since they have persuaded me to jettison a position I had long defended, I want to emphasize my conversion. I've changed my mind about how to handle the homunculus temptation: the almost irresistible urge to install a "little man in the brain" to be the Boss, the Central Meaner, the Enjoyer of pleasures and the Sufferer of pains. In *Brainstorms* (1978), I described and defended the classic GOFAI strategy that came to be known, thanks to Lycan (1987), as "homuncular functionalism": *replacing the little man with a committee.*

> The AI programmer begins with an intentionally characterized problem, and thus frankly views the computer anthropomorphically: if he solves the problem he will say he has designed a computer that can [e.g.] understand questions in English. His first and highest level of design breaks the computer down into subsystems, each of which is given intentionally characterized tasks; he composes a flow chart of evaluators, rememberers, discriminators, overseers and the like. These are homunculi with a vengeance. . . . Each homunculus in turn is analyzed into smaller homunculi, but, more important, into less clever homunculi. When the level is reached where the homunculi are no more than adders and subtracters, by the time they need only the intelligence to pick the larger of two numbers when directed to, they have been reduced to functionaries "who can be replaced by a machine." (Dennett 1978, p. 80)

I still think this is on the right track, but I have come to regret—and reject—some of the connotations of two of the terms I used: "committee" and "machine." The cooperative bureaucracy suggested by the former, with its clear reporting relationships (an image enhanced by the no-nonsense flow charts of classical cognitive science models) captured the dream of top-down GOFAI, but it suggested a profoundly unbiological sort of efficiency. Turing's strange inversion of reasoning is still intact: *eventually* in our decompositional cascade we arrive at elements whose tasks are so rigid and routinized that they "can be replaced by a machine," just like Turing's diligent human computer. The simplest moving parts *within* neurons, the motor proteins and microtubules and the like, really are motiveless automata, like the marching broomsticks in *The Sorcerer's Apprentice*, but neurons themselves, in their billions, play more enterprising and idiosyncratic roles than the obedient clerks I was imagining them to be, and that fact has major implications for the computational architecture of brains.

Tecumseh Fitch (2008) coined the term "nano-intentionality" to describe the agency exhibited in neurons, and Sebastian Seung gave a keynote address to the Society for Neuroscience (2010) on "selfish neurons," and earlier (2003) wrote of "hedonistic synapses." What could a neuron "want"? The energy and raw materials it needs to thrive—just like its unicellular eukaryote ancestors and more distant cousins, the bacteria and archaea. Neurons are sorta robots; they are certainly not conscious in any rich sense—remember, they are eukaryotic cells, akin to yeast cells or fungi. If individual neurons are conscious, then so is athlete's foot. But neurons are, like yeast and fungi, highly competent agents in a life-or-death struggle, not in the environment between your toes but in the demanding environment between your ears, where the victories go to those cells that can network more effectively, contributing to more influential trends at the levels where large-scale human purposes and urges are discernible.

Turing, Shannon, and von Neumann, tackling a spectacularly difficult and novel engineering project, intelligently designed computers to keep needs and job performance almost entirely independent. Down in the hardware, the electric power is doled out evenhandedly and abundantly; no circuit risks starving. At the software level, a benevolent scheduler doles out machine cycles to whatever process has highest priority, and although there may be a bidding mechanism of one sort or another that determines which processes get priority, this is an orderly queue, not a struggle for life. (It is a dim appreciation of this fact that perhaps underlies the common folk intuition that a computer could never "care" about anything. Not because it is made out of the wrong materials—why should silicon be any less suitable a substrate for caring than organic molecules?—but because its internal economy has no built-in risks or opportunities, so its parts don't have to care.)

The top-down hierarchical architecture of computer software, supported by an operating system replete with schedulers and other traffic cops, nicely implements Marx's dictum: "To each according to his needs, from each according to his talents." No adder circuit or flip-flop needs to "worry" about where it's going to get the electric power it needs to execute its duty, and there is no room for "advancement." A neuron, in contrast, is always hungry for work; it reaches out exploratory dendritic branches, seeking to network with its neighbors in ways that will be beneficial *to it*. Neurons are thus capable of self-organizing into teams that can take over important information-handling work, ready and willing to be given new tasks which they master with a modicum of trial-and-error rehearsal. This is *rather* like the thousands of unemployed women who showed up at the gates of Oak Ridge to take on the tasks they didn't need to understand, but with a big difference: there is no General Leslie Groves to design the brain; it needs to be designed bottom-up.

It is time to take stock of what these considerations imply. The top-down intelligent design of the classic computer created a hyper-competent but deeply unbiological marvel. This is not because computers are made of the wrong kind of stuff but because the stuff they are made of is organized into the wrong kind of hierarchies: the kind of planned bureaucracies that may be "well-oiled machines" but depend on regimentation of the moving parts, suppressing both exploration and improvisation at every level. It is important to stress again that *in principle* this conclusion leaves open the door for silicon-based (or Turing-based, whatever the physical medium) brains; *simulated* or *virtual* neurons are possible in principle but computationally very expensive; each individual neuron would have to be modeled in all its idiosyncrasies, habits, preferences, weaknesses. To put the idea in perspective, imagine that intergalactic scientists discovered our planet and proceeded to study, from afar, the "behavior" of our cities and towns, our highways and railways and communication systems, as if these were organisms of some sort. Being enthusiastic developers of computer models, these scientists decide to make a simulation of New York City, in all its activities. Newyorkabot, they call it. "Go ahead and try," say the skeptics, "but do realize that you will have to model all the millions of citizens (those squishy little moving things) in considerable detail if you want to have much hope of making a good, predictive model. They are none of them exactly alike, and they are both curious and enterprising."

Top-down intelligent designs depend on foresight, which evolution utterly lacks. The designs generated by natural selection are all, in a way, retrospective: "this is what has worked in the past." Whether it will work in the future depends on whether the regularities on which the design is founded persist. Moth-design could rely on the regularity of the sun and the moon being the only light sources in the environment, so "fly at constant angle n to the light" was a winning habit to hard-wire into moths until candle flames and electric light bulbs entered the scene. This is where too much regularity in the selective environment can be a trap, committing a lineage to a locked-in technique or disposition that can turn toxic when conditions change. As noted in chapter 5, variable selective environments, precisely because of their unpredictability, favor the selection of incomplete designs, along with mechanisms to tune the design to suit the circumstances, exploitable plasticity or, to use the everyday term, learning. Top-down GOFAI-style systems can be equipped with tunable parameters and learning capabilities, of course, but the hierarchical control structure tends to commit to a fixed range of plasticity, based on a "worst case scenario" estimate of what is likely to happen.

Foresightless, backward-"looking" evolution by natural selection is not intelligent design but still powerful R&D, sufficiently discerning so that the general division of labor in the brain can be laid down in the genetic recipes that have been accumulated and refined over the billion years or so of mobile life. And, as we have seen, the refinements and variations in the genes can set enough parameters to ensure that elaborate instinctive behaviors, such as nest building, can survive generation after generation. Fixing *all* the wiring details by genetic instruction would be impossible; it would require many times more bits of Shannon information than can be transmitted in a genome, even a three-billion-nucleotide-long genome like ours. This is the undeniable message of "evo-devo" (evolutionary-developmental biology), which stresses that the production of the next generation of any organism is not a straightforward matter of following a blueprint or recipe laid down in the genes; it is a construction job that must rely on local R&D that depends on the activity of (myopic) local agents that conduct more or less random trials as they interact with their surroundings in the course of development. Neurons, for instance, are akin to Skinnerian agents, taking their chances and making the most of the opportunities that come their way, exploiting their own plasticity to improve the performances for which they are rewarded or reinforced. Their efforts are, however, guided by some helpful landmarks laid down by the genes, giving them paths to follow. *Brains are more like termite colonies than intelligently designed corporations or armies.*

How do brains pick up affordances?

A recurring theme so far has been that all organisms, from bacteria to us, are designed to deal with a set of affordances, the "things" that matter (in a broad sense of "thing"), and that this catalogue, an organism's *Umwelt*, is populated by two R&D processes: evolution by natural selection and individual learning of one sort or another. Gibson was notoriously silent on how organisms pick up the information needed to notice, identify, and track their affordances, and I have been postponing the question as well, up to now.

Here's what we have figured out about the predicament of the organism: It is floating in an ocean of differences, a scant few of which might make a difference to it. Having been born to a long lineage of successful copers, it comes pre-equipped with gear and biases for filtering out and refining the most valuable differences, separating the semantic information from the noise. In other words, it is prepared to cope in some regards; it has built-in expectations that have served its ancestors well but may need revision at any time. To say it has these expectations is to say that it comes equipped with partially predesigned appropriate responses all ready to fire. It doesn't have to waste precious time figuring out from first principles what to do about an A or a B or a C. These are familiar, already solved problems of relating input to output, perception to action. These responses to incoming stimulation of its sensory systems may be external behaviors: a nipple affords suckling, limbs afford moving, a painful collision affords retreating. Or they may be entirely covert, internal responses, shaping up the neural armies into more effective teams for future tasks.

How does this training happen? Here it will help to resurrect a somewhat discredited distinction from the early days of cognitive science: competence models versus performance models. A competence model (such as the grammar of a language) says how the system *ought* to work—it gives the specs, as in the elevator example of chapter 4—while leaving the details of how to accomplish those requirements as a design problem with potentially many different solutions, different performance models. A performance model for *implementing* a grammar, for making a speaker of grammatical English, for instance, would be a job for neurolinguistics far in the future. In the early days of cognitive science, theoretical linguists argued over grammars without concerning themselves with how on earth a brain might "follow the rule book"; they just wanted to get the rule book right first.

Meanwhile, psycholinguists devised clever experiments that showed patterns in grammatical errors that children made, sources of interference, failures of grammatical judgment, and the like, and when the theorists couldn't explain these patterns, they could handily excuse themselves, saying it was too early to try to account for grubby details of performance. People make mistakes, people have faulty memories, people jump to conclusions, but they still have the basic competence enshrined in their grammars, however sloppy their performance on this or that occasion. Performance models would come later.

This was not always a useful division of labor, and it created a gulf in cognitive science that is still creating tension and miscommunication today. The linguists were right to insist that until you have a *fairly* clear idea of what a brain *can and ought to do* in the way of perceiving and uttering language, your efforts to reverse engineer the brain's language-manipulating machinery are apt to be misguided, tackling misconceived problems. But part of the specs for any reverse-engineering exploration has to be the limitations and strengths of the available machinery, and by postponing questions about the brain, and underestimating the importance of the telltale performances elicited by the psycholinguists, the theoretical linguists conducted some wild goose chases of their own.

With that caution in hand, we can address the idea that is now sweeping through cognitive science as a very promising answer to how the brain picks up, and uses, the available semantic information: Bayesian hierarchical predictive coding. (For excellent accounts see Hinton 2007; Clark 2013; and the commentary on Clark, Hohwy 2013.) The basic idea is delicious. The Reverend Thomas Bayes (1701–1761) developed a method of calculating probabilities based on one's *prior expectations*. Each problem is couched thus: Given that your expectations based on past experience (including, we may add, the experience of your ancestors as passed down to you) are such and such (expressed as probabilities for each alternative), what effect on your future expectations should the following new data have? What adjustments in your probabilities would it be *rational* for you to make? Bayesian statistics, then, is a normative discipline, purportedly prescribing the *right* way to think about probabilities.[41] So it is a good candidate for a competence model of the brain: it works as an expectation-generating organ, creating new affordances on the fly.

Consider the task of identifying handwritten symbols (letters and digits). It is no accident that such a task is often used by Internet sites as a test to distinguish real human beings from bots programmed to invade websites: handwriting perception, like speech perception, has proven to be an easy task for humans but an exceptionally challenging task for computers. Now, however, working computer models have been developed that can do a good job identifying handwritten—scribbled, really—digits, involving a cascade of layers in which the higher layers make Bayesian predictions about what the next layer down in the system will "see" next; when the predictions prove false, they then generate error signals in response that lead to Bayesian revisions, which are then fed back down toward the input again and again, until the system settles on an identification (Hinton 2007). Practice makes perfect, and over time these systems get better and better at the job, the same way we do—only better (see chapter 15).

Hierarchical, Bayesian predictive coding is a method for generating affordances galore: we expect solid objects to have backs that will come into view as we walk around them; we expect doors to open, stairs to afford climbing, and cups to hold liquid. These and

---

41   There are anti-Bayesian statisticians who are troubled by its reliance on a "subjective" starting point: given *your* experience to date, what ought you conclude on this occasion? Where do you get your priors, and what if mine disagree? Gelman 2008 is a clear and accessible survey of this and other objections. This objection is irrelevant to our application of Bayesian thinking, which is specifically designed to address the predicament of an agent blessed (or cursed) with a set of priors and wondering what to do next. The subjectivity of the agent is an integral part of the design problem being faced.

all manner of other anticipations fall out of a network that doesn't sit passively waiting to be informed but constantly makes probabilistic guesses about what it is about to receive in the way of input from the level below it, based on what it has just received, and then treating feedback about the errors in its guesses as the chief source of new information, as a way to adjust its prior expectations for the next round of guessing.

Not least of the attractive properties of these applications of Bayesian thinking to the problem of how brains learn is that they provide a simple and natural explanation of an otherwise perplexing fact of neuroanatomy: in visual pathways, for instance, there are more *downward* than *upward* pathways, more *outbound* than *incoming* signals. On this view, the brain's strategy is continuously to create "forward models," or probabilistic anticipations, and use the incoming signals to prune them for accuracy—if needed. When the organism is on a roll, in deeply familiar territory, the inbound corrections diminish to a trickle and the brain's guesses, unchallenged, give it a head start on what to do next.

These Bayesian models are descended from the "analysis by synthesis" models of early cognitive science, in which top-down curiosity ("Is it a deer?" "Is it a moose?") guide the formation of hypotheses for testing against the incoming data. (Your brain *analyzes* the data by making a guess, *synthesizing* a version of what you're looking for and matching it against the data.) In *Consciousness Explained* (Dennett 1991, p. 10ff), I offered a speculative model of dreams and hallucinations based on analysis by synthesis, arguing that the elaboration of content in these phenomena could be due to nothing more than a "disordered or random or arbitrary round of confirmation and disconfirmation" (p. 12). This has been spectacularly supported and refined by recent work at Google Research (e.g., Mordvintsev, Olah, and Tyka 2015). Today I could simplify my account: in a Bayesian network silence counts as confirmation. Whatever the higher levels guess counts as reality by default in the absence of disconfirmation.

Another virtue of Bayesian models from our vantage point

is that an organism can be blessed by natural selection with a high-powered statistical analysis engine without having to install a mathematician-homunculus in a special office in the bureaucracy. These are, if you like, expectation-generating *fabrics* with a remarkable competence they don't need to understand. In earlier work I have warned against postulating "wonder tissue" in brains to do the hard work that has to be done somehow without saying how the wonder tissue manages to be so splendid. These Bayesian fabrics are demonstrably not magic; they work just fine running on nonmysterious off-the-shelf computers. The details of neural implementation of such hierarchical predictive coding networks are still obscure, but they don't seem out of reach.[42]

This is all very well for animal cognition, and a number of the commentators on Clark's fine survey venture the tentative conclusion that animal minds *are* Bayesian expectation machines relying on probabilities eked out of the passing show to guide all their behavior. But if so, then animal minds are, once again, termite colonies, not intelligent designers. They are so competent that we can concede them all kinds of behavioral comprehension, but still, in some important way yet to be articulated, they are missing one of our tricks: *having* reasons and not just *acting* for reasons. Bayesian anticipators don't need to express or represent the reasons they track; like evolution itself, they "blindly" separate the informational wheat from the chaff and act on it. Reasons are not things in their ontologies, not salient items in their manifest images.

But reasons are things for us. They are the very tools and objects of top-down intelligent design. Where do they come from? How do they get installed in our brains? They come, I am now at last ready

---

42   There are skeptics. Domingos (2015) points out that the implementation of Bayesian expectation-generation depends heavily on the MCMC (Markov Chain Monte Carlo) algorithm, a brute force algorithm that does the heavy lifting; it gets the job done, but "does not simulate any real process" (p. 164). "Bayesians, for one, probably have MCMC to thank for the rising popularity of their methods more than anything else" (p. 165). Perhaps, but since the models simulate massively parallel architectures, their use of MCMC is almost a forced move.

to argue in some detail, via cultural evolution, a whole new process of R&D—less than a million years old—that designs, disseminates, and installs thinking tools by the thousands in our brains (and only our brains), turning them into minds—not "minds" or sorta minds but proper minds.

Feral neurons?

We're going to leave the level of neurons behind in subsequent chapters, but before we do, I can't resist a speculation that ties some of our themes together and might turn out to play a role in accounting for the fact that *H. sapiens* is, so far, the only species on Earth with proper minds, encultured minds full of thinking tools. Neurons, I have claimed, are in effect the domesticated descendants of very ancient eukaryotes that faced life for eons as self-employed microorganisms. To distinguish such eukaryotes from bacteria and archaea they are often called *protists,* of the kingdom *Protista.* They managed well, of course, or we wouldn't be here, and the protists of today, such as amoebas and single-celled algae, still thrive, thanks to their many and varied talents. When protists first began forming colonies, and eventually multi-celled organisms, they brought with them their genomes, instructions for all the competences they had acquired by natural selection over billions of generations. Many of these talents were no longer needed in their new, sheltered neighborhoods, so they were lost—or at least these *expressions of the relevant genes* were lost, since they no longer paid for themselves. But it proves to be cheaper and quicker to leave the genes alone and turn off (some of) their expression in development than to erase genes altogether. (Over time, they will erase themselves, if never used.)

This is nicely parallel to a tactic in software development: "legacy code." When software is revised, as it often is (I am writing this book in Word 14.4.8, which wears its genealogy on its sleeve, being the eighth minor revision of the fourth major revision of the fourteenth New and Improved Version), the programmers would be foolish to discard all the hard work of earlier generations of programmers—who knows when you might need it for something?—so they have the practice of leaving obsolete code intact, right beside the code that replaces it, but just "commented out." Enclosing the legacy code in brackets or asterisks or some other convention recognized by the compiler program guarantees that it will be passed over by the compiler and hence not represented in the code to be executed by the computer.[43] Remember that there is no source-code/compiled-code distinction in genomes, but the same silencing of code can occur thanks to a simple mutation in a regulatory gene (a gene for controlling the "expression" of genes for proteins).

When animals are domesticated, their domesticators select, either unconsciously or methodically, desirable traits. The differences in a litter (or herd) that are large and visible are likely to be the result of relatively minor genetic differences, and these may be differences in the expression of a gene shared by all. The domesticators need not have any understanding of, or intention to produce, the effects they are having on the genes of their livestock, but silencing genes while leaving them unchanged is a likely path to the desired results. Then, when descendants of the domesticated animals escape and go feral, all that has to happen, genetically, to restore the wild characteristics that have been suppressed is the removal of the "brackets" that enclose the legacy code, which then springs back into action. This may well explain the strikingly quick reversion to wild traits that occurs in feral lineages. In a few generations feral pigs, for instance, will recover many traits of their wild boar relatives, in both appearance and behavior, and feral horses—called mustangs in the United States—are only a few hundred years away from their domesticated forebears but differ significantly in disposition.

---

43   There are several distinct meanings of the term; in another sense, legacy code *is* compiled but restricted for special applications (for instance, to permit the program to run on an outdated platform).

Your average neuron is apparently docile enough to spend its long lifetime doing the same job while still maintaining a spark of autonomy, a modest ability and disposition to try to improve its circumstances when opportunity knocks. We can easily imagine that some environments in which neurons find themselves favor more enterprising, risky exploration of options, breeding more versatile, aggressive neurons. In such conditions, selection could favor the renewed expression of long-silenced genes, at least in some subpopulations of neurons in critical regions of the brain. These would be feral neurons, in effect; a bit less stable, more selfish, a bit more likely to morph themselves into novel connections with neighbors.

What sort of condition might favor this trend? How about the arrival of invaders? New kinds of entities need local supporters to further their own replication and help them compete against each other for hegemony in the brain. What do invading armies do when attempting to assert control over newly occupied territory? They open the prisons, releasing the prisoners in order to staff security forces with hardened natives who know their way around. This oft-reinvented good idea could be the free-floating rationale of a parallel development in brains. We are about to embark on a detailed account and defense of *memes*, informational viruses that govern infectious habits. When they spread through a population, they need neural resources to replicate in much the way viruses need to commandeer the copy machinery in individual cells to get new copies of themselves made. Viruses obviously can't understand the rationale for their quest; it is just their nature to have this effect on cells they encounter. They are macromolecules, after all, not even living things. And memes, as viruses of the mind, made of nothing but information, have to get into minds, and then have to get rehearsed and rehearsed and rehearsed, but they don't need to comprehend this or anything else.

This strange idea of mindless informational things that provoke coalitions of rebellious neurons to support them, the idea of memes infecting brains, will take several chapters to articulate and defend.[44] Here I am pointing out that if anything like that invasion happened, it would no doubt involve a coevolutionary process in which human brains—the only brains so far seriously infected with memes—were selected for handling memes, sheltering them and aiding in their reproduction, in much the way fig trees have coevolved with the parasitic wasps who return the favor by helping pollinate the fig trees. Neurons come in different shapes and sizes, and one kind, the von Economo neuron, or spindle cell, is found only in animals with very large brains and complex social lives: humans and other great apes, elephants, and cetaceans (whales and dolphins). Since the many other species that share common ancestors with these lineages lack von Economo cells, these neurons have apparently evolved quite recently and independently (convergent evolution) to play similar roles. Perhaps they are just long-distance communicators in those large brains, but they also tend to be concentrated in brain areas known to be involved in self-monitoring, decision-making, and social interaction, to put it very generally. The possibility that their presence in these different highly social lineages is no coincidence suggests that they are likely candidates for further exploitation by meme-invaders. But we may discover others, once we have a better idea what to look for.

We have set the stage for the invasion. Brains are computers, but they are very unlike the computers in use today. Composed of billions of idiosyncratic neurons that evolved to fend for themselves, the brain's functional architecture is more like a free market than a "politburo" hierarchy where all tasks are assigned from on high. The fundamental architecture of animal brains (including human brains) is probably composed of Bayesian networks

44   One of the many virtues of the detailed model proposed by Eliasmith in his groundbreaking *How to Build a Brain* (2013) is that his "semantic pointer architecture" accounts for behavioral comprehension (perception and control) of the animal variety, while leaving plenty of plasticity—knobs to turn—for invaders—words and other memes—to exploit. It is not a language-centric model of intelligence from the outset. The details of how this might work would fill another book.

that are highly competent expectation-generators that don't have to comprehend what they are doing. Comprehension—*our* kind of comprehension—is only made possible by the arrival on the scene quite recently of a new kind of evolutionary replicator—culturally transmitted informational entities: memes.

The Role of Words in Cultural Evolution

The evolution of words

The survival or preservation of certain favoured words in the
struggle for existence is natural selection.
—Charles Darwin, *Descent of Man*

n chapter 7, I briefly previewed the theme of this chapter with figures 7.4 and 7.5, the Darwinian Spaces in which we can map a few
dimensions of cultural evolution. Figure 7.4 (p. 146) shows words,
like viruses, as *relatively* simple things (compared with organisms
and religions, for instance) that don't just wax and wane (like the
oxygen level in the atmosphere, like trust) but *replicate*—they have
offspring. Figure 7.5 (p. 148) shows the evolution *of the evolution* of
culture, from profoundly Darwinian processes (involving minimal
comprehension, bottom-up generation of novel products by random
search processes) to processes of intelligent design (comprehending, top-down generation of novel products by directed search). It is
time to provide, and argue for, the details of this evolution. Words,
I will argue, are the best example of *memes*, culturally transmitted
items that evolve by differential replication—that is, by natural
selection.

Other species have some rudiments of cultural evolution.

Chimpanzees have a few traditions: cracking nuts with stones, fishing for ants with sticks or straws, and courtship gestures (and a few others) that are not transmitted genetically between the generations but rather are *ways of behaving* that depend on the offspring's perception of the elders' behavior. Birds have a well-studied variety of ways of acquiring their species-specific songs: the vocalizations of seagulls and chickens, for instance, are entirely instinctual and need no acoustic model to develop, while the fledglings of most other species have an "instinct to learn" (as ethologist Peter Marler put it) but need to hear their parents sing their species-specific song. Experiments in *cross fostering* place eggs from two different species in the "wrong" nests and reveal that the hatchlings copy their foster parents' song as best they can. A wide variety of animal dispositions (ways of behaving) that were long thought to be genetically transmitted "instincts" have proven to be "traditions" transmitted between parent and offspring via perceptual channels, not genes (Avital and Jablonka 2000), but in none of these species are there more than a handful of such acquired ways of behaving.

We *Homo sapiens* are the only species (so far) with richly cumulative culture, and the key ingredient of culture that makes this possible is language. Thanks to our accumulating human culture, we have engulfed the planet, transforming the environment and also triggering an extinction event that may soon rival the mass extinctions of earlier eons if we don't take steps to reverse some of the trends. Our species has had a sustained population growth unprecedented in any vertebrate species (aside from our chickens, cows, and pigs). In the last two centuries our numbers have swelled from a billion to over seven billion, and while the rate is now slowing significantly, we will reach eight billion in less than a decade, according to the latest estimates from the United Nations.

Our genes haven't changed very much in the last 50,000 years, and the changes that have occurred were probably all driven directly or indirectly by new selection pressures created by human cultural innovations, such as cooking, agriculture, transportation, religion, and science. The widespread adoption of a new way of behaving creates a one-way ratchet: once almost everybody is eating cooked food, the human digestion system evolves, genetically, to make it no longer practical—and then no longer possible—for humans to live on a raw diet. Once transportation moves people around the world, first populating all the islands and continents, and then visiting/ invading all the islands and continents, it is no longer possible for human beings to live without a host of defenses against the diseases brought by travelers.

One of the facts of life, both genetic and cultural, is that *options become obligatory.* A clever new trick that gives its users a distinct advantage over their peers soon "spreads to fixation," at which point those who don't acquire it are doomed. If hiding from predators in a hole is a frequent enough lifesaver, eventually having a hole to hide in becomes no longer an eccentricity of a few members of a population but a species necessity, embodied in an instinct. Most mammal species can synthesize their own vitamin C, but primates can't; their growing reliance on fruit in their diet over the millennia led to the loss of the innate ability to make vitamin C—use it or lose it. Without vitamin C, you get scurvy and other illnesses, a discovery first made possible by human long-distance navigation, which (in effect) isolated test populations whose health and diet were important enough for navies and ship's surgeons to attend to. This discovery (along with hundreds of others) led over a few centuries—an evolutionary eye-blink—to today's exquisitely detailed scientific knowledge of the biochemistry of human metabolism. We are now *obligate* ingesters of vitamin C, but no longer obligate frugivores, since we can take vitamins as necessary. There is no law today obliging people to have either a credit card or a cell phone, which just a few years ago were luxurious options. Today we are so dependent on them, there doesn't need to be a law requiring them. Soon cash may disappear, and we will become *obligate* users of the computer technology of credit cards or their descendants.

We are approximately as dependent on words, today, as we are on vitamin C. Words are the *lifeblood* of cultural evolution. (Or should we say that language is the *backbone* of cultural evolution or that words are the *DNA* of cultural evolution? These biological metaphors, and others, are hard to resist, and as long as we recognize that they must be carefully stripped of spurious implications, they have valuable roles to play.) Words certainly play a central and ineliminable role in our explosive cultural evolution, and inquiring into the evolution of words will serve as a feasible entry into daunting questions about cultural evolution and its role in shaping our minds. It is hard to know where to start. It is obvious that we now have culture galore, and that there was a time when our ancestors didn't have culture, and hence that it must have evolved, but how? The answer to that question is far from obvious. As Richerson and Boyd (2004) observe:

> A little scientific theorizing is necessary to convince us that the existence of human culture is a deep evolutionary mystery on a par with the origin of life itself. (p. 126)

Darwin wisely began his "one long argument" in the middle. The *Origin of Species* has nothing to say on the topic of how it all began, leaving the murky details about the ultimate origin of life for another occasion. A dozen years later, in a famous letter to Joseph Hooker in 1871, he entertained the surmise that perhaps the conditions in a "warm little pond" would be just right for getting life going, and more than a dozen decades later we're still exploring that warm pond hypothesis and other alternatives. The origin of human language is another unsolved puzzle, and ever since Darwin it has provoked curiosity and controversy for much the same reasons. First, there are few "fossil traces" of its beginnings, and those that can be confidently identified are subject to multiple interpretations. Moreover, both puzzles concern momentous events that may well have occurred only once on this planet.

In both cases—we must be careful how we put this—it is *possible* that life, and language, arose several times or even many times, but if so, those extra beginnings have been extinguished without leaving any traces we can detect. (But who knows what tomorrow may bring?) There has to have been at least one origin for each, and there is no compelling evidence for more than one origin, so let's start with the assumption that there was just one in each case. We are then left with the task of composing scientifically plausible and indirectly testable hypotheses for how these two histories of successful evolutionary R&D might have arisen. Empirical work in both areas has made enough progress in recent decades to encourage further inquiry, taking on board the default (and tentative) assumption that the "trees" of existing lineages we can trace back eventually have single trunks.

Phylogenetic diagrams, or *cladograms*, such as the Great Tree of Life (which appears as figure 9.1 of the color insert following p. 238) showing all the species, or more limited trees of descent in particular lineages, are getting clearer and clearer as bio-informatics research on the accumulation of differences in DNA sequences plug the gaps and correct the mistakes of earlier anatomical and physiological sleuthing.[45] Glossogenetic trees, lineages of languages (figure 9.2), are also popular thinking tools, laying out the relations of descent among language families (and individual words) over many centuries.

Biologists drawing phylogenetic trees run into difficulties when they must represent *anastomosis*, the *joining together* of what had been distinct lineages, a phenomenon now understood to have been prevalent in the early days of life (witness the endosymbiotic origin of eukaryotes). Historical linguists drawing glossogenetic trees must confront the widespread anastomoses of merging languages (creating *pidgins* and *creoles* for instance) and the even more frequent jumping of individual words from one language to another. It is obvious that Inuit words like *igloo* and *kayak* are going to be adopted

---

45   The fact that a Tree of Life diagram is in one regard upside-down, with the later *descend*ants shown *higher* in the branches than their ancestors, is not such a strange inversion, but if it bothers you, turn the diagram sideways. I have never seen a horizontal cladogram where time runs from right to left, but perhaps they are used in Arabic and Hebrew texts.

FIGURE 9.2: Glossogenetic tree of all languages. © *Nature magazine.*

in many languages as soon as acquaintance with igloos and kayaks is common, and *computer* and *goal* from English are almost universally understood by people who don't speak English.

Dawkins (2004, pp. 39–55) points out that, in many instances, tree diagrams showing the lineages of individual *genes* are more reliably and informatively traced than the traditional trees showing the descent of *species*, because of "horizontal gene transfer"—instances in which genes jump from one species or lineage to another. It is now becoming clear that especially in bacteria and other unicellular organisms, genes are often traded or shared by a variety of processes independent of reproduction, which is *vertical* gene descent.

Similarly, etymologies (descent lineages) for *words* are more secure than the descent of the *languages* in which they are found, because of horizontal word transfer between languages.

The idea that languages evolve, that words today are the descendants in some fashion of words in the past, is actually older than Darwin's theory of evolution of species. Texts of Homer's *Iliad* and *Odyssey*, for instance, were known to descend by copying from texts descended from texts descended from texts going back to their oral ancestors in Homeric times. Philologists and paleographers had been reconstructing lineages of languages and manuscripts (e.g., the various extant copies of Plato's *Dialogues*) since the Renaissance, and some of the latest bio-informatic techniques used today to determine relationships between genomes are themselves refined descendants of techniques developed to trace patterns of errors (mutations) in ancient texts. As Darwin noted, "The formation of different languages and of distinct species, and the proofs that both have been developed through a gradual process, are curiously the same" (1871, p. 59). We will follow Darwin's example and begin in the middle, postponing the primordial origin of language until later in the chapter, and looking at what can be confidently said about the ongoing evolution of language, and in particular, words.

Looking more closely at words

What are words? A useful bit of philosophical jargon first formulated by Charles Sanders Peirce in 1906 is the *type/token* distinction.

> (1) "Word" is a word, and there are three tokens of that word in this sentence.

How many words in sentence (1)? Fifteen if we're counting tokens (like the word counter in Microsoft Word), but only thirteen if we're counting types. If you say sentence (1) aloud, there will be three different acoustic events, tokens of the type "word," occurring in succession over several seconds. On the printed page there are three different ink patterns that look very much alike (the first is capitalized, but still a token of the type "word"). Tokens can be audible events, ink patterns, skywriting patterns, grooves carved in stone, or bit-strings in computers.[46]

They can also be silent events in your brain. When you read sentence (1) silently, in addition to the visible tokens on the page there are tokens in your brain, and, like tokens spoken aloud, these tokens have physical properties and clockable times of occurrence. (Written tokens can have long lifetimes, starting when the ink is not yet dry and lasting until the paper decays or is burned.) We don't yet know how to identify brain-tokens of words by their physical properties—*brain "reading"* hasn't been figured out yet, though we're getting closer—but we can be sure that the brain-tokens of the type "word" will be as physically different from the written and spoken tokens as those are from each other. The brain-tokens will not look like "word" or sound like "word" (they're *brain* events, and it's dark and quiet in there), but they will no doubt be physically similar to some of the events that normally occur in your brain when you see or hear "word."

This curious fact is often overlooked by theorists of consciousness (as we will see in chapter 14), with devastating effects. Read sentence (1) *silently* but in a high-pitched voice. Now read it *silently* with a Norwegian accent (lots of ups and downs of pitch). Did it "sound Norwegian" in your mind's ear? Pronouncing a word in your mind is extra work, not necessary for a practiced language-user, but a use-

---

46 And to honor a sometimes important technical distinction, we should note that in the sentence of the *type* "the cat is on the mat" there are two *occurrences* of the type "the" and four *occurrences* of the type "t." Occurrences are, like types, abstractions, not concrete entities. A similar distinction must be made (but is often ignored, with resulting confusion) when we speak of genes. To a first approximation there is at least one *token* of each of your gene *types* in each of your cells but also multiple *occurrences* of the same sequence of codons in your genome. For instance, gene *duplication*, which is crucially involved in many of the major innovations in evolution, is the amplification of a single multiple copying of a gene *token* into trillions of duplicate tokens in the cells of offspring. For a lucid account see Dawkins (2004).

ful crutch when learning a new language—like moving your lips when you read. Your brain is equipped with auditory "machinery" that specializes in distinguishing speech sounds and with articulatory "machinery" that specializes in controlling your tongue, lips, and larynx. These neural systems were trained in the course of learning a language, and they play a major role in "saying words in your mind," but this activation of brain areas doesn't itself make any noise—don't hunt for any vibrations that could be picked up by a suitably sensitive microphone implanted in your brain. Now close your eyes and "*see*" the first four words of sentence (1) in your mind's eye—first in yellow against a black background and then in black against a white background. Judging by introspection (a tricky and unreliable method, but still indispensable at times like this, the outset of investigation), when we "have words running through our heads" we don't normally go to all this trouble of "hearing" them with a full complement of auditory features such as tone of voice, punctiliousness of pronunciation, and emphasis, or "seeing" them (in a particular font, color, and size). When you can't get her name out of your mind, is it always capitalized, or pronounced with a sigh?

So there seem to be lots of intermediate cases of words—definite, specific words—tokened in our minds without going into the distinction between spoken or written, heard or seen. And there also seems to be "wordless" thinking where we don't even go to the trouble of "finding" all the words, but just *tokening their bare meanings*. In the "tip-of-the-tongue" phenomenon, for instance, we may know a lot about the word we're "looking for" but are unable to retrieve it until suddenly it "comes to us." It ends in "ize" and has three syllables and means sort of the opposite of *ignore*. Ah! *Scrutinize!* Sometimes, failure to find the word we are groping for hangs us up, prevents us from thinking through to the solution of a problem, but at other times, we can wordlessly make do with unclothed meanings, ideas that don't have to be in English or French or any other language. An interesting question: Could we do this if we didn't have the neural systems, the "mental discipline," trained up in the course of learning our mother tongue? Might wordless thought be like barefoot waterskiing, dependent on gear you can kick off once you get going? Psycholinguists and neurolinguists have made a good beginning on supplementing (and often replacing) this casual armchair introspection with controlled experiments and models of speech production and perception. (Jackendoff 2002, especially chapter 7, "Implications for Processing," is a valuable introduction.)

The problem with introspection is that it *acquiesces* in the illusion that there is an inner eye that sees and an inner ear that hears—and an inner mind that just thinks—these intimately familiar items, the objects of consciousness that parade across the stage in the Cartesian Theater (Dennett 1991). There is no Cartesian Theater; there just seems to be a Cartesian Theater. There do indeed *seem* to be these items, audible and visible and thinkable, and—setting magic aside—it couldn't seem so unless there really were physical tokens in the brain *playing those roles*, but how, physically, anything in the brain manages to play any of these roles is a topic for future scientific investigation, *not* introspection.

We're getting ahead of ourselves; there will be an occasion for gathering the implications of these observations for a theory of consciousness in chapter 14. For now, I simply want to acknowledge the obvious fact that in addition to public tokens of words, there are private, internal tokens of words and draw attention to the much less obvious fact that we don't yet know anything much about their physical properties. Internal tokens *seem to resemble* external tokens, but this is because they make use of the very same neural circuitry we use to detect the resemblances and differences between external tokens, not because this neural circuitry *renders copies* of what it identifies.[47]

---

47   It may help to imagine this if you compare the phenomenon to Shazam, the smartphone app that can identify music playing on the radio. The signal transduced by the microphone doesn't have to get "turned back into sounds" in order to be identified by the program.

Even if we someday succeed in mastering the art of identifying the words being entertained—tokened—by people in their brains, this will not by itself constitute an ability to "read their minds" in the sense of discovering their beliefs or even their thoughts. To see this, please silently utter "down with democracy!" five times. Suppose I could read your brain-tokens, I could discover what you were covertly saying, but it wouldn't show me that you believed what you silently rehearsed, would it? Mind reading of this sort may someday be a reality, and if so, mind-word-token identifying would play approximately the same evidential role as wiretapping does today: telltale but far from conclusive grounds for attributions of belief.

There is scant likelihood, in any case, that your brain-tokens of "word" are physically similar (in shape, or location, or other physical properties[48]) to my brain-tokens of "word," and I wouldn't bet against the proposition that you yourself have many physically *different* kinds of brain-tokens of "word," involving coalitions of neurons scattered through the language-processing parts of your brain, more different in shape and other physical properties than these written tokens:

word word WORD WORD *word* **WORD** word

What makes something a token of a type is not simple resemblance at all: the written token "cat" and the sound you make when you read it aloud are both tokens of the type "cat" and don't resemble each other at all. Even spoken tokens can have strikingly different properties: a *basso profundo* uttering "cat" creates a sound that shares hardly any physical properties with a whispered rendering of the word by a five-year-old girl, but both are *easily* recognizable by

48   We know that there are brain regions where knowledge of different types of words—for food, for musical instruments, for tools—is stored (if they get knocked out by a stroke or other cerebral accident, there is selective loss of knowledge of meanings), so your tokens of "violin" and my tokens of "violin" may both involve activity in the same place in our brains, but it would be beyond optimistic to hope that having located "violin," "trumpet" and "banjo" in my brain would help researchers pinpoint the other terms in yours.

English speakers as tokens of *cat*. Suppose Alan utters "chocolate" to Beth, who writes "chocolate" on a slip of paper and hands it to Cory, who whispers "chocolate" to Dave, who types "chocolate" on his iPhone and shows it to Emily, and so on. Any process that makes a new token of a type from an existing token of a type counts as a replication, whether or not the tokens are physically identical or even very similar. Tokens of words are all physical things of one sort or another, but words are, one might say, made of information, like software, and are individuated by types, not tokens, in most instances. The version of Microsoft Word currently installed on your laptop is one token, but when we talk about Microsoft Word—even Microsoft Word 14.4.1—we generally mean the type, not the token.

The idea that words have evolved by natural selection was obvious to Darwin, and one would think that Noam Chomsky, who pioneered the idea of an innate Language Acquisition Device in the brain, would look with favor on evolutionary accounts of how languages in general, and words in particular, came to have their remarkable features, but he has disparaged evolutionary thinking in linguistics in almost all regards (more on this in chapter 12). Following Chomsky, many linguists and philosophers of language have staunchly resisted evolutionary thinking, with a few notable exceptions. In her pioneering book *Language, Thought and Other Biological Categories* (1984), philosopher Ruth Millikan surmised that words were *descendants* of "vocal tract gestures." Philosopher David Kaplan (1990) introduced a model of words as *continuances* and specific utterances or inscriptions (or occurrences in the head) as *stages*, a "naturalistic" model that was clearly inspired by Darwin, especially in its rejection of unchangeable Platonic essences, and its recognition that tokens can vary widely in physical properties while still being of the same type. (In fact, Kaplan disparaged the token-type terminology—long a staple in philosophy of language—because of its residual whiff of essentialism but relented in the end: "I don't mind if you want to continue to call utterances and inscriptions 'tokens,' although I'd prefer 'utterance' and 'inscription,' so long as we do not get caught up in the metaphysics of the token/type model," p. 101.) Kaplan's flirtation with evolution met resistance from the Chomsky camp at MIT, with philosopher Silvain Bromberger (2011) reacting in typical MIT style:

> The indisputable fact remains that talk of words changing is at best a convenient shortcut, and a way of being noncommittal about empirical details. But the moral, once again, is that anyone seriously interested in the ontology of language ought not to take *façons de parler* about change at face value. . . . People changed, not "words"! (pp. 496–497)

Bromberger added a footnote: "If they are abstract entities, how could they change?" As if this were decisive. What must Bromberger think of genes? That they are also a mere *façon de parler*? The opposition to evolutionary thinking of Chomsky and some of his colleagues and followers around the world used to discourage philosophers of language from looking at the possibilities, but the times are changing. In addition to Millikan, Kaplan, and me, Daniel Cloud (2015) has taken up the perspective, and Mark Richard (forthcoming) explores the analogy between biological species and linguistic meanings (more on this in chapter 11).

My colleague the linguist Ray Jackendoff (2002) has developed an evolutionarily informed theory of language intended to join forces with recent neuroscience, and on his account, words are *structures in memory* that are autonomous in the sense that they must be independently acquired (learned).[49] They are items of *information*, as defined in chapter 6. Here are some other *informational structures*: stories, poems, songs, slogans, catchphrases, myths, techniques, "best practices," schools of thought, creeds, superstitions, operating systems, web browsers, and Java applets, among other things. Informational structures come in various sizes (measuring the number

---

49   Problematic connotations of the everyday term "word" led Jackendoff to propose and define a technically more precise substitute, "lexical item." Not all lexical items have pronunciations, for instance.

of parts, not the physical size of some of the tokens, such as the HOLLYWOOD sign). Novels are quite large, poems usually much smaller, traffic signs (KEEP RIGHT) smaller still, trademarks often single shapes (of any material).

Words have, in addition to the visible or audible parts of their tokens, a host of informational parts (making them nouns and verbs, comparatives and plurals, etc.). Words are autonomous in some regards; they can migrate from language to language and occur in many different roles, public and private. A word, like a virus, is a minimal kind of *agent:* it *wants to get itself said* (Dennett 1991, pp. 227–252). Why? Because if it doesn't, it will soon go extinct. A word is *selfish* in exactly the same way a gene is selfish (Dawkins 1976). This metaphorical usage has proven to be very effective in directing our minds to illuminating perspectives on evolution. (Don't be afraid of a little metaphor; it won't bite you, but you should always make sure you know how to cash it in for unvarnished fact when you feel the urge.)

An informational thing doesn't have a mind, of course, any more than a virus does, but, like a virus, it is designed (by evolution, mainly) to provoke and enhance its own replication, and *every token it generates is one of its offspring.* The set of tokens descended from an ancestor token form a *type,* which is thus like a *species.* Now we can appreciate Kaplan's reluctance to keep the type-token distinction, since tokens shade off gradually until a new *type* emerges, much as the chain of offspring of some dinosaur eventually emerged as members of some new bird species. Some of a word-token's offspring will be private utterances: its human host is talking to herself, maybe even obsessively rehearsing the word in her mind, over and over, a population explosion of tokens, building an ever more robust niche for itself in a brain. (And it may well be that many more internal tokenings—offspring—are born outside our conscious attention altogether. At this very moment, words may be replicating competitively in your head as inconspicuously as microbes replicate in your gut. More on this later.) Some of its offspring will be uttered aloud, or written, and maybe even printed in books, and a scant few of these offspring will take up residence in other brains, where they will either find a home already prepared for them—they are recognized by their new hosts—or they will have to start a new niche.

How do words reproduce?

As with viruses, there is safety in numbers. A single copy of a virus can infect a new host, but a flood is much more likely to gain a foothold and set up a reproducing colony of sorts. Similarly, a single hearing of a word can be enough, especially in an interested adult listener, to establish a new vocabulary item, a new generator of word-tokens, but multiple hearings are more likely to make an impression on an infant, almost literally.

How do words get themselves installed in infant brains? Children learn about seven words a day, on average, from birth to age six. (How do we know? Simple: measure the vocabulary of a six-year-old—in the range of 15,000 words—and divide by the number of days she's been alive—2,190 on her birthday. She learns about 200 words the first two years of her life, and then the acquisition process picks up speed for a few years, before tapering off sharply. How many words have you learned this week?) If we zero in on her earliest days of word learning, we discover that it takes, on average, about six tokenings of a word in the presence of the child to generate her first clear efforts to say the word, to utter a copy (Roy 2013; see also Roy's brilliant TED talk at http://www.ted.com/talks/deb_roy_the_birth_of_a_word). So you might say that unlike a virus, a word has to have multiple parents before it can be born, but they don't have to come together at the same time. Most of the words infants hear are not directed to them; the infants are overhearing the speech of parents and caregivers. The first occurrence of a word is just a novel auditory event in some complex, largely inscrutable perceptual context, but it makes an impression of sorts on the brain. The second occurrence adds to that impression (if the child could talk, it might say, "Oh, there's *that sound* again"), and it occurs in a context that may (or may not) share some features of the first context. The third occurrence is a smidgen more familiar, and the context in which it occurs may begin to come into focus in some ways. The fourth, fifth, and sixth occurrence drive home the auditory signature, the *phonology* as linguists say, and create an anchor of sorts in the brain. At that point the child has a target for action: *say it!* (We can surmise that the disposition to form this goal has by now been genetically installed in the human genome, but in the earliest days of hominin vocalization/communication,[50] this was probably a variable idiosyncrasy, a fortuitous extension of a general tendency in favor of imitation of, and interaction with, parents and caretakers. (Much more on the origins of language in chapter 12.)

Once the child starts trying to say words, the adults respond by slowing down and simplifying the vocalizations they make in the child's company (Roy 2015). The child is no longer being treated as an inadvertent overhearer, but as a candidate participant in conversation. After only a few more generations of tokens, with the child using feedback from his or her own performance to improve execution, recognizable copies begin to emerge: "baby's first words," but the baby doesn't have to understand them at all or even understand that they are words. The baby has simply developed a habit of uttering these structured sounds in ever more specific contexts, and she sometimes gets an immediate reward, in much the way her "instinctual" crying tends to lead to food, or cuddling, or cessation of discomfort. The sounds of specific words become more and more familiar, recognizable, identifiable, as the repetitions pile up in both uttering and hearing. A pronounceable word has taken up residence in the infant's brain, and at this point it is good for . . . nothing but reproducing itself. Soon it begins to acquire uses discernible (unconsciously) by the infant, and gradually it begins to mean something to the child.

---

50   *Hominids* are the category of primates that includes the chimpanzees and other apes, while the more narrow category of *hominins* is the six-million-year-old branch that includes only humans and their nearest ancestors.

These generations of offspring may be unconscious since the infant probably hasn't yet developed a competent consciousness out of its torrent of stimulation and reaction. Let's approach this contentious issue cautiously. At this point, we needn't try to "draw the line" distinguishing conscious from unconscious activity even if there is, in the end, a sharp line to be drawn. Just as waking from a deep sleep is often accomplished gradually, with no moment standing out as the first moment of resumed consciousness, an infant's development of consciousness in a form worthy of the term (something beyond mere sensitivity, responsivity to stimuli, which is ubiquitous in plants and bacteria) is probably gradual, like all other great transformations in life.

Some people cling to the view that consciousness is the big exception, an all-or-nothing property that divides the universe into two utterly disjoint classes: those things it is *like something to be,* and those that it is not like anything to be (Nagel 1974; Searle 1992; Chalmers 1996; McGinn 1999). I have never encountered a persuasive argument for why this should be so. It strikes me as a dubious descendant of vitalism, the now all-but-abandoned doctrine that being alive is a metaphysically special property that requires some sort of infusion of wonder-stuff, *élan vital.* Since this metaphysically extravagant view of consciousness is still popular among some very thoughtful people, including both philosophers and scientists, I will acknowledge that at this stage infants are only *probably* not (really, very) conscious of the word-offspring that are beginning to thrive in their brains. If those very thoughtful people prove someday to be right—there really is a cosmic divide—I will revise my claim. If the magic spark is identified tomorrow, I will agree that consciousness arrives in the infant (or fetus) at *Magnificent Time T,* as we now know, but I will then insist that consciousness does not develop its familiar suite of talents, the conscious agent's ability to *do things because it is conscious,* until that agent gradually gets occupied by thousands of memes—not just words—that (re-)organize the neural connections on which these talents depend.

So we are looking at that brief period in an infant's life when random vocalizations (or maybe they are attempts to copy a heard sound) can become meaningless baby words that never get promoted beyond this functionless thriving. These vocal microhabits manage to get themselves rehearsed for months until they either go extinct in the face of competition with more effective vocalizations or acquire a local meaning thanks to eagerly cooperating interlocutors. For most people, their baby words are the only words they will successfully "coin" in their whole lives. If doting parents and siblings fall into echoing the nonsense words back to baby, without ever attaching a meaning to them, they are complicit in the creation of a genuine pronounceable meme with no function other than reproduction, a commensal mind-virus that neither helps nor hinders but prospers for a while, nonetheless.[51]

The more effective words, the mutualist memes that become the indispensable toolkit or vocabulary of every person, acquire their semantics, their syntax, and their connotations by a gradual, quasi-Darwinian coevolutionary process of bottom-up learning, exploiting the brain's pattern-detecting prowess and taking advantage of the brain's predesigned set of affordance detectors (Gorniak 2005; Gorniak and Roy 2006). This bald assertion, in the face of all the known complexities of the semantics and syntax of natural languages, may strike theoretical linguists as irresponsibly speculative, but consider how modest a claim it makes: phonology comes first, creating a node or focus in the brain for the auditory signature of the word, and then this becomes the basis, the anchor or collection point, for the semantics and syntax to develop around the sound, along with its articulatory profile—how to say it. I'm not taking sides in the controversies, even where I have convictions; I have more than enough to defend without defending a verdict on how much bias, how much "Universal Grammar," as the Chomskyans say,

---

51 Symbionts are classified into three kinds: *parasites*, which are deleterious to the host's genetic fitness; *commensals*, which are neutral (but "eat at the same table"); and *mutualists*, which benefit the host, enhancing its genetic fitness.

must be genetically installed in the "Language Acquisition Device" nor do I stipulate in any way what form this biasing must or might take. One way or another, without deliberate (let alone conscious) "theory construction" of any kind, the infant brain homes in on habits that it will share with its parents, responding sensitively to semantic information present to its senses.[52]

Today, the installation of a native language in a human infant is a well-designed process, benefiting, no doubt, from the selective impacts on thousands of generations of human-language learners. Many shortcuts may have been discovered by evolutionary processes along the way, making languages easier to acquire, words easier to pronounce and recognize, assertions and questions and requests and commands easier to formulate and express. These evolutionary processes would not initially have involved differential replication of *genes*, but rather the much swifter differential replication of *memes*. Language evolved to fit the brain before the brain evolved to better accommodate language.[53]

No doubt pronounceable memes are to some degree constrained by the physics and physiology of articulation and human hearing. However much variation exists in a particular language, children normally acquire adult competence in recognition and articulation with little or no instruction. And no doubt the semantic properties their newly minted words share, once well installed, are heavily dependent on the structures of the nonlinguistic part of the human *Umwelt*, the manifest image of affordances and the actions that deal

52   Here I mean semantic information in the broad sense developed in chapter 6. I am referring to the (semantic) information *about* the syntax, the phonology (in the case of hearing infants), the pragmatics, the contexts, and the meanings (semantics) of the items being acquired.

53   Deacon (1997) offers a good account of this. See also Deacon (2003), and Dennett (2003b) both in Depew and Weber, ed., *Evolution and Learning: The Baldwin Effect Reconsidered*. More recently, Christiansen and Chater (2008) present an ambitious defense of (an exaggerated version of) this thesis in a *Behavioral and Brain Sciences* Target Article, surveying the alternatives and generating a particularly valuable collection of responses.

with them. Here too, children acquire the semantics with almost no instruction. Doting parents in most cultures love to teach their children the names for things, but however enthusiastic they are about this project, their children acquire the meanings of most of their words gradually, without explicit attention, and without any deliberate parental guidance. Repeated exposure to thousands of words in contexts provides almost all the information needed to come to a reliable sense of what they must mean, requiring just a little fine-tuning as occasion demands later on. (Is a pony a kind of horse? Is a fox a kind of dog? Am I *ashamed* or *embarrassed*?)

Socrates was puzzled about how people could figure out a *definition* of a word from just talking it over with each other. If they didn't already know what the word meant, what good would be served by querying each other, and if they already did know what it meant, why bother defining it—and why should defining it be so difficult? Part of the dissolution of this conundrum is the recognition that understanding a word is *not* the same as having acquired a definition of it.[54]

Perhaps, over and above these semantic and phonological constraints there are specific, genetically designed *syntactic* constraints and attractors imposed by structural properties of our brains. Perhaps these are frozen accidents of earlier evolutionary history that now limit the class of learnable languages in ways that could have been otherwise had history taken a slightly different course. It might help to imagine an example: suppose, as some theorists do, that language had its origins in gesture, not vocalization; it might be that some ordering principle that was optimal *for a gesture language* (because, say, it minimized abrupt shifts in the position of the arms that would have strained the muscles of loquacious ges-

---

54 Here is a place where the "top-down," intelligent design aspect of GOFAI is particularly vivid: if you think of an agent's knowledge as a set of axioms and theorems, like Euclidean geometry, you're going to need to "define your terms." The fact that CYC has many thousands of definitions hand-coded into its database is a testimony to the conviction among some people that top-down AI is still the way to go.

turers) was somehow carried over as a constraint on vocalized languages, not because it made them more efficient, but because it *used* to make gesture languages more efficient. This would be like the QWERTYUIOP pattern on our keyboards, which had a good reason back in the days of mechanical linkages between key press levers and type levers. Letter combinations that are frequent in English, such as "th" and "st," were assigned to relatively distant keys, to prevent jamming of type levers on their way to and from the paper. Or perhaps the syntactic constraints imposed by evolved brain structures are *globally* optimal; given the neural architecture of mammalian brains here on Earth at the time language first evolved, it is no accident that terrestrial languages have these structural features. Or perhaps they are Good Tricks that we would predict to appear in any natural language anywhere in the universe. If the tricks are that good everywhere and always, there is a good chance that (by now) they have been installed in our genome, by the Baldwin Effect. (It is possible for a behavioral innovation X to prove so useful that anybody who doesn't X is at a disadvantage, and any variation in the population for *acquiring* or *adopting* behavior X will be selected for genetically, so that after some generations, the descendants prove to be "natural born Xers" who need hardly any rehearsal; Xing has been moved, by the Baldwin Effect, into their genomes, as an instinct.) Or perhaps they must be learned anew by each generation but readily learned since they are such obviously Good Tricks.[55] Nobody would suppose that the fact that human beings everywhere always throw their spears pointy-end first shows that there must be a gene for a pointy-end-first instinct.

In any case, the epistemological problem confronting the newborn infant may *resemble* the epistemological problem of an adult field linguist confronting an alien language and trying to adduce a grammar and lexicon, but the methods each uses to close the gap between igno-

---

55   Note: A Good Trick could be "obvious" to natural selection while quite inscrutable to inquiring human investigators. Never forget Orgel's Second Rule: Evolution is cleverer than you are.

rance and knowledge are strikingly different. The adult frames and tests hypotheses, seeking evidence that could confirm or disconfirm these informed guesses, refining the generalizations, noting the irregular exceptions to rules, and so forth. The infant just babbles away, trying to assuage her hungers and curiosities, and gradually bootstraps herself into comprehension by an unconscious process of massive trial and error, accomplishing the same epistemic feat, but without the benefit of theory, the way evolution by natural selection accomplishes the feat of designing the wings of different bird species without the benefit of a theory of aerodynamics. This is reminiscent, of course, of the contrast between top-down intelligent design—Gaudí and Turing and Picasso being our exemplars—and Darwinian R&D. Children acquire their native languages by a quasi-Darwinian process, achieving the competences that are the foundation for comprehension by a process that is competent without comprehension.

First words—both in infants today and in the deep history of language in our species—are thus best seen as synanthropic species (see chapter 6), evolved by natural selection (both genetic and cultural) to thrive in human company, with all its peculiarities of physiology, habitat, and needs. Synanthropy is probably the route to domestication taken by most of our domesticated species. For instance, as Coppinger and Coppinger (2001) argue, the myth of wolves being turned into dogs via the deliberate removal of wild wolf pups from their dens by intrepid domesticators is hardly plausible. Almost certainly once human settlements began to generate concentrations of discarded edible food, these dumps became attractive to wild wolves who varied, initially, in their ability to tolerate proximity to dangerous human beings. Those who could bear to be that close became geographically and reproductively isolated from their more wary cousins, who kept their distance, and what emerged over time was the junkyard dog, belonging to nobody, nobody's pet or companion, but a familiar presence in the community, like house mice, rats, and squirrels. Dogs in effect *domesticated themselves* over many generations until their human neighbors became their owners, their companions, their guardians, their masters.

Similarly, we can imagine that it only gradually dawns on infants that they *have* all these words they can start to *use*. Indeed they start using them, and enjoying the benefits of using them, long before they realize that they are doing this. Eventually, they reach a state where the words in their manifest image become *their own* words, affordances that belong to their kit, like a club or a spear, rather than affordances that just appear in Nature, like a stone to throw or a cave to hide in. Their words have become domesticated, in the process that Darwin outlined in the opening chapter of *Origin of Species* (1859) and developed in his magisterial work *The Variation of Animals and Plants under Domestication* (1868). It starts with what Darwin calls "unconscious" selection, with people willy-nilly or inadvertently favoring some offspring at the expense of others, thereby creating a selective force that later becomes more focused and more directed. Eventually we get to "methodical" selection, in which pigeon fanciers, or rose growers, or horse or cattle breeders, for instance, have specific goals in mind, specific targets of features they are trying to select for, and this is a major step toward top-down intelligent design, where the domesticators *have* reasons (good or bad) for what they are doing and what they hope the outcome will be. In the case of the domestication of words, this emerges when individuals begin to become reflective or self-conscious about their use of language, shunning words they find unimpressive or offensive or old-fashioned (or too slangy and new).

When people start editing their language production (literally in the case of writing, or figuratively in the way they silently try out sentences in their heads before uttering them aloud), they are fulfilling Darwin's definitive requirement for domestication: controlling the reproduction of the members of the species—at least those that they "own."[56] Notice that there may *be* reasons why people "uncon-

---

56   See Cloud (2015) for a development of this issue. I view Cloud's account as a substantial friendly amendment to my earlier brief remarks about the domestication of words, not a rebuttal, but there are some residual disagreements between us to be sorted out later.

sciously" shun certain words or favor others, but the result is not fully domesticated language until they *have* reasons. Synanthropic words, like synanthropic organisms, have to fend for themselves when it comes to reproduction; they don't get any stewardship from owners. Once domesticated, they can relax somewhat: their reproduction is all but assured, thanks to the care of their guardians. Carefully chosen technical terms in a scientific field, for instance, are prime examples of fully domesticated memes, patiently taught to the young, with enforced rehearsal and testing. They don't *have* to be amusing, exciting, tinged with the seductive aura of taboo, like the most successful synanthropic reproducers. Domesticated words have the backing of the Establishment and get reproduced with the help of whatever methods have proven to work, but we are wise to follow Darwin's lead and recognize that there is something of a continuum between these full-blown cases of domestication and the patterns of preference that lead to unconscious selection of words for particular contexts.

Phonemes are perhaps the most important design features of human language, and they probably were largely in place when words were synanthropic, barely noticed features of the memetic environment, not valued possessions of human beings. Phonemes accomplish the *digitization* of the auditory medium for speech. Each spoken language has a finite auditory alphabet of sorts, the phonemes from which words are composed. The difference between *cat* and *bat* and *pat* and *sat* is the initial phoneme, and the difference between *slop* and *slot* is the terminal phoneme. (When one learns to read by the "phonics" method, one learns how written forms map onto the phones [sounds] that make up the phonemes [distinctive sound units] of a language. The "t" sound in "tuck" is in fact slightly different from the "t" sound in "truck" but these are different phones, not different phonemes, because no two words [in English] differ by just this difference [imperceptible to most English speakers] the way "tuck" and "duck" do, for instance.)

What is brilliant about phonemes is that they are largely impervious to differences in accent, tone of voice, and other variations common in everyday speech: There are hundreds of different ways of uttering "please pass the butter" (cowboy twang, Virginia drawl, Valley girl whine, Scottish whisper . . .) that are all effortlessly identified by almost all English speakers as instances of "please pass the butter." There is well-nigh continuous variation in the physical phenomena that get pushed through the procrustean filters that "correct to the norm" and "fish-or-cut-bait," turning every borderline case into one phoneme or another. This is the heart of digitization, obliging continuous phenomena to sort themselves out into discontinuous, all-or-nothing phenomena. Phonemes protect the boundaries of *types* so that they can have wildly variant *tokens*, a requirement for reliable replication. Our auditory systems have been tuned since before birth (yes, in the womb) to classify a wide and continuously varying spectrum of sounds as instances of the phonemes of our native or first language, and if we wait until we are teenagers to learn another language we will probably always have difficulty not only with our accents but with effortless perception of the words spoken by others, since we will have unreliable and imperfect phoneme detectors.

Without a digitization scheme, audible sounds are hard to remember, hard to reproduce, likely to drift imperceptibly away from their ancestors, leading to the "error catastrophe" phenomenon, in which natural selection cannot work because mutations accumulate faster than they can be selected for or against, destroying the semantic information they might otherwise carry. Whether or not an audible sound is *worth* remembering, reproducing, or copying, if it is not digitizable, its prospects of being well remembered are dim. So synanthropic audible memes, whether mutualist, commensal, or parasitical, will persist and spread widely only if they happen to have phonemic parts: *fiddle-de-dee, razzamatazz, yada yada*.

This is the same digitization that gives computers their striking reliability. As Turing noted, nothing in nature is truly digital; everywhere there is continuous variation; the great design move is making devices that *treat* all signals as digital, discarding instead of copying the idiosyncrasies of particular tokens. That is how it is possible to make practically *perfect* copies of text files and music files. Compare a CD copier with a photocopier: if you photocopy a photograph, for instance, and then copy the photocopy, and then copy the copy of the photocopy and so forth, you will gradually accumulate blurring and other visible "noise," but a JPEG picture file on a CD can be downloaded to a computer and then used to make another copy on another CD, and so forth, indefinitely with no loss of clarity or fidelity at all, because at each copying step whatever tiny variations there are in the individual tokens of 0 and 1 are ignored during the copying process. The same is true in written language, of course, with its finite alphabets normalizing essentially infinite variability in letter shape and size. What this provides for spoken (and written) language is extremely reliable transmission of information *in the absence of comprehension.*

TAE CAT

FIGURE 9.3: Selfridge's automatic CAT.

A famous written example is due to Oliver Selfridge (figure 9.3). An English speaker will read this as "THE CAT" even though the second and fifth symbol are exactly the same intermediate shape. Spoken words are also automatically shoehorned into phonemic sequences, depending on the language of the hearer. English-speaking audiences have no difficulty reproducing with perfect accuracy "mundify the epigastrium" on a single hearing, even when they have no inkling what it could mean (soothe the lining of the stomach—a slang term for "have a drink" in some quarters), but they are unable to reproduce accurately the sounds that might be transliterated as "fnurglzhnyum djyukh psajj." No matter how loudly and clearly articulated, this sequence of vocal sounds has no automatic decomposition into phonemes of English. Even nonsense ("the slithy toves did gyre and gimble in the wabe") can be readily perceived and accurately transmitted, thanks to this system of norms.

Phonemes are not just a brilliant way of organizing auditory stimuli for reliable transmission; they are also a kind of benign illusion, like the ingenious user-illusion of click-and-drag icons, little tan folders into which files may be dropped, and the rest of the ever more familiar items on your computer's desktop. What is actually going on behind the desktop is mind-numbingly complicated, but users don't need to know about it, so intelligent interface designers have simplified the affordances, making them particularly salient for human eyes, and adding sound effects to help direct attention. Nothing compact and salient inside the computer corresponds to that little tan file-folder on the desktop screen. And nothing compact and salient can be discerned among the physical properties of different occurrences of the phonemes that distinguish *cat* from *bat* from *bad* from *bed* and *ball* from *bill* and *fall* from *full*. The differences between these words seem simple and obvious, but that is an illusion engendered by our inbuilt competence, not by underlying simplicity in the signal. After decades of research and development, speech-recognition software is finally almost as competent as a five-year-old child at extracting the phonemes of casual speech from the acoustic maelstrom that arrives at an ear or a microphone.

The digitization of phonemes has a profound implication: words play a role in cultural evolution that is *similar* to the role of DNA in genetic evolution, but, unlike the physically identical ladder rungs in the double helix made of Adenine, Cytosine, Guanine, and Thymine, words are not physically identical replicators; they are "identical" only at the user-illusion level of the manifest image. Words, one might say, are a kind of *virtual DNA*, a largely digitized medium that exists only in the manifest image.

Millions of acorns and apples are broadcast into the world by oak trees and apple trees "in hopes of" starting new descendants. When we broadcast millions of words the "hopes" are not for descendants of *us* but descendants of the words themselves, like the cold virus descendants we sneeze on our companions. When they arrive in other brains they may "go in one ear and out the other" or they may, occasionally, take root. One of my grade school teachers used to tell us "use a word three times and it's yours!" as a way of increasing our vocabularies, and she wasn't far off the mark.

Some philosophers will frown at this point, and worry that I am gliding over some treacherous thin ice. Do words even exist? Are they part of your ontology? Should they be? This talk of words being "made of information" is pretty dicey, isn't it? Just a lot of hand-waving? Some philosophers will bite the bullet at this point and insist that words *don't exist*, strictly speaking. They have no mass, no energy, no chemical composition; they are not part of the scientific image, which they say should be considered the ultimate arbiter of ontology. But words are very prominent denizens of our manifest image, and even if science doesn't have to *refer* to them or *mention* them, you couldn't do science without *using* them, so they should perhaps be included in our ontology. They loom large for us, readily occupying our attention.[57]

The contrast between us and chimpanzees here is striking: by now, thousands of chimpanzees have spent their entire lives in human captivity, and such chimps have heard almost as many words as human children hear, but they seldom pay any attention. Human speech is to them pretty much like the rustling of leaves in the trees, even though speech contains vast amounts of semantic information that could be of use to them, if only they tumbled to it. Think of how much easier it would be for chimps to escape captivity or foil experimenters if they could overhear and understand the conversa-

---

57 This is perhaps too strong; there may be languages in which there is no word for *word*, and the speakers of such languages may not have paid attention to the fact that their languages can be broken down into their parts! In much the way children have to be taught how to "sound out" words when learning to spell, everybody has to learn that what they are so naturally speaking is something composed of reusable parts. This part of folk linguistics may not be as obvious to *everybody* as it is to us self-conscious word-mongers.

tions of their keepers. It takes a prodigious training regime to get a chimpanzee to acquire the habit of attending to words, spoken or signed or tokened in plastic shapes. Human infants, in contrast, are hungry for verbal experience from birth. Words are affordances that *our* brains are designed (by evolutionary processes) to pick up, as Gibson said, and they afford all manner of uses.

The Meme's-Eye
Point of View

Words and other memes

I think that a new kind of replicator has recently emerged on this very planet. It is staring us in the face. It is still in its infancy, still drifting clumsily in its primeval soup, but already it is achieving evolutionary change at a rate which leaves the old gene panting far behind. . . . The new soup is the soup of human culture. We need a name for the new replicator, a noun which conveys the idea of a unit of cultural transmission, or a unit of *imitation*.
>   —Richard Dawkins, *The Selfish Gene*

I am convinced that comparisons between biological evolution and human cultural or technological change have done vastly more harm than good—and examples abound of this most common of intellectual traps. . . . Biological evolution is powered by natural selection, cultural evolution by a different set of principles that I understand but dimly.
>   —Stephen Jay Gould, *Bully for Brontosaurus*

Words exist in the manifest image, but what are they? Dogs are a kind of mammal or a kind of pet. What are words a kind of? They're a kind of *meme*, a term that

Richard Dawkins coined in his book *The Selfish Gene* and is defined in the epigraph. Which kind of meme are words? *The kind that can be pronounced*. In addition to words, there are other *lexical items* (Jackendoff 2002) such as irregular plurals and exceptions to the "rules" of a language that also have to be independently stored in memory. *What is transmitted* from other speakers to a novice speaker who picks up the irregular plural of *child*, for instance, is a lexical item. These are all paradigmatic *memes*, not *innate* features of language or grammar, but optional, culture-bound features that have to spread through linguistic communities, and they are often in competition with variants. Other memes, such as the meme for wearing your baseball cap backward or gesturing *just so* to indicate approval or for building an arch that looks like *this*, also have no pronunciation, and hence aren't words.

What are memes a kind of? They are a kind of *way of behaving* (roughly) that can be copied, transmitted, remembered, taught, shunned, denounced, brandished, ridiculed, parodied, censored, hallowed. There is no term readily available in the technical language of the scientific image that aptly encapsulates what kind of a thing a meme is. Leaning on the ordinary language of the manifest image, we might say that *memes are ways*: ways of doing something, or making something, but not *instincts* (which are a different kind of ways of doing something or making something). The difference is that memes are transmitted perceptually, not genetically.

They are semantic information, design worth stealing or copying, except when they are misinformation, which, like counterfeit money, is something that is transmitted or saved under the mistaken presumption that it is valuable, useful.[58] And as we have seen, *disinformation* is design worth copying. Internet spam, like propaganda, is designed, after all. Who benefits? Sometimes the hosts, sometimes the authors, and sometimes the memes themselves, who may have

---

58   That presumption will often be part of a free-floating rationale, not anything the recipient consciously presumes. A gullible child may be victimized by liars without having yet developed an appreciation for truth versus falsehood.

no authors but who nevertheless, like viruses (who also have no authors) have their own fitness, their own relative replicative ability.

Words are the best examples of memes. They are quite salient and well individualized as items in our manifest image. They have clear histories of descent with modification of both pronunciation and meaning that can be traced back thousands of years in many cases. They are countable (think of vocabulary sizes), and their presence or absence in individual human vectors or hosts is detectable by simple test. Their dissemination can be observed, and now, thanks to the Internet, we have a fine laboratory in which more data can be gathered. As always in a lab, there are prices to pay in restricting the phenomena to a circumscribed artificial environment and in the risk of deep bias in the population of our experimental subjects (not all language users are Internet users, obviously). It is no accident that the population of tokens of the "meme" species (the term), founded by Dawkins in 1976, languished somewhat until the Internet provided an ideal niche for their use.

If words are the best memes, why didn't Dawkins feature them prominently in his exposition? In fact, he began his chapter on cultural evolution by citing the accumulation of changes in English from Chaucer to the present. "Language seems to 'evolve' by non-genetic means, and at a rate which is orders of magnitude faster than genetic evolution" (1976, p. 203). His first list of meme examples was "tunes, ideas, catch-phrases, clothes fashions, ways of making pots or of building arches" (p. 206). Catch-phrases are lexical items made of words, but Dawkins didn't stress the role of words in transmitting "ideas," no doubt because ideas can be wordlessly shared and transmitted (showing something, not describing or defining something, for instance), and Dawkins wanted to highlight the unobvious *extension* of the concept of memes beyond words to other kinds of cultural things: tunes, not necessarily songs with lyrics, fashions in clothing *whether or not there are names for them*, and so forth. Terms like *miniskirt* and *bow tie* get spread, along with their exemplars, but so do unattended features that are only noted, and subsequently named, after they have multiplied many-fold:

wearing your baseball cap backward, jeans with rips in them, socks with sandals.

The main enabling technology of genomics is PCR, the polymerase chain reaction that takes any sample of DNA and copies it so voluminously that it can be readily detected and distinguished and manipulated. Multiple copies of advertisements lined up on the walls of the escalators in the London Tube, or plastered side by side on the temporary wooden walls around urban construction sites—"hoardings" in British English—catch the observer's eye by making an irresistible pattern. As always, repetition is a key ingredient in creating new affordances, whether it is arrayed synchronously in space like the advertisements or identical fragments of nucleic acid; or sequenced in time, like rehearsing a tune or a word; or noticing, again and again and again, the Eiffel Tower in the distance in Paris. Multiple copies of anything tend to enable your pattern-recognition machinery to make yet another copy, in the recognizer, and thus a meme can get spread.

In figure 7.4 (p. 146), words are shown as high on reproduction versus growth, high on culture versus genetic, and low on complexity. Perhaps if Dawkins had developed the theme of words as paradigmatic memes, he might have forestalled or softened some of the resistance his proposal encountered from other researchers of culture: historians, literary theorists, philosophers, linguists, anthropologists, and sociologists, for instance. They have never supposed that words—except, perhaps, for a handful of coined words—were deliberately designed creations. Since there has never been a myth of *words* as intelligently designed cultural artifacts, they would have had no vested interest in protecting such a myth, and might well have been more open-minded to the suggestion that rather more elements of culture were *like words* than they had realized. In any event, we can make the point now and go on to note that once words are secured as the dominant medium of cultural innovation and transmission, they do begin to transform the evolutionary process itself, giving rise to new varieties of R&D much closer to the traditional, mythical ideal of intelligent design.

What's good about memes?

The price we pay for our promiscuous lust for adaptive information is playing host to sometimes spectacularly pathological
cultural variants.
            —Richerson and Boyd, *Not by Genes Alone*

Among scientists researching cultural evolution, there has been
something of an embargo against using Dawkins's (1976) term
"meme" to refer to the ways of doing and making things that spread
through cultures, but every theorist I can think of countenances
such items under one term or another: ideas, practices, methods,
beliefs, traditions, rituals, terms, and so on. These are all *informational things* that spread among human beings more or less the way
germs or viruses do. Information is what cultural evolution, like
genetic evolution, deals in, and as Richerson and Boyd (2005) note,

> we need some expedient agreement about what to call the
> information stored in people's brains. This problem is not triv-
> ial, because psychologists have deep disagreements about the
> nature of cognition and social learning. (p. 63)

When we talk about culturally evolved information we seldom if
ever are talking about *bits* (Shannon style) of information, and it
would be good to have a general term for a salient hunk or morsel
of information, a term to go alongside—and contrast with—*gene.*
Since the word *meme* has secured a foothold in the English language, appearing in the most recent edition of the *Oxford English
Dictionary* defined as "an element of culture that may be considered
to be passed on by non-genetic means," we may conveniently settle
on it as the general term for any culturally based *way.* Those who
are squeamish about using a term whose identity conditions are still
embattled should remind themselves that similar controversies continue to swirl around how to define its counterpart, *gene,* a term
that few would recommend abandoning. So, bucking the tradition

of shunning the word, I will persist in calling (many of) the items of culture *memes*, and will defend that terminology as we go along, since I think the contributions Dawkins's concept makes to our understanding of cultural evolution far outweigh any unfortunate connotations the term has acquired.

Why did the concept of a *meme* acquire a bad reputation? It was thanks in part to some exaggerations put forward by some would-be memeticists who didn't do their homework, and in part to theorists who had already figured out some of the aspects of cultural evolution on their own and didn't want to give Dawkins undue credit by adopting his coinage when talking about their own theoretical work. There were also a few careful criticisms of aspects of Dawkins's account of memes, which we will look at in due course. But more influential, in my opinion, has been a frantic campaign of misbegotten criticism, an allergic reaction of sorts from scholars in the humanities and social sciences who were unnerved to see dread biology invading their sacred precincts.[59] I will respond to the standard objections from this quarter in chapter 11, after recounting some of the benefits of adopting the meme's-eye point of view, and also respond to the serious objections.

Chief among the insights that Dawkins's memes contribute to the study of cultural evolution are three conceptions:

1. *Competence without comprehension.* The *designedness* of some cultural items is attributable to no author or syndicate of authors, no architects, no intelligent designers at all; the undeniable cleverness or aptness of the arrangement of the parts is due entirely

---

59 For years this hostility was encouraged by the late Stephen Jay Gould, whose campaign against "Darwinian Fundamentalism" in general and Dawkins in particular persuaded many nonbiologists that they should ally themselves with Gould's purportedly more ecumenical and humanities-friendly version of Darwinism, when in fact his version paid a steep price in obfuscation and confusion for its aura of gentleness. Gould himself claimed to "understand but dimly" how human culture fits in the evolving world. If you want to understand more vividly, abandon Gould and start down the trail blazed by Dawkins.

to natural selection: the differential replication of informational symbionts whose hosts *can* be as clueless about their cleverness as butterflies are of the eye-spots on their wings. (Human beings often do appreciate the virtues of the ideas they eagerly spread, but this is, as we shall see, a mostly recent and entirely optional feature of much of cultural evolution. Human comprehension—and approval—is neither necessary nor sufficient for the fixation of a meme in a culture.)

2. *The fitness of memes.* Memes thus have their own reproductive fitness, just like viruses. As Dawkins put it, "What we have not previously considered is that the cultural trait may have evolved in the way that it has, simply because it is *advantageous to itself*" (1976 [1989], p. 200). Whether or not a meme enhances the reproductive fitness of its hosts—the human beings who adopt and use the designs they carry—natural selection, of the memes themselves, is doing the work. Memes can spread to fixation in a single decade, and go nearly extinct in the same time frame, an evolutionary process much too swift to be discernible in any increase or decrease in human offspring. (Even if all memes were *mutualist* symbionts, importable adaptations that do indeed enhance the reproductive fitness of their hosts, the speed of their spread could not be explained by this minuscule effect.)

3. *Memes are informational things.* They are "prescriptions" for ways of doing things that can be transmitted, stored, and mutated without being executed or expressed (rather like recessive genes traveling silently in a genome). Marco Polo is thought to have brought the pasta meme to Europe from China; he didn't have to become a pasta chef to do this; all he had to do was disperse copies of the meme in environments where other human beings could become infected by it, and express it in their ways of behaving.

Traditional theories of cultural change in the humanities and social sciences have often struggled without these ideas. Old-fashioned Durkheimian *functionalism*, for instance, uncovered many highly plausible functions and purposes of social arrangements—taboos, practices, traditions, social distinctions, and so forth—but was unable to explain how these effective arrangements came into existence. Was it due to the cleverness of the kings, chieftains, or priests; a gift from God; or just luck? Or was it some mysterious "group mind" or shared "genius of the people" that appreciated these features and thereupon led to their adoption? You don't get "teleology" for free, and the Durkheimians had no defensible way of paying their debts. But no magical mechanism need be postulated; natural selection of memes can do the design work without any obligatory boost from human, divine, or group comprehension. As David Sloan Wilson (2002) points out, functionalism died out for lack of a credible mechanism. The theory of meme evolution by natural selection can provide the needed mechanism just as plate tectonics provided the needed mechanism for the hypothesis of continental drift to be taken seriously.[60]

Another difficulty with traditional theories of culture is that they tend to settle on "psychological" categories for their informational entities: *ideas* or *beliefs*, in particular. But although ideas and beliefs (however we want to characterize them as mental or psychological states or episodes) obviously play a big role in human culture, not all cultural transmission and evolution depends on anything like *conscious uptake*. A group may *imperceptibly* shift its pronunciation of a word, or execution of a gesture, or way of applying plaster to walls, taking no notice of the drift. Even the meanings of words

---

60   The evolutionary account Wilson champions isn't memetic evolution but rather his theory of "multilevel group selection." As I have pointed out (2006, pp. 181–188), the conditions required for his group selection to occur are both too infrequent and require too many "generations" (of groups, not individuals) to be effective over historical time. Moreover, since Wilson needs memes anyway for his own account, there is no good reason not to acknowledge memetic evolution as just the right phenomenon to account for the unwitting Darwinian evolution of functional elements of social groups.

can evolve by processes quite outside the ken of those using the words, thanks to differential replication. Today when somebody says, enthusiastically, "That was an *incredible* lecture, simply *terrific!*" they don't mean they didn't believe it, but nonetheless found it terrifying. Nobody decided to adjust these meanings, or even approved the trend; it just happened, by a shift in the population of tokens produced in the culture. (See Richard, forthcoming, for more on this.)

The fact that changes in cultural features can spread *without notice* is hard to account for—and hence likely to be overlooked—when one adopts the traditional psychological perspective of ideas and beliefs. The memes perspective provides a valuable corrective to this oversight, but much more important is the way memes can provide an alternative vision of how culture-borne information gets installed in brains *without being understood*. The problem with the standard view is that it relies *uncritically* on the assumption of rationality that is built into the intentional stance: the default presumption of folk psychology is that people, and even "higher" animals, will *understand* whatever is put before them. Novel ideas, beliefs, concepts are, almost "by definition," *understood* ideas, beliefs, concepts. To have an idea is to know what idea you have, is to conceive of it clearly and distinctly, as Descartes said. We hardly notice that whenever we try to identify an idea to discuss or consider, we identify it by its content (not, for instance, as *the first idea mentioned by Tom*, or *the idea I keep right next to my idea of baseball*). This is a ubiquitous application of the pre-Darwinian notion that comprehension is the source of competence, and it is important that we invert the idea so that we can at least sketch, for the first time, an account of *how* uncomprehending competences can gradually yield comprehension (a task for chapter 13).

Traditional theorists of culture tend to overestimate the contribution of individual artificers, imagining that they must have had more comprehension of the designs they invent, transmit, and improve than need be the case. Comprehension is not even necessary for design *improvement*. In a study of the evolution of the Poly-nesian canoe, Rogers and Ehrlich (2008) cite a passage by Alain ([1908] 1956), a French philosopher (writing about fishing boats in Brittany, not Polynesian canoes[61]):

> Every boat is copied from another boat. . . . Let's reason as fol-lows in the manner of Darwin. It is clear that a very badly made boat will end up at the bottom after one or two voyages and thus never be copied. . . . One could then say, with complete rigor, that it is the sea herself who fashions the boats, choosing those which function and destroying the others.

The boat builders may have theories, good or bad, about what would be an improvement in the craft their fathers and grandfa-thers taught them to build. When they incorporate these revisions in their own creations, they will sometimes be right and some-times wrong. Other innovations may simply be copying errors, like genetic mutations, unlikely to be serendipitous improvements, but once in a blue moon advancing the local art of boatbuilding. The boats that come back get copied; the rest are consigned to obliv-ion, a classic case of natural selection at work. The copied boats may pick up gratuitous but harmless elements—decorations, in effect—in addition to their genuinely functional adaptations. At a later date these merely decorative elements may acquire a function after all, or they may not. The memetics perspective can thus accommodate both functional and merely traditional embel-lishments to artifacts, rituals, and other human practices, and account for our inability to "draw a bright line" distinguishing between the two categories without any recourse to the inscruta-bility of the "folk genius" that created them. No comprehension is

---

61   I mistakenly took Alain to be writing about the Polynesian canoes when I cited this passage in Dennett 2013. Roger DePledge pointed out that Alain (pseudonym of Émile August Chartier) was writing "not about Polynesian but Breton fisher-men on the Île de Groix (47° 38' N, 3° 28' W) off Lorient."

required even if it probably accelerates R&D processes more often than it retards them.[62]

Yet another shortcoming of much traditional thinking about culture is the tendency to concentrate on the good stuff and ignore the junk. Adopting an *economic model* of culture is appropriate *as a first approximation* of the obviously useful (or just benign) elements of a culture: its technology, science, art, architecture, and literature— its "high culture," in short. These are indeed a society's treasures, and it makes good economic sense (it is rational) to preserve and maintain these things, to spend whatever it takes to sustain them, and make sure that they get passed along to the next generation as a handsome legacy, even if it takes years of arduous instruction and rehearsal to keep them "alive and well" over the centuries.[63] But what about the malignant, useless, burdensome, fitness-reducing elements of a culture? What about those elements not appreciated by anybody, really, but too costly to eradicate, like the cold virus and malaria? The meme perspective covers both extremes and the middle, in fact, via the tripartite distinction from biology between mutualist, commensal, and parasitical symbionts (see ch. 9, fn. 51, p. 193).

As I noted in chapter 7, many people are repelled by the idea that memes are like viruses because they believe, mistakenly, that viruses are always bad for us. Not at all, in fact only a tiny minority of the trillions of viruses that inhabit each of us right now are toxic in any way. It is not yet known for sure if some viruses are actually good for us or even obligate mutualists like the microflora in our gut without which we would die.

Do we *need* some viruses in order to thrive? Perhaps. We certainly need lots of our memes. Daniel Defoe's *Robinson Crusoe* (1719) is a

---

62   The history of invention and engineering is replete with examples of bad theory blocking the path to innovation for decades or even centuries. For instance, if it "stands to reason" that you can't sail upwind, you won't bother experimenting with designs that might achieve this "impossibility."

63   Richerson and Boyd (2005) note that "the rational-choice model is a limiting case of cultural evolution," p. 175.

veritable encyclopedia of the memes one needs to get by, and—a nice historical twist—the average eighteenth-century adult probably was better equipped with memes for such a self-reliant life than the average twenty-first-century adult from the developed world. (How many of us know how to gather seeds for next year's planting, kindle fires without butane lighters, or fell trees?)[64]

Can genuinely *parasitic* memes really thrive? A positive answer to this question was one of the most arresting points in Dawkins's introduction of memes, and probably this accounts for much of the antipathy for memes. Dawkins seemed to be saying (to many readers) that *all* culture is some horrible sickness, an infection of the brain, provoking self-destructive activity. That was not what he was saying. An egregious case of this mistaken presumption is found in David Sloan Wilson (2002), who clearly sees the need for an evolutionary foundation for culture:

> The success of some [social experiments] and the failure of others on the basis of their respective properties constitutes a process of blind variation and selective retention operating during the course of recent human history in addition to the distant evolutionary past. (p. 122)

But he peremptorily dismisses memes because he thinks they are "parasitic" (p. 53), so he never considers the possibility that memes might provide an evolutionary account for some cultural phenomena—religions, in particular—more effectively and testably than his own "multilevel group selection" hypothesis. In fact, he cannot see that he himself on occasion alludes to memes—for instance, he finds it useful to characterize catechisms "as 'cultural genomes,'

---

64   Obviously no one individual needs to harbor all the memes that are required to sustain human life these days, since we have long ago established a division of labor and expertise that permits individuals to specialize. Robinson Crusoe is a fascinating thought experiment in many ways, one being that it demonstrates just how dependent we have become, even in the eighteenth century, on a wide variety of competences that few if any can claim to have mastered.

containing in easily replicated form the information required to develop an adaptive community" (p. 93). And on other occasions (e.g., pp. 118–189, 141, 144) he simply fails to notice that memetic alternatives to his hypotheses (such as my account in *Breaking the Spell*, 2006) are not only available, but more plausible.[65]

Many memes, maybe most memes, are mutualists, fitness-enhancing prosthetic enhancements of our existing adaptations (such as our perceptual systems, our memories, our locomotive and manipulative abilities). If there weren't mutualist memes from the outset, it is hard to see how culture could have originated at all (more on that origin later). But once the infrastructure for culture has been designed and installed (by an interplay between cultural and genetic evolution, as we shall see), the possibility of parasitical memes exploiting that infrastructure is more or less guaranteed. Richerson and Boyd (2005, p. 156) call such memes "rogue cultural variants." To take the obvious recent example of such a phenomenon, the Internet is a very complex and costly artifact, intelligently designed and built for a most practical or vital purpose: today's Internet is the direct descendant of the Arpanet, funded by ARPA (now DARPA, the Defense Advanced Research Projects Agency), created by the Pentagon in 1958 in response to the Russians beating the United States into space with its Sputnik satellite, and its purpose was to facilitate the R&D of military technology. Today, spam and porn (and cat photos and other Internet memes) dwarf any high-tech design-sharing by Pentagon-funded labs.

Before settling on the handy stereotype of such junk as examples of memes that don't enhance human fitness, we should remind ourselves of what "fitness" means in the context of evolutionary biology: not health or happiness or intelligence or com-

---

65 For instance, Wilson insists that "the *details* [my emphasis] of Calvinism are better explained as a group-level adaptation than by any other competing hypothesis" (p. 119), but only a few of these many details have any bearing on group solidarity.

fort or security but *procreative prowess.* Which memes actually make their bearers more likely to leave more grandchildren than average? Very few, it seems. And many of our most cherished memes are demonstrably fitness-*reducing* in the biological sense. *Getting a college education,* for instance, has such a striking fitness-reducing effect that if, say, eating broccoli had a similar-sized effect it should probably carry a public health warning: "CAUTION! Eating broccoli is likely to reduce your number of surviving grandchildren to below average."[66] When I ask my students if this striking fact alarms them, they deny it, and I believe them. They think there are more important things in life than out-reproducing their conspecifics, and their attitude is, if not quite universal, the prevailing view of all humanity. That fact, by itself, sets us apart from all other species. We are the only species that has managed to occupy a perspective that displaces genetic fitness as the highest purpose, the *summum bonum* of life.[67] The quest for progeny is why birds build nests and beavers build dams; it's why whales migrate tens of thousands of miles and some male spiders rush to their deaths in the embrace of their mates. No salmon fighting their way upstream can reconsider and contemplate a life of learning to play the violin instead. We can.

We are the only species that has discovered other things to die for (and to kill for): freedom, democracy, truth, communism, Roman Catholicism, Islam, and many other meme complexes (memes

---

66   Or as Richerson and Boyd quip: "If you want to improve your kids' genetic fitness, for goodness sake don't help them with their homework!" (p. 178).

67   There are many tales of dogs risking, or even giving up, their lives for their masters, so they are an important exception to this generalization. Dogs are much more like human beings than any other species of domesticated animals (including cats), and this is no accident; they have been unconsciously selected for these similarities for thousands of years. And ironically, another striking fact about both *Homo sapiens* and *Canis familiaris* is that in spite of their shared ability to subordinate procreation to a higher call, they vastly outnumber their closest undomesticated kin.

made of memes). Like all living things, we are born with strong biases in favor of staying alive and procreating, manifest in the all but irresistible urges of our "animal nature" and a host of more subtle habits and dispositions. But we are also the persuadable species, not just learners, like birds and monkeys, and not just trainable, like dogs or horses (and lab animals under strict regimes), but also capable of being moved by reasons, reasons *represented to us*, not free-floating. We have already seen, in the castle-building termites and stotting antelopes, examples where there are reasons but not the organism's reasons. When animals do things for reasons, their comprehension of those reasons—of why they are doing what they are doing—is either nonexistent or quite limited, as revealed in a host of experiments demonstrating their inability to generalize, to use what they at first appear to understand in novel variations or wider contexts.

We, in contrast, don't just do things for reasons; we often *have* reasons for what we do, in this sense: we have articulated them to ourselves and have endorsed them after due consideration. Our understanding of the reasons we can cite for acting as we do may often be imperfect, confused, or even self-deceived, but the fact that we can *own* these reasons (wisely or foolishly) makes us susceptible to being talked out of them and talked into others. When we have our minds changed, as we say, by some persuasion (possibly self-administered), there is always the possibility that our acknowledgment and acquiescence in the revision of our reasons won't really be "taken to heart," in spite of what we say on the spur of the moment. The lesson may not have the long-term effect on our convictions and attitudes that was sought by the persuader or teacher. That is why we rehearse and revise and re-explain and retest and review our reasons, in the practice of reason-giving and reason-taking (and rejecting) highlighted by Wilfrid Sellars (1962). (See chapter 3 on the "space of reasons".)

Aristotle distinguished our species as the *rational* animal, and Descartes and many others, up to the present day, have attributed our talent as reasoners to a special *res cogitans*, or *thinking thing*, implanted in our brains by God.[68] If our rationality isn't God-given, how could it evolve? Mercier and Sperber (2011) argue that the individual human being's capacity to reason, to express and evaluate logical arguments, arises out of the social practice of persuasion, and it shows fossil traces, in effect, of this origin. The notorious *confirmation bias* is our tendency to highlight the positive evidence for our current beliefs and theories while ignoring the negative evidence. This and other well-studied patterns of errors in human reasoning suggest that our skills were honed for taking sides, persuading others in debate, not necessarily getting things right. Our underlying talent "favors decisions that are easy to justify but not necessarily better" (p. 57). The evolutionary process of R&D that could design such a skill would necessarily depend on some prior, if inchoate, skill in language use, so we may think of it as a *co*evolutionary process, part cultural evolution and part genetic evolution, with the cultural evolution of pronounceable memes, words, leading the way.

---

68 "Technology is a gift of God. After the gift of life it is perhaps the greatest of God's gifts. It is the mother of civilizations, of arts and of sciences." Freeman Dyson (1988). At least Dyson sees our talent as growing out of technology, not its ultimate source.

11

What's Wrong with Memes? Objections and Replies

Memes don't exist!

On a number of occasions when I have encountered hostility to memes, I have been treated to a haughty sneer: "Memes? Memes? Can you prove to me that memes even exist?" Claims about nonexistence are almost always slippery, especially when scientists wax philosophical or philosophers wax scientific. We can all agree—can't we?—that mermaids and poltergeists and phlogiston and élan vital don't exist, but controversies simmer in some quarters over genes, strings (of string theory), public languages (as contrasted with idiolects), numbers, colors, free will, qualia, and even dreams. Sometimes, the negative claims are based on a highly austere doctrine of reality according to which nothing *really* exists except "atoms and the void" (or subatomic particles and physical fields, or just one subatomic particle dashing back and forth between the beginning of time and the end of time, weaving a fabric with itself—a proposal by the physicist John Archibald Wheeler). Both philosophers and scientists have long been attracted to such minimalist visions. As a student of mine once noted:

Parmenides is the philosopher who said, "There is just one thing, and I'm not it."

Sometimes the sweeping negative claims encompass all of the manifest image: the items in the official ontology of the scientific image really exist, but solid objects, colors, sunsets, rainbows, love, hate, dollars, home runs, lawyers, songs, words, and so on, don't really exist. They are useful illusions, perhaps, like the user-illusion of the desktop icons. The patterns of colored pixels on the computer screen are real, but they portray entities that are as fictional as Bugs Bunny and Mickey Mouse. Similarly, some would say, the manifest image has some reality as a collection of images—the movie we live in, perhaps—but it is a mistake to think of the "things" we interact with and manipulate, and fall in love with, as reality.

That's a defensible position, I suppose. In fact, it's a *version* of what I have said about the manifest image of each species: a user-illusion brilliantly designed by evolution to fit the needs of its users. My version differs only in being willing and eager to endorse these ontologies as ways of carving up *reality*, not *mere* fictions but different versions of what actually exists: real patterns. The more shocking way of saying it—"we live in a fictional world, in a dream world, in unreality"—tends to cause confusion because it suggests we are somehow victims, duped by some evil force. Some theorists deny that, say, dollars exist, but have no difficulty accepting love, hate, and rainbows; the reality of love, hate, and rainbows is never mentioned, however, since it is "obvious" to them that these exist (but not obvious to them that dollars exist!).

Setting aside such radical metaphysical doctrines, usually what is meant by denying that Xs exist is that some theory or other of Xs is a bad theory. When the theory is a "folk theory" drawn from the manifest image, this is often undeniable. For instance, the folk theory of color is rife with error and confusion. But does that mean that colors don't really exist? Are we as deluded about colors as we would be about mermaids if we believed in them? Some would say that. Others are less willing to go down that path. Should Sony be sued for false advertising for claiming to sell color television sets? An intermediate position declares that colors don't *really* exist but acting as if they do is a "useful fiction."

Yet another option insists that colors do indeed exist; they just aren't what the folk think they are. I have followed that path throughout my career, insisting that not only are colors real but also consciousness, free will, and dollars. Dollars are an interesting case. Probably one of the most powerful subliminal supports for the belief in the reality of dollars comes from the undoubted existence of dollar *bills* and *coins*, legal tender you can see and weigh and manipulate and carry around. Bitcoin, in contrast, seems much more illusory to most folks, but if they reflect on it, they will discover that the palpable, foldable dollar bills that are physical objects are ontological crutches of sorts, to be thrown away once you've learned how to walk, leaving in place only the mutual expectations and habits that can in principle support bitcoin as well as dollars. Well into the twentieth century there were heated political debates on the necessity of maintaining such crutches—the gold standard and silver certificates, and the like—but tomorrow's children may grow up with only credit cards to hold and manipulate and be none the worse off for it. Dollars are real; they just aren't what you may think they are.

Consciousness exists, but just isn't what some folks think it is; and free will exists, but it is also not what many think it must be. I have learned, however, that some people are so sure that they know what consciousness or free will *would have to be* to be *real* that they dismiss my claims as disingenuous: They claim I'm trying to palm off a cheap substitute for the real thing. For instance:

> Of course the problem here is with the claim that consciousness is "identical" to physical brain states. The more Dennett et al. try to explain to me what they mean by this, the more convinced I become that what they really mean is that consciousness doesn't exist. (Wright 2000, ch. 21, fn. 14)

[Dennett] doesn't establish the kind of absolute free will and moral responsibility that most people want to believe in and do believe in. That can't be done, and he knows it. (Strawson 2003)

According to my theories, consciousness is *not* a nonphysical phenomenon and free will is *not* a phenomenon isolated from causation, and by the lights of Wright and Strawson (and others), I *ought* to have the courage then to admit that neither consciousness nor free will really exists. (Perhaps I could soften the blow by being a "fictionalist," insisting that they don't really exist but it is remarkably useful to act as they did.) I don't see why my critics think their understanding about what *really* exists is superior to mine, so I demur.

What I should say about memes is in any case easier. The theory I am defending declares, nonmetaphorically and literally and without caveat, that words are memes that can be pronounced. Memes exist because words are memes, and words exist, and so do other *ways of doing things* that are transmitted nongenetically. If you are one of those who wants to deny that words exist, you will have to look elsewhere for a rebuttal, since I am content, without further argument, to include all the furniture of the manifest image—words very much included—in my ontology. Now let's consider the substantive objections to the theory of memes I am defending.

Memes are described as "discrete" and "faithfully transmitted," but much in cultural change is neither

Richerson and Boyd (2005), while sympathetic to many of Dawkins's claims, resist calling any cultural information item a meme because of its connotations as a "discrete, faithfully transmitted genelike entity, and we have good reason to believe that a lot of culturally transmitted information is neither discrete nor faithfully transmitted" (p. 63). We've already seen that words really are quite "discrete, faithfully transmitted" entities, and they are at least somewhat "genelike"; or better, genes are *somewhat* "wordlike," as we saw in chapter 8, in Dawkins's (2004) analogy with the Mac toolbox of subroutines:

> In one respect the analogy with words in misleading. Words are shorter than genes, and some writers have likened each gene to a sentence. But sentences aren't a good analogy, for a different reason. Different books are not put together by permuting a fixed repertoire of sentences. Most sentences are unique. Genes, like words but unlike sentences, are used over and over again in different contexts. A better analogy for a gene than either a word or a sentence is a toolbox subroutine in a computer. (pp. 155–116)

Words are like genes, then, in being informational structures (like subroutines) that determine ways of doing things. Richerson and Boyd are right that "a lot of culturally transmitted information is neither discrete nor faithfully transmitted," as we shall see, so perhaps their objection, applied to the view here, is that memes are only a small, maybe a minor, part of the information that cultural evolution discovers and refines. Suppose words were the *only* memes, or almost the only memes; then memetics or the meme's-eye perspective could hardly be considered a *general* theory of cultural evolution. Similarly, Dan Sperber (2000), another serious critic of memes, defines memes as "cultural replicators propagated through imitation" and acknowledges that while a few such entities exist—he mentions chain letters, but ignores words as examples—they represent a minuscule fraction of cultural transmissions:

> For memetics to be a reasonable research program, it should be the case that copying, and differential success in causing the multiplication of copies, overwhelmingly plays the major role in shaping all or at least most of the contents of culture. (p. 172)

I think that this is too strong a demand, and I anticipate that Sperber might well agree, given his vacillation between "overwhelmingly" and "or at least most" of the contents of culture. In any case, are words the only good memes? Even if they were, we might still find that the reliance on these memes for *cumulative* culture like ours gives the meme's-eye view a dominant place in theories of cultural evolution, however small a portion of transmitted information was in the form of words. But in fact there are more high-fidelity replicators in human culture than these critics have acknowledged, and they all depend, in one way or another, on some variety of digitization rather like phonemes, which, you will recall, create tokens that are *subjectively similar* (immediately identifiable, uncontroversially categorizable) in spite of their huge physical differences, thanks to evolved features of the human auditory system and vocal tract. The descendant tokens of "cat" are not *physical* replicas of their ancestors, but they are—we might say—*virtual* replicas, dependent on a finite system of norms to which speakers unconsciously correct their perceptions and their utterances, and this—not physical replication—is what is required for high-fidelity transmission of information.

We see the same sort of abstraction, and the same correction to the norm, in music. (I'll consider Western tonal music, the theory of which is better known to me and most of the likely readers of this book.) Nobody invented tonal music—*do, re, mi, fa, sol, la, ti*—but many musicians and music theorists contributed to codifying it and choosing the syllables to sing for each tone and perfecting a system of musical notation; a fine mixture of Darwinian cultural evolution and intelligent design over hundreds of years beginning in the eleventh century. Tonal music is a good example of a digitized alphabet that allows correction to the norm. (You're singing that note a bit sharp. Fix it!) Many musical innovations involve bending, sliding, deliberately flatting the notes (for instance in the blues), but standing behind these deviations are the canonical tones. Can you sing or hum the melody of "Greensleeves" (the same tune as the Christmas carol "What child is this?")? If you can, you are making another in a very long lineage of tokens of that melody, which is not in any particular key or at any particular tempo. "Greensleeves" played at breakneck speed by a bebop saxophonist in C-minor is a token of the same type as "Greensleeves" played pensively on a guitar in E-minor. If the ancient melody were not in the public domain, its copyright would cover it in any key at any tempo by any voice or on any instrument—a collection of audible events with precious little *physical* similarity but readily detectable identity in melody-world, an important part of our manifest image. And, of course, this melody can now readily be written down in musical notation, which preserves it today much the way writing preserves the ancient orally transmitted sagas. But still, independently of musical notation, the *unwritten* systems of scales—like the systems of language before writing was invented—was itself a sufficient digitization to preserve simple melodies and harmonies "by ear."

In literature, there are still higher levels of abstraction that yield high-fidelity tokening of types. Not only do two different printings (in different fonts, perhaps) of *Moby Dick* count as copies of the same novel but so do translations into other languages. (If not, could you claim to have read Homer's *Iliad* or Tolstoy's *War and Peace*?) Translation is a process that *requires comprehension* (at least until Google Translate was developed) but that doesn't prevent reliable transmission from leaping over the language barrier, permitting global spread of cultural items that preserve many important features with high fidelity. Such examples show that beyond simple physical replicahood, there are a variety of higher levels of abstract similarity that can preserve information through selection events, permitting cultural evolution by natural selection to occur without succumbing to the "error catastrophe" described in chapter 7 (see p. 141). *West Side Story* is a descendant of *Romeo and Juliet*, not because the sequence of inkmarks or sounds that compose Shakespeare's play are physically replicated, but because the sequence of circumstances and relationships portrayed are copied, a purely semantic-level replication.

To paraphrase the familiar disclaimer: any resemblance between the characters in *West Side Story* and the characters in *Romeo and Juliet* is *not* purely coincidental. If you wonder if such semantic replication "counts" as meme replication, recall that memes are informational structures that are normally valuable—they are *worth* copying—and copyright laws have been devised and refined to protect that value. Not only translations, faithful or not, but also abridgments, cinematic treatments, plays and operas based on novels, and even video games can count as meme replications. Given copyright laws, you can't take a novel about fictional folks in nineteenth-century Boston and change all the names of the characters and set the novel in eighteenth-century Paris and expect to get away with it. That is plagiarism, theft of design, even though the physical object that counts as the copy need share no distinctive physical properties at all with the source object.

What is particularly important in this exploration of memes is that some of these higher levels really do depend on *comprehension*, not just copying *competence*, even though they are based on, and rely on, systems of copying competence that do *not* require comprehension. In fact, from this vantage point we can see that the high-fidelity copying of DNA, our chief model for replication, stands out as an extreme case of mindlessness in replication. To get high-fidelity replication at the molecular level, using macromolecular structures (polymerase molecules and ribosomes, for instance) as the "readers" or "copyists," you have to rely on dead-simple copying—pretty much atom-for-atom "recognition" and subsequent duplication. At higher levels, with more sophisticated, more competent "readers," you can create systems that can tolerate more physical variation. Spoken words are the chief example here, but there are others. Scamrlbed wrdos rae aesliy nusracmbedl. Mst ppl wll hv lttl trbl rdng ths sntnc. Turing saw the importance of basing his great invention on as mindless a recognition system as he could imagine—binary choices between 0/1, hole/no-hole in a piece of paper, or high-voltage/low-voltage between two points in a circuit. In principle, high-fidelity transmission could be accomplished in a binary code that used *love letters* for 0 and *death threats* for 1, but you'd need readers capable of comprehension to get reliable transmission.[69]

This is possible in principle, and so are countless other ditigization schemes, but do they actually play much of a role in human culture? In fact they do. Computer programmers have an excellent jargon term, the *thinko*. A thinko is like a typo—a typographical error—but at a higher, semantic level—misthinking not miswriting. Typing "PRITN" instead of "PRINT" is a typo; forgetting to put the brackets or asterisks (or whatever the programming language calls for) around a comment is a thinko, and so is defining a three-place function where you need a four-place function. The notorious Y2K bug, which didn't leave room in data structures for dates that didn't begin with "19," was no typo; it was a thinko. A thinko is a *clear mistake* in any endeavor where the assumed goals of the enterprise require certain identifiable "best practices." "Bugs" in computer programs may be attributable to typos in the source code, but more often they are thinkos. (Most typos are spotted by the compiler program and sent back to the programmer for correction before the construction of executable code.) If a falling short is not uncontroversially identifiable by the experts, it is perhaps a lamentable lapse but not a thinko. (A thinko is like an *error* in baseball: you get "charged" with an error, a substandard execution of a play; when there is uncertainty about what the best play would have been or when a play is regrettable because it wasn't as spectacular as it might have been, it is not an error.) What matters for high-fidelity replication is the existence of canons of practice relative to which thinkos can be routinely corrected. Truman Capote once dismissed a draft short story submitted for his approval with "This isn't writing; this is typing!," but he was not, presumably, identifying a thinko; he was

69 Such a system would have an interesting digital property: although the reader-writers would have to be comprehenders of the texts—the love letters and death threats—used as binary digits, they might be completely uncomprehending of the message being transmitted in binary code. It could be encrypted, or, say, a plain e-mail in Swedish.

displaying a standard of excellence that could not be routinely met by best practices.

Routines are themselves memes, honed by differential replication over the generations and composable into larger practices that can be "read" and "written" by experts. Making arrows and axes, tending fires, cooking, sewing, weaving, making pots and doors and wheels and boats, and setting out fishnets are ways that can be corrected over many generations by the combined action of simple physical requirements and local traditions. When these local traditions evolve into simple "alphabets" of moves, they provide a host of thinkos, cookos, weavos, and other readily correctable errors so that the traditions can be transmitted reliably—perhaps without language (this is an issue for chapter 13). And, as usual, nobody has to understand that such "alphabets" have this benefit, certainly not the memes that are the direct beneficiaries. A tradition composed of bad habits—parasitic reducers of human fitness—that happens to get digitized in this way will outlive its competition in the replication contest, other things being equal.

Dance provides some interesting variations on these themes. Compare "folk dancing" with sophisticated choreography. Folk dancing typically has an alphabet well known to the folk who do the dances. In square-dancing and contra-dancing, for instance, there are a small number of *named* basic moves that everybody knows; the caller can create a new dance on the spot by just changing the order of the moves or their "values": honor your *partner*, honor your *corner*, gents to the center with a right-hand *star*, . . . *swing* your partner and *promenade* her home. It's a dead simple programming language for contra-dance construction. As a result of this digitized system, people who haven't danced a Virginia Reel for decades, none of whom can remember just how it goes, can nevertheless partner up and recreate the dance with only a few rehearsals (replications), since majority rule will rinse out the typos or thinkos (missteps), restoring the original "spelling" of the dance. No two couples will do the *allemande left* in exactly the same way, but their variants will be acceptable fillers for the same role in the dance. Such "voting" has been reinvented again and again in human culture, and it is a reliable way of enhancing fidelity of transmission via unreliable, low-fidelity individual memories.

Unison chanting is ubiquitous in traditional religions and other ceremonies, and it similarly serves to repair the memories of the chanters, none of whom could provide a faithful copy of last year's rendition unaccompanied. When chronometers were invented in the eighteenth century, navigators knew to take three, not two, chronometers on their long voyages, so that the majority rule trick—by then a recognized, domesticated meme—could be used. The replicative power of tribal chanting didn't have to be recognized or appreciated by the chanters, though some reflective folks may well have caught on. Free-floating rationales have been as ubiquitous in cultural evolution as in genetic evolution, and as informative to us as theorists, once we manage to suppress our traditional habit of attributing *comprehension* to any organisms—or memes—observed to be doing something clever.

Sophisticated choreography, in contrast with folk dancing, requires subtleties that demand a more expressive system of recipes, and while cinema and video recording has "solved" the problem by brute force, much the way recorded music can "solve" the problem of a missing musical score, choreography, in spite of its name, to date has lacked a system of *graphing* dance moves that is effective enough to go to fixation among dancers and choreographers. Labanotation, invented in the 1920s by Rudolf Laban, and refined and supplemented over the years, has its loyal corps of aficionados, but has not yet established itself as the lingua franca of dance.

It is tempting to see a gradual transition from

1. "infectious" rhythmic entrainment among tribespeople dancing and noisemaking and vocalizing in impromptu gatherings, repeating their favorite moves and imitating each other—the *synanthropic* origins of dance, requiring no leaders, no callers, no choreographers

2. through more self-conscious rituals (with rehearsal required and deliberate teaching and correcting)—the *domestication* of dance with careful control of reproduction

3. to today's professional choreographers—*memetic engineers*, intelligently designing their art objects and launching them hopefully into the cultural world.

The original *ways* of dancing, on this view, were memes that nobody "owned," mindlessly evolving to exploit human idiosyncrasies of skeleton, gait, perception, and emotional arousal. What were they good for? For thriving in human company, habits that spread because they could spread, like the common cold.

It would help, of course, if the original dance memes offered some benefit to the genes of the proto-dancers, but that mutualism could evolve gradually, out of a commensal or even parasitic relationship, as competing memes fought for time in the available bodies. Infectious bad habits can be hard to eradicate, but if they can morph into useful habits, their reproductive prospects are enhanced. Once recognized, at first dimly (Darwin's "unconscious selection") and then consciously (Darwin's "methodical selection"), their reproduction would be more or less ensured by their hosts, so the memes could relax, become less exciting, less irresistible, less captivating, less vivid and unforgettable because they had become so useful. (The brains of domesticated animals are always smaller than the brains of their nearest wild kin; use it or lose it, and domesticated animals have a relatively unchallenging life, being protected from predators and starvation, and provided with mates at procreation time.) The corollary of this, of course, is that for something boring to spread, it has to be deemed by its hosts to be particularly useful, or particularly valuable, and hence worth *breeding*: inculcating via extensive training. Double-entry bookkeeping and trigonometry come to mind. Being-deemed-valuable-by-influential-hosts is itself a frequent adaptation among memes, and among the more extreme examples are soporific atonal "serious" music and some lineages of contemporary conceptual art that—like laying hens—are so domesticated that they would go extinct in a generation without the assistance of their well-heeled stewards.

So even if it is true, as Richerson and Boyd and Sperber and others have noted, that many important areas of human culture exhibit change over time that is not *just* a direct effect of high-fidelity transmission systems of "discrete, genelike" informational entities, such systems—not only words, but music, dance, crafts, and other traditions of affordances—provide a variety of paths that can carry cultural information from generation to generation with enough fidelity for *relatively* mindless gradual mutation to accumulate improvements with minimal comprehension required. And as we saw in the case of the canoes, artifacts themselves provide norms of correction. As a general rule of thumb, any artifact found in abundance and showing signs of use is a good whatever-it-is; following this rule, you can often tell the good ones from the not so good ones without knowing exactly why the good ones are good. Copy the good ones, of course. Darwin's brilliant idea of *unconscious selection* as the gradualist segue into domestication gets put to important use in cultural evolution as well. Our ancestors "automatically" ignored the runts of the litter, and the lemons of the fleet, and the result in each case was the gradual improvement (relative to human tastes and needs) of the offspring.

Memes, unlike genes, don't have competing alleles at a locus

The way genes are represented in DNA is as long sequences of nucleotides—A,C,G,T—which make it possible to compare different genomes and identify the places—*loci*—where they are alike and where they differ. There is nothing like this (yet) in the physical representation of memes. There is no common underlying code for representing a particular meme or its competition wherever their tokens occur. For instance, Benjamin Franklin invented bifocals, and no doubt had some neural structure(s) in his brain that embodied or represented his brainchild, the concept of bifocals. When he publicized his invention, others picked up the idea, but there is no good reason to suppose the neural structures in their brains "spelled" the *bifocals* concept the same way he did. For some critics, this is tantamount to admitting that there is no *bifocal* meme, that memes don't exist, that meme theory is a bad idea. For one thing, it means that we can't use a common code to identify *loci* in different brains and see if, at the *bifocals locus*, there is the same, or a different, spelling of the meme, the way genetic testing can reveal if you have the *allele* (variant) for Huntington's chorea or Tay-Sachs disease.

But as we have seen, there are "alphabetic" systems for words, music, and many other families of memes, and they create roles or pigeonholes or loci for somewhat different sorts of allelic competition. In words, there are competitions between pronunciation (*CONtroversy/conTROVersy; dayta/datta*) and foreign-word borrowings often exhibit a competition between (an approximation of) the foreign pronunciation and a more rudely naturalized version. You can have some *Oh juice* on your beef *au jus*, while a French *chaise longue* becomes an American *chaise lounge*; and *lingerie (lanjeree*, roughly*)* evolves into *lawnjeray*. Or we can hold pronunciation constant and look at competing meanings at a phonological locus. There are the opposite meanings of *terrific* and *incredible*, where the negative senses are almost extinct, and *homely*, which has come to have hugely different meanings on opposite sides of the Atlantic. For a while, *beg the question* was hanging on to its original, technical meaning in philosophy and logic, but many (most?) of its tokens today mean *raise the question* (roughly). This meaning is now issuing from the mouths and pens of influential authors, so the charge of misuse is beginning to sound pedantic and antiquarian.

As with genes, mutations are transmission *errors*, but on occasion such an error is a serendipitous improvement; what started out as a bug has become a feature. Then there are cases like *skyline*, which originally meant "the line in a visually perceivable scene that demarcates the terrestrial from the sky" (s.v. *OED*) but has imperceptibly shrunk to urban applications only (Richard, forthcoming). A mountain ridge seen in the wilderness is no longer aptly called a skyline.[70] Or we can hold meaning constant and look at semantic *loci* where the war of the synonyms is waged, with regional strongholds and nostalgia-inducing antique terms on their way to obsolescence. (A soda or soft drink was a *tonic* in my 1950s Massachusetts neighborhood, and a milkshake was a *frappe*—pronounced *frap*.) It is not that these shifts in meaning and pronunciation (and grammar, too) went unnoticed until memes came along. In fact they have been studied rigorously for several centuries by historical linguists and others, but these scholars have sometimes been hobbled by presumptions about *essential* meanings, creating artifactual controversies that look strikingly like the taxonomists' battles over species, genera, varieties, and subspecies that relaxed after Darwin came along.[71] ("The word "incredible" *really* means *unbelievable*, of course—just look at its etymology! Yes, and birds are *really* dinosaurs and dogs are *really* wolves.)

In music, we can look at each popular song as a locus, with the *covers* by different bands duking it out for domination. From this perspective we can see that often the hit version of a song, which goes nearly to fixation, is a mutation, a rival allele of the originally released version. Some singer-songwriters are well known for having their original versions supplanted by later covers. (Have you ever heard Kris Kristofferson's original version of "Me and Bobby McGee?") They still get royalties on the other singers' versions, even if the value added by later interpreters earns them the lion's share of record sales. Here, the existence of the technology of sound recording provides fixed, canonical versions of a song, hardening the physical identity of each version. Experiments by Daniel Levitin (1994, 1996) used this regularity to test human subjects on their memory

70   Reading, Pennsylvania, has a Skyline Drive along a ridge that overlooks the city, a fossil trace of the earlier meaning. Think of all the riverside towns whose names end in "ford" where there is no longer a ford to use when crossing the river.
71   Richard (forthcoming) shows how W. V. O. Quine's (1951) assault on the notorious analytic-synthetic distinction in philosophy of language can be seen in a good light as extending Darwin's anti-essentialism to philosophy. As Ernst Mayr urged, population thinking should replace the rigid taxonomies of all natural kinds.

for pitch and tempo of popular songs. Whereas "Greensleeves" has no canonical pitch (key) or tempo, "Hey, Jude" and "Satisfaction" do, and fans of these records were surprisingly good at humming/singing these songs from memory with close to the right pitch and tempo.

When we consider whether memes are replicated faithfully enough to permit the accumulation of adaptations, we should include the enhancing technology in culture just as we do with genes. A lot of evolutionary R&D went into improving the replication machinery of DNA during the first billion or so years of life. The invention of writing has similarly boosted the fidelity of linguistic transmission, and it was the product of many minds in many places over several millennia. Few if any of the "inventors" of writing had—or needed to have—a clear vision of the "specs" of the machine they were inventing, the "problem" they were "solving" so elegantly. (Compare their work to Turing inventing the computer or the "top-down" designing of the elevator controller program.) As noted in chapter 6, copyright requires the existence of a relatively permanent record (writing or audio recording), for obvious evidential reasons. But the ability to translate or transduce an invention into such a medium has a more profound impact than mere legal protection. "The written medium allows more complexity because the words on a page don't die on the air like speech, but can be rescanned until you figure out what the writer intended" (Hurford 2014, p. 149). This gives us the ability to *offload* a key phrase or sentence (or verse or saga or promulgation) into the world, taking a burden off our limited working memory, and allowing us to *take our time* going over and over a bit of text without worrying about whether it will slowly morph in our rehearsals or dissolve from our memory altogether.[72] (It is possible, with much practice, to do three-

---

72 Susan Sontag, *On Photography* (1977), discusses the huge epistemological advance (rivaling microscopes and telescopes) that high-speed photography provided for science. "Freezing" time is just as valuable as magnifying space. The recent development of image-stabilizing binoculars exhibits the same advantage: a boat name that is utterly unreadable because it is bouncing in the binocular view becomes easily read when it is locked down briefly.

digit multiplication in your head, but think how much easier it is with a pencil and a scrap of paper.)

Thanks to further innovations we may now be on the verge of a major transition in cultural evolution. There is no DNA for culture now, but HTML (the underlying language of content representation on the Internet) or its descendants may become so dominant in the future that few memes can compete for "eyeballs" and "ears" in this crowded world of information without getting themselves represented in HTML. There are now bots and apps that can search for content and recognize or discriminate content (e.g., Shazam, the smartphone app). If their descendants begin making what amount to value judgments (what Richerson and Boyd, 2004, call biased transmission), encouraging the differential replication of memes without any reliance on human brains, eyes, or ears, then the memes of the near future may thrive without direct human intervention, still synanthropic, like barn swallows and chimney swifts, but dependent on the amenities of the technological niche constructed in the twenty-first century by human beings. These digitally copied memes had been called "temes" by Susan Blackmore (2010), but she has more recently decided to call them *tremes* (personal correspondence, 2016). (More on them in the final chapter.)

Not all processes of change in culture exhibit features analogous to competition of alleles at loci, but that feature is in any case just one of the dimensions along which evolutionary processes can vary, some more "Darwinian" than others. Darwin didn't have the concept of a locus with alleles, and he didn't need it to establish in general outline his theory of evolution by natural selection. (We'll look at some of the other dimensions of variation, arrayed on one of Godfrey-Smith's Darwinian Spaces in chapter 13.)

Memes add nothing to what we already know about culture

The claim is that the would-be memeticist has simply taken over wholesale all the categories, discovered relationships, and phenomena already well described and explained by cultural theorists of more traditional, nonbiological schools of thought and rebranded the whole enterprise as memetics. According to this objection, there is no value added by switching from idea-talk and belief-talk and tradition-talk and institution-talk to meme-talk (with its pseudo-biological, pseudo-scientific sheen). Where are the novel insights, the corrections, the clarifications that might motivate or warrant such an imperialistic takeover? At best, memeticists are reinventing the wheel. (I hope I have done justice to the adamant outrage with which this objection is often expressed.)

There is some truth to this charge, and, in particular, memeticists (myself included) have on occasion offered presumed novel insights to traditional cultural theorists that they had long ago formulated and digested. It must be particularly irritating for a culture theorist—a historian or anthropologist or sociologist, for instance—to endure an elementary memetics-based explanation of some phenomenon that unwittingly echoes some much more detailed and supported explanation they had authored some years earlier. It would be shocking to discover that earlier cultural theorists were all wrong in their theories or that they had been blind to all the distinctions noted by memeticists, so it should not embarrass the memeticist to acknowledge that *in some regards* culture theorists had gotten along quite well in their explanatory projects without memetics.

In fact, memeticists should seek out and treasure the insights garnered by traditional investigators of culture following the good example of Darwin, who harvested the wealth of natural history he learned from his large coterie of correspondents. The data gathered and tabulated on all aspects of plants and animals by *pre*-Darwinian investigators had the seldom-sung virtue of being *theoretically untainted*, not gathered by ardent Darwinians (or ardent anti-Darwinians) whose biases might cloud their judgment. And non-Darwinian, pre-memetics studies and theories of culture should be valued for the same reason: the data gathering—always at risk of being tainted by observer bias in the social sciences—has at least not been gathered

FIGURE 3.3: Australian termite castle.

© *Photograph by Fiona Stewart.*

FIGURE 3.4: Gaudí, La Sagrada Familia.

© *Diariodiviaggio.org.*

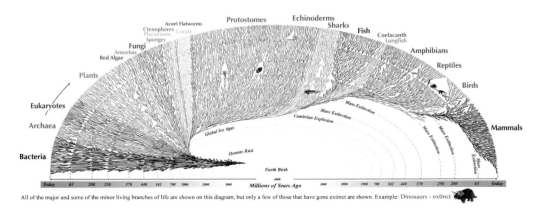

FIGURE 9.1: The Great Tree of Life. © *Leonard Eisenberg.*

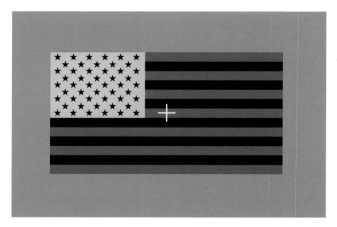

FIGURE 12.1: Claidière et al., random patterns (on left) evolve into memorable tetrominos (on right). © Nicolas Claidière.

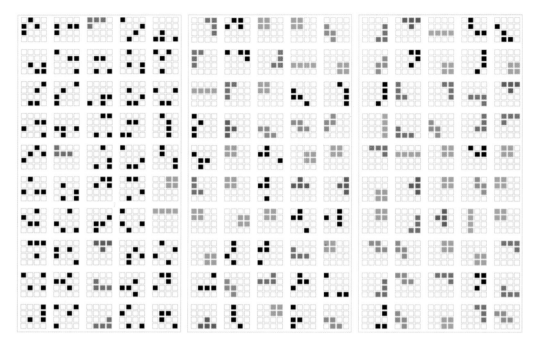

FIGURE 14.1: Complementary color afterimage.
Fixate on the white cross in the top panel for about ten seconds and then switch to the white cross in the bottom panel.

with intent to support memetics! Memetics would be a dubious candidate for a good theory of cultural evolution if it *didn't* take over large swathes of earlier theory, making minor terminological adjustments by recasting the points in meme-talk.

But memetics should also offer genuine advances in outlook, and I have already sketched what I take to be its main contributions, along with the corresponding weaknesses in traditional theories of culture. Cultures are full of well-designed elements, and traditional theorists of culture have either overendowed people with comprehension, with inventive skill, with genius, or—seeing that people, like butterflies, like antelopes, can be beneficiaries of arrangements that they needn't understand—have simply abdicated the responsibility of accounting for how the requisite R&D might have come to pass. Genetic evolution ("instincts") can't operate fast enough to do the job, leaving a yawning gap to be filled by memetics, and no positive ideas of anything else coming from traditional approaches to culture that could do the job.

The memetics perspective is also valuable in *depsychologizing* the spread of innovations (good and bad). Traditional approaches to cultural evolution (in the neutral sense of cultural change over time) such as "history of ideas" and cultural anthropology take people to be, in the first place, perceivers, believers, rememberers, intenders, knowers, understanders, noticers. A comatose or sleeping person is not a transducer/transmitter of culture under any conditions, so the natural assumption is that cultural innovations are *noticed* and then (often) *adopted*. The innovations are typically held to have been adopted *because* they were *appreciated* or *valued* or just *desired*. (They may be adopted by mistake, of course.) This all fits smoothly—too smoothly—with the default vision of people as rational agents, intentional systems whose behavior can be predicted in principle by attributing beliefs, desires, and rationality to them. This in turn yields one version or another of the economic model of cultural preservation and transmission: those cultural "goods" deemed valuable are preserved, maintained, and either bequeathed to the next generation or sold to the highest bidder. But much cultural innovation happens by what might be called subliminal adjustments over long stretches of time, without needing to be noticed or consciously approved at all. Like viruses, memes can spread from host to host without the hosts taking notice. Population-level effects can accumulate unheeded by the members of the population. These accumulated shifts can often be recognized in retrospect, as when an expatriate community is joined by a new person from the old country whose way of speaking is both strangely familiar and strangely unfamiliar; *Aha! I remember we used to talk like that too!* Not just pronunciation and word meaning can subliminally shift. In principle, attitudes, moral values, the most emblematic idiosyncrasies of a culture can soften, harden, erode, or become brittle at a pace too slow to perceive. Cultural evolution is lightning fast, compared to genetic evolution, but it can also be much too gradual for casual observation to discern.

Then there are the "pathological cultural variants." Many persistent cultural phenomena are maladaptive, and no theory that aligns cultural innovations with genetic innovations ("adaptations transmitted by other means") has any way of accounting for these. Richerson and Boyd's (2005) chapter, "Culture is Maladaptive," is not presented as a defense of memetics, but it adopts the memetic perspective to account for a variety of otherwise puzzling phenomena, such as the thriving (so far) of Anabaptists such as the Amish. They grant, in spite of their misgivings about some aspects of memes (see, e.g., p. 6), that "the selfish meme effect is quite robust" (p. 154).

> So far, only the Anabaptists and a few similar groups, like ultra-Orthodox Jews, seem to have much resistance to modernity's infections. Anabaptism is like a tightly made kayak navigating the turbulent modernist sea. It looks so fragile but survives because it doesn't leak despite the enormous stresses it faces. One serious cultural leak anywhere, and it's gone. Anabaptism's evolutionary future or futures are impossible to predict. In the meantime, you can't help but admire the beauty of the design! (p. 186)

The would-be science of memetics is not predictive

In one sense, this is true but not an objection. The theory of genetic evolution is also not predictive in the sense that it permits us to describe what the future of *H. sapiens* (or codfish or polar bears or spruce trees) will be with any degree of certainty. There is a deep reason for this: Darwinian evolutionary processes are *amplifiers of noise*. As I noted in chapter 1, a curious feature of evolution by natural selection is that it depends crucially on events that "almost never" happen. So evolutionary theory, not being able to predict the once-in-a-billion events that in due course get amplified into new species, new genes, new adaptations, can't predict the future except very conditionally: *If* such and such happens, then (unless some *other* unpredictable event gets amplified into a clobberer), such and such will happen. But this sort of real-time prediction is not a necessary requirement for serious science; prediction of what we will and won't find if we dig for fossils—to take an obvious case—is still prediction, even though it's ultimately about events that occurred millions of years ago. Evolutionary biology predicts that you will never find a bird with fur instead of feathers, no matter how isolated it is on some faraway island, and that you will find quite specific DNA sequences in quite specific places in the genomes of any heretofore unknown insect that is captured anywhere in the world.

So the fact that memeticists cannot reliably say which songs will make the top twenty or whether hemlines will rise in 2017 is not a serious objection to a science of memetics. A better question is whether memetics can provide unified explanations of patterns that have been recorded by earlier researchers of culture but not unified or explained at all. I have claimed that the meme's-eye view fills the large and awkward gap between genetically transmitted instincts and comprehended inventions, between competent animals and intelligent designers, and it fills it with the only kind of theoretical framework that can nonmiraculously account for the accumulation of *good* design: differential replication of descendants. Whether some other evolutionary theory of cultural change can fill this gap remains to be seen.

Memes can't *explain* cultural features, while traditional social sciences can

This objection misses the point of memes, as we can see by mounting a similar "objection" to genes. *Genes can't explain adaptations (structures, organs, instincts, etc.).* That's true; that's why we need molecular biology, physiology, embryology, ethology, island biogeography, and all the other specializations of biology if we are going to explain how particular adaptations work and why they are adaptations. We also need these other fields to explain how parasites exploit their hosts, how spider webs are cost-effective food traps, how beavers build their dams, why whales vocalize, and all the rest. Similarly, we need psychology, anthropology, economics, political science, history, philosophy, and literary theory to explain how and why cultural features (good and bad) work the way they do.

Theodosius Dobzhansky was right when he so quotably said, "Nothing in biology makes sense except in the light of evolution," but he didn't say that the light of evolution makes sense of everything in biology. There is nothing in memetics that can replace the hard-won data gathering and pattern proving of the "behavioral ecology" of people in music and the arts, religions, corporations, armies, clubs, families, teams, schools, kitchens, and all the other contexts of human life. What memetics promises to do is provide the framework for making sense of all this in *some* of its aspects. Nobody is born a priest or a plumber or a prostitute, and how they "got that way" is not going to be explained by their genes alone or just by the memes that infest them. Pre-Darwinian natural history at its best was well-developed, systematic science, with hypotheses to test and explanations galore, but Darwin enriched it with a host of questions that put all its generalizations in a new light. My overarching claim in this book is that the evolutionary perspective in general and the memetic perspective with regard to culture transform many of the apparently eternal puzzles of life, that is, meaning and consciousness, in ways inaccessible to those who never look beyond the manifest image that they grew up with and the disciplines they are trained in.

Cultural evolution is Lamarckian

This is one of the most popular "explanations" of why Darwinian approaches to cultural evolution are misguided, and it wears its confusion (and desperation) on its sleeve. Presumably what those who make this "charge" have in mind is the doctrine most famously advanced by Jean-Baptiste Lamarck, a predecessor of Darwin, according to which characteristics acquired by an individual organism in its lifetime can be transmitted genetically to its offspring: the blacksmith's bulging biceps, acquired by strenuous use, are passed on to his progeny *through his genes* (not through his enforcement of rigorous exercise habits in his children). Similarly, on this Lamarckian view, the fear of human beings inculcated in a dog as a result of having an encounter with a cruel master can become an *instinctual* fear in the dog's puppies, which never have to experience cruelty to anticipate and fear it. Not so. There is no way for an acquired trait to adjust an organism's genes so that the trait gets passed along to the next generation genetically. (The organism's body would have to reverse engineer the somatic [bodily] change and work out what the genetic recipe for making that novelty in normal development would be, and then revise the DNA sequence in the sperm or eggs accordingly.) This truly heretical idea was not only espoused by Lamarck, but variations on it were also quite popular in the nineteenth century. Darwin was tempted by a version of it, but it is indeed discredited in standard neo-Darwinian evolutionary theory. There are various real phenomena that look rather "Lamarckian" because they do involve something like transmission of acquired characteristics to offspring but not genetically. For instance, a fearful dog might well transmit fear to her puppies through a pheromone ("the smell of fear") or via a hormone she shared through the placenta, but not through her genes. There's also the Baldwin Effect, which *looks* Lamarckian in some regards, since behaviors acquired by one generation can create a selection pressure favoring offspring who have a heightened tendency or talent for acquiring that behavior, eventually making it "instinctual." And then there are the current hot topics: the complexities of development uncovered by "evo-devo" biologists, including especially "epigenetic" traits. A consensus definition of an epigenetic trait was achieved in 2008: "a stably heritable phenotype resulting from changes in a chromosome without alterations in the DNA sequence" (Berger et al. 2009). The topics are hot for several reasons: the researchers have uncovered some fascinating new wrinkles at the molecular level in the unfolding saga of evolutionary theory, some of the researchers have trumpeted their discoveries as truly revolutionary, calling into question neo-Darwinian orthodoxy, and as a result, Darwin-dreaders outside biology have perked up their ears, inflated what they've heard, and spread the false alarm that evolution by natural selection has been "refuted" by the discovery of Lamarckian inheritance. Nonsense.

Presumably certain critics of memes operate under the impression that if cultural evolution is Lamarckian, then it cannot be Darwinian (whew!). The critics are right that cultural transmission permits any traits that are acquired by the parent to be inculcated in the young (by setting an example, by training, by admonition), but this does not show that cultural evolution is not accomplished by Darwinian natural selection. Far from it. After all, parents can also pass on their acquired germs and microbes and viruses to their young, and all of these phenomena are uncontroversially subject to (non-Lamarckian) Darwinian natural selection. Maybe these critics are forgetting that in memetic evolution it is the fitness of the memes themselves that is at stake, not the fitness of their hosts. The question of Lamarckianism should be whether features *acquired by the memes* can be transmitted to *their* offspring. And here, since the genotype/phenotype distinction that makes Lamarckian transmission all but impossible is absent, Lamarckian transmission would not be a heresy, but just an alternative variety of natural selection—

after all, memes don't have genes. (See, for instance, figure 7.2, p. 144, the Darwinian Space with the G dimension, standing for the germ/soma distinction; different positions on the G dimension admit of Lamarckian inheritance, so understood.) In a virus, for instance, there is no clear distinction between germ line mutation and acquired trait. The replicator and the interactor are—for most purposes—one and the same thing.

There is another, more interesting way cultural evolution might be considered Lamarckian, and it is easiest to see if we consider our paradigm memes, words. As we saw in chapter 9, before an infant will begin trying to imitate a word, it has to have heard about six tokens—six "parents," not the usual two (or one, in the case of asexually reproducing organisms). But suppose I, as an adult, learn a new word, hearing you say it in a context that makes it clear what it means. I store a copy of your token as a new lexical item in my brain. The informational structure in my head is a descendant of your token, and every time I rehearse it these newly made tokens are further descendants—grandchildren and great grandchildren, in effect—of the single token you uttered. But suppose before I utter the word aloud I hear you say it again, or hear it on the lips of several other speakers. Now I have multiple models for my new word, multiple parents, in effect.

We might then consider that my new word, like the infant's, has multiple parents that *all* contributed to fixing the informational structure descended from them. Or we might consider the first utterance heard the *sole* parent (just like a virus, which reproduces asexually) and consider all the other tokens that helped fix the stable features of my version of the word to be not parents but just *influencers*. After all, having heard the first token from you, my memory and perceptual system is set up to recognize further tokens, and when I hear the later tokens from others, this may induce me to make further tokens of my own, but this won't be *imitating* the others; it will be "*triggered reproduction*" (Sperber 2000). But triggering reproduction is itself, if not a parenting at least a bit of midwifery, making a contribution—and perhaps a biased contribution—to the further reproduction of the type. My informational structure for the word may be adjusted by encountering these later tokens of the word, and these could lead to changes in features that could then be inherited by any further offspring of my version. This would be Lamarckism of a sort at the meme level, one of many possible variants of natural selection in Darwinian populations. Alternatively, we could consider words, and memes more generally, to be the result of variable, temporally extended processes of reproduction (as if father's contribution was not made "at conception" but at some later time, after mother had already given birth), an imaginable variation on our normal mode of sexual reproduction.

So there are several ways we could consider cultural evolution to be Lamarckian without thereby erecting a barrier against the imperialistic forays of dread Darwinism into the sacred precincts of culture. Steven Pinker—no friend of memes—has candidly acknowledged: "To say that cultural evolution is Lamarckian is to confess that one has no idea how it works." He goes on, however, to say quite a bit about how he thinks it works:

> The striking features of cultural products, namely their ingenuity, beauty, and truth (analogous to organisms' complex adaptive design), come from the mental computations that "direct"—that is, invent—the "mutations," and that "acquire"—that is, understand—the "characteristics." (1997, p. 209)

This perfectly expresses the traditional view that it is comprehension, by inventors, by intelligent designers, that accounts for the improvements—the "ingenuity, beauty, and truth"—we observe in cultural items. Yes, *some* of the marvels of culture can be attributed to the genius of their inventors, but much less than is commonly imagined, and all of it rests on the foundations of good design built up over millennia by uncomprehending hosts of memes competing with each other for rehearsal time in brains.

Several themes emerge from this consideration of objections to memes. First, the issue should not be whether Dawkins, when introducing the concept, managed to articulate the best definition, or whether I or anybody else has yet succeeded in coming up with a bullet-proof formulation. The issue is—as it usually is in science—whether there are valuable concepts and perspectives emerging from the explorations to date. The analogy with genes is rich but partial, and perhaps the chief benefit of the meme's-eye point of view is that it suggests questions about cultural phenomena that we might not otherwise think of asking, such as: Is x the result of intelligent design? Is x a good worth preserving and bequeathing or a bit of parasitic junk? Are there alternatives (alleles) to x that have been encountered and vanquished?

Second, memes are in competition with traditional accounts of cultural change only when those accounts attribute comprehension to people (or mysterious social forces) for which there is no evidence other than the excellence of the designs. Thus, as we will see in chapter 13, the memetic perspective provides a level playing field where all degrees and kinds of human comprehension can be located.

There is not yet a formalized science of memetics, and there may never be, though various pioneering forays have been attempted. But then, the parts of evolutionary theory in biology that have blossomed under mathematical treatments are only parts, and Darwin managed without any of them. Darwin also wisely postponed inquiry into how evolution got started until he had carefully described its midcourse operation: "descent with modification." Following his lead, now that we have seen how words and other memes can descend, with modifications, from their ancestors, we can return to the difficult question of how cultural evolution in general, and language in particular, originated in our species.

12

The Origins of Language

I cannot doubt that language owes its origins to the imitation
and modification of various natural sounds, and man's own distinctive cries, aided by signs and gestures.
—Darwin, *The Descent of Man*

The chicken-egg problem

Now that we've seen how memes in general, and words in particular, create the environment in which we grow our minds,
it is time to go back to the origins of language, and culture,
and see what might be figured out about how this happened. As
noted in chapter 9, the topic of the origin of language is like the
topic of the origin of life itself; both are (probably) unique events
on this planet that left only a few tantalizing clues about how they
occurred. The evolution of language has been deemed "the hardest
problem in science" (Christiansen and Kirby 2003, p. 1), and it has
certainly been controversial since shortly after Darwin published
Origin of Species in 1859.

In fact, one of the most popular memes in linguistics (a meme I
have been complicit in spreading, until now) is that the topic proved
to be so speculative that the Linguistic Society of Paris banned all
discussion of language evolution in 1866, a policy also followed
by the Philological Society of London in 1872 (see, e.g., Corballis
2009). For decades this story has decorated the battle flag of many
Chomskyans who want to dismiss all evolutionary approaches to
language as mere "just-so stories" unfit for scientific assessment.

It turns out, however, that the Linguistic Society of Paris was not upholding sober science but just the opposite: taking a dogmatic stand against the materialists of the Anthropological Society, and devoting itself to advancing the opposing views of monarchists and Catholics! As I recently learned from Murray Smith:

> The Société de Linguistique de Paris still exists. Its website has a history page, which says that in 1866 by-laws were drawn up that specified (article 2) that "the society receives no communications concerning either the origin [not "evolution"—MS] of language nor the creation of a universal language." In establishing this rule, the history continues, the society sought to distinguish itself from "positivist and republican" circles. The society was not outlawing a controversial topic; it was taking a position on it. (Personal correspondence, 2015)

For decades the factoid about the ban—too good to be true, as it turns out—has helped discourage evolutionary approaches to language, but recent advances across the relevant fields have tempted intrepid researchers in linguistics, evolutionary biology, anthropology, neuroscience, and psychology into the open, with a panoply of prospects. A bird's-eye view is all we need for our purposes, fascinating though the various positions are. As we consider the possibilities, we must be alert to the danger of gambit blindness (see ch. 2, p. 30), the besetting blockade on adaptationist imagination. As noted earlier, the first living, reproducing thing may well have been an ungainly hodgepodge of inefficient parts, only later streamlined by further evolutionary competition into the lean, effective bacteria we know today. What might the ancestors of today's well-designed languages have been? They were probably inefficient, hard-to-learn behavioral patterns that seldom "worked." What conditions had to be in place to make those early versions worth investing in? They may not even have "paid for" the expense of using them. They may have been parasitic habits that were infectious and hard to shake. So we need to ask: Who were the principal beneficiaries, the words or the speakers? It has seemed obvious to many that the answer must be the speakers, but this may be because they never considered the meme's-eye point of view. We should be on the lookout for a circuitous route, with gambits galore, and while there can be little doubt that language *today* serves us human hosts remarkably well, the early days of language might have been more of an imposition than a gift. Wherever and however language got started, here is a tentative list of the functions all languages eventually served:

Communicative Utility. The power of languages to command, request, inform, inquire, instruct, insult, inspire, intimidate, placate, seduce, amuse, entertain.

Productivity. The power of languages to generate a Vast (see ch. 6, fn. 36, p. 130) number of different meanings (sentences, utterances) composed from a finite stock of lexical items. Thinking formally, there is no end to the number of grammatical sentences of English—no rule that says, for instance, that no sentence can have more than *n* words—but if we were to restrict ourselves to sentences of no more than, say, twenty words, the number of clearly grammatical, sensible sentences, readily understandable on first hearing by normal adults is still Vast. (The previous sentence has almost certainly never before been expressed by anyone, and it has sixty-three words in it, but was not a struggle, was it?)

Digitality. The ability, as we have seen, of language receivers/transmitters to "correct to the norms," thereby rinsing much of the noise out of the signal, even in the face of incomprehension.

Displaced Reference. The power of language to refer to things not present in the environment of the communicators, out of sight, in the past, imaginary, or hypothetical.

Ease of Acquisition. The remarkable swiftness (compared to reading, writing, and arithmetic, for instance) with which spoken or signed language is picked up by children.

It would seem that other social mammals—apes, wolves, elephants, whales, bats, for instance—could make spectacular use of language if they had it, and some of them have well-known communicative talents, but no other species has a faculty remotely like human language in its power.[73] Somehow our languageless ancestors stumbled onto a rare path to this treasure and myopically followed it, getting some advantage along the way, or at least not losing ground, until they hit the jackpot. How and why? And why haven't other species found the same Good Trick?

I can't think of anyone who denies that a necessary precursor of language had to be *some* kind of prelinguistic cultural transmission supported by some kind of genetic adjustment. For instance, an instinct to *cooperate with one's extended family of conspecifics* is found in many languageless social species, from prairie dogs to elephants and whales; and if we add to that only an imperious instinctual habit of *imitation* (of one's parents, one's elders, one's peers), this would perhaps be enough to create a supportive environment for the runaway cultural evolution that eventually emerged in us, riding on language. In other words, words may be the best memes, but they weren't the first memes. Such instincts would not blossom and thrive without an environmental/behavioral innovation that raised the bar for subsequent members of a population, creating a selection pressure for *enhanced* dispositions to cooperate—to cooperate sooner, more readily, more attentively. Similarly, an instinct to copy one's elders could be a fitness booster under many but not all conditions. Did genetic evolution revise one lineage of hominids, giving them instincts that in one way or another allowed them

---

73 I find that some of my students are offended by this complacent generalization, and insist that we simply cannot know that dolphins, say, don't have productive languages like ours. I will concede this much: Yes, dolphins may conceivably have language and intelligence the equal of ours, and if so, they have done a heroic and brilliant job of concealing that fact from us, sacrificing themselves on countless occasions when a little dolphin chitchat could have saved their lives or the lives of some of their conspecifics. (There are a few experiments that suggest that dolphins have some limited capacity to inform each other about out-of-sight affordances.)

to share cultural innovations more than the other lineages? What new selection pressure, restricted to one group, might set that off? Did group cooperation evolve before language? Chimpanzees often engage in *somewhat* cooperative hunting of colobus monkeys. Did a new challenge in the environment oblige one group of hominids (our ancestors) to rely more heavily on cooperation, and did this give rise to novel habits of attention? (Tomasello 2014). What benefit got our ancestors' children so *interested* in the vocalizations of their group and so eager to imitate them?

Richerson and Boyd, in *Not by Genes Alone* (2004), are deliberately nearly silent about language. Halfway through the book they say, "So far we have said nothing about language, and the reason is simple: paleoanthropologists have no idea when human language evolved" (pp. 143–144). There is a two-*million*-year window of controversy, with some neuroanatomical details (as deduced from the fossilized skulls of hominins) suggesting that elements of language have been present for millions of years, and other evidence suggesting that even 50,000 years ago language was rudimentary at best.

Is language required for transmitting skill in knapping stone tools? Trading goods (materials, food, tools)? Tending a fire? Can we imagine young hominins acquiring the self-control and foresight to tend a fire effectively without verbal instruction (and later self-admonition)? According to Sue Savage-Rumbaugh (personal communication, 1998), bonobos in captivity love to sit around a campfire, but could they be trained by us to gather firewood, and to tend a fire indefinitely, conserving fuel and avoiding bonfires? Whatever the answer to that question, it would not actually shed much light on whether our ancestors could train each other to tend fires before they had at least some crude version of language. Why not? Because bonobos have been evolving away from the common ancestors we share for as long as we have—about six million years— and there is an untabulated and unsurveyable world of differences between the perspective and talents of a language-equipped, educated human trainer who can enforce training regimes on a captive bonobo and the perspective of an adult hominin raising one of its offspring. As I once observed (1978), "Bears can ride bicycles, a surprising fact of elusive theoretical significance." Could the cave paintings of Lascaux (between 20,000 and 30,000 years old) have been painted by *H. sapiens* artists without language? (See Humphrey 1998.) These and other key questions remain unanswered and perhaps unanswerable.

A benefit of postponing any hypotheses about when functioning language entered the scene is that it allows Richerson and Boyd to explore models that are minimalistic in an important sense: they presuppose almost no comprehension because they presuppose no *communication*. (I find that it helps my students, when reading their book, to ask themselves: "What would it take to get cultural transmission to thrive among turtles or seagulls or sheep?" This is one way of keeping premature assumptions of comprehension on the part of the actors from clouding one's imagination.) For instance, Richerson and Boyd explore models in which there is *biased transmission*, and the bias can be *anything at all* that makes some memes more likely to be passed on to others.[74] Biased transmission counts as a selective force, whether or not it is a good or wise or understood bias. A bias that is apt to be more valuable than "copy anything that moves" or "copy the first adult you see" is "copy the majority" (conformist bias) "copy the successful" or "copy the prestigious." And while information moves from A to B whenever B copies A, this is not always communication, any more than B catching a cold from A is communication (even though we talk of *communicable* diseases).

---

74 As noted, Richerson and Boyd prefer not to use the word "meme" because of its connotations as a "discrete, faithfully transmitted genelike entity" (p. 63), but they recognize the "need for some expedient agreement" for what to call "the information stored in people's brains." Since we are allowing memes to vary along these dimensions (see this book pp. 146–149), we are not taking sides in these controversies so, acknowledging their resistance, I lean on *OED*'s definition and recast their points in meme-talk. A useful exercise would be to go through *Not by Genes Alone* and recast all their points in terms of memes to discover just which disagreements still persist. I take the substantial issues to have been covered in the defense of memes in chapter 11.

Under such minimalistic conditions, when does a habit of (basically clueless) copying do better than engaging in your own trial-and-error learning? Better for whom? *Cui bono?* Better for the individual copiers, better for the population of adopters, or better for the memes themselves? Richerson and Boyd are well aware that maladaptive culture can spread, but they maintain that until *adaptive* culture (culture that enhances the genetic fitness of those who have it) has established a foothold, the genetically maintained habits and organs on which culture depends would soon be selected against. The parallel between memes and viruses makes the reason clear. Viruses can't reproduce on their own; they depend on commandeering the reliable copy machinery in the nucleus of living cells, and that copy machinery was the product of a billion years of R&D. Memes, helpful or not, must above all get themselves copied, so at least that much genetically established infrastructure—dispositions to attend to others, and to copy (some of) the *ways* perceived—is the only ground in which memes could take root and bloom.

A lot of design work is required to create *and optimize* a user-friendly medium of information exchange, and "someone has to pay for it," but once a rudimentary copy system is in place, it can be hijacked by selfish interlopers. Perhaps we are just apes with brains being manipulated by memes in much the way we are manipulated by the cold virus. Instead of looking only at the prerequisite *competences* our ancestors needed to have in order for language to get under way, perhaps we should also consider unusual *vulnerabilities* that might make our ancestors the ideal hosts for infectious but nonvirulent habits (memes) that allowed us to live and stay mobile long enough for them to replicate through our populations.

The upshot of cultural evolution—so far—has been a spectacular increase in human population, and there are hundreds of thousands of memes to be found in every corner of the Earth, including many environments that could not be occupied by *H. sapiens* without their assistance. But just as the ant climbing the blade of grass is the lancet fluke's way of getting into the belly of a cow or sheep, so perhaps we should think of astronauts going to the moon as the memes' way of getting into the next generation of science nerds.[75]

The meme's-eye point of view permits a minor but useful amendment to the various *dual-inheritance* or *dual-highway* models of the relationship between genetics and cultural evolution (Cavalli-Sforza and Feldman 1981; Boyd and Richerson 1985, among others). Adaptations (fitness enhancers) can be either genetically or culturally transmitted. The *genetic* information highway has been optimized over billions of years, an engineering marvel of breathtaking complexity, with DNA copying machines, editing machines, and systems for dealing with genomic parasites (rogue DNA that gets established in genomes, where the damage they do is more or less controlled and minimized).[76] The *cultural* highway, over a much shorter time period, has also evolved a host of design features to facilitate reliable transmission of information, and much of the R&D creating and honing these features has been the evolution of memes themselves, to "fit the brain" while genetic adjustments to the brain followed in their wake, a coevolutionary process in which the "research" is mainly done by the memes and the later "development" is mainly done by the genes. That is, innovations *in memes* that made them more effective reproducers in brains that were not *yet* well designed for dealing with them could provide the early "proof of concept" that would underwrite, in effect, the more expensive and time-consuming genetic adjustments in brain hardware that would improve the working conditions for both memes and their hosts.

This basic pattern has been echoed hundreds of times since the

75 "The nucleic acids invented human beings in order to be able to reproduce themselves even on the Moon" (Sol Spiegelman, quoted in Eigen 1992, p. 124).

76 Some of the curious complexities found in the copy system for DNA are the effects of billions of years of war—an arms race between selfish DNA, which just "wants" to get itself copied without contributing anything to the maintenance and survival of the organism, and policing systems that have evolved to keep this uncooperative DNA from overrunning the genetic system. A fascinating analysis can be found in Burt and Trivers, *Genes in Conflict: The Biology of Selfish Genetic Elements* (2008).

dawn of the computer age, with software innovations leading the way and hardware redesigns following, once the software versions have been proven to work. If you compare today's computer chips with their ancestors of fifty years ago, you will see many innovations that were first designed as software systems, as *simulations* of new computers running on existing hardware computers. Once their virtues were proven and their defects eliminated or minimized, they could serve as specifications for making new processing chips, much faster versions of the simulations. As novel environments for computers emerged, behavioral competences were first explored in the form of programs running on general-purpose computers, where revision is quick and cheap, and then extensively field-tested against the competition, with the best designs eventually being incorporated into "dedicated" hardware. For instance, today's smartphones have, in addition to layers and layers of software running on software running on software, special-purpose graphics and speech-synthesis and recognition *hardware* in their microprocessors, the descendants of software systems that explored the Design Space first.

All of this computer R&D has been top-down intelligent design, of course, with extensive analysis of the problem spaces, the acoustics, optics, and other relevant aspects of the physics involved, and guided by explicit applications of cost-benefit analysis, but it still has uncovered many of the same paths to good design blindly located by bottom-up Darwinian design over longer periods of time. For instance, your smartphone has special-purpose hardware for speech processing but *not* for speaking English or Chinese; those competences can be added with software once the smartphone has landed in a home environment. Why? For the same reason (free-floating rationale) an infant brain is language neutral: versatility widens the "market" for the design. This could change; if a single human language went more or less to global fixation (a process portended by the extinction of hundreds of languages every year), infant brains that happened to have built-in biases for learning that language might be positively selected for, leading over many generations to an eventual loss of linguistic versatility in the underlying human hardware. This would be a vivid example of the Baldwin Effect, showing how it *reduces* genetic variance and versatility by driving a behavior (or developmental option) into a "best practices" straitjacket controlled genetically, turning options into obligate behaviors, as we saw in chapter 9 (p. 196).

Once the cultural highway is in place, in any stable collaboration between meme features and brain features, software and hardware, cultural parasites (rogue cultural variants, in Boyd and Richerson's terms, parasitic memes, in Dawkins's terms) can thrive, in spite of countermeasures, in an arms race much like any other in evolution.

Not all environments are fertile ground for the evolution of cultural evolution. If we think of copiers as information *scroungers* and learners as information *producers* (Richerson and Boyd p. 112, citing Kameda and Nakanishi 2002), we can see that there should be a trade-off, with the possibility of an equilibrium in which some innovators bear the costs of R&D (not of the information highway but of the invention/discovery of whatever they produce) and the rest help themselves to the information at lower costs.[77] The fact that simple models exhibit this dynamic equilibrium suggests that it could exist in a population lacking language but having sufficient variation in inquisitiveness (favoring R&D, even though it is expensive) and conformity (favoring copying, even though it may be outdated information).[78] This feature depends, however, on the variability of the environment: if the environment is either too predictable or too unpredictable, cultural transmission (copying) doesn't gain a foothold. This gives us an example of a possible threshold condition: cultural transmission won't evolve except in a Goldilocks environment that is neither too hot—chaotic—nor too cold—unchanging—for long enough to provide evolution a chance to create some

77 *Mindlessly copy the majority* turns out to be a remarkably effective strategy. The individuals with competence (or behavioral comprehension) soon lose their advantage to the copiers (Rendell et al. 2010)

78 Richerson and Boyd point out that some experiments with people and pigeons (Kameda and Nakanishi 2002, for instance) exhibit the same division of labor with no communication between participants.

new habits and fix them in a population. If such long periods are rare, populations that would otherwise be well equipped to take advantage of cultural transmission will be unlikely to take the necessary steps.

Culture has been a spectacularly successful Good Trick for *H. sapiens*. Any theory of the origin of culture without at least one threshold, one barrier to adoption that has somehow been overcome, is vulnerable to the objection that if acquiring cultural transmission were that easy, many species—mammals, birds, fish—would have it by now. An environment with a history of just the right amount of variation, neither too monotonous nor too abrupt, is one possible threshold. Other thresholds have been proposed. Consider *bipedality*, walking upright on two legs, a perennial favorite since Plato described man as a featherless biped. Among hominids, only the hominins exhibit this feature, and it certainly freed up our ancestors' arms for making artifacts, and carrying them and the raw materials needed from place to place. But evolution has no foresight, so the original free-floating rationale of this excellent design innovation may have been something else. Here is yet another chicken-egg problem: Did rudimentary tool making (of the sort seen by chimpanzees) create a selection pressure for the ability to carry raw materials or finished tools for long distances, or did upright walking, evolved for other reasons, open up the Design Space for effective tool making? There are many hypotheses. The savannah hypothesis proposes that once a drier climate drove our ancestors out of the trees and onto the grasslands, upright locomotion was favored since it permitted seeing farther over the tall grass (and/or minimized exposure to the blazing sun, and/or minimized energy expended in locomotion), while variations on the wading (or aquatic ape) hypothesis (A. Hardy 1960; Morgan 1982, 1997) propose that harvesting shallow-water shellfish was the ecological innovation. Or perhaps aquatic plants (Wrangham et al. 2009) became a critical "fallback food" during difficult times, promoting wading in ever deeper water, breath holding, and perhaps other physiological revisions. All very controversial and likely to be controversial for some time. In any case, could bipedality and its ensuing suite of enabled competences open the floodgates for language and culture?

Another proposed threshold is *social intelligence* (Jolly 1966; Humphrey 1976): the competence to interpret others as intentional systems whose actions can be anticipated by observing what these others observe and figuring out what they want (food, escape, to predate you, a mating opportunity, to be left alone). This competence is often called TOM (an acronym for theory of mind), which is an ill-chosen term because it invites us to imagine those having this competence as comprehending theoreticians, astute evidence gatherers and hypothesis considerers, instead of flying-by-the-seat-of-their-pants agent anticipators, blessed with an interpretive talent they don't have to understand at all. It could be, in any case, that such perspective taking is required for complex cultural information to be transmitted, just as bipedality is apparently required for complex material culture, such as tools, weapons, housing, containers, clothing, and boats.

According to Michael Tomasello (2014), another leading researcher in the area of the evolution of human cognition, such perspective taking, initially evolved for enhancing *competition* (between conspecifics and in hunting prey), could evolve into an instinct to *cooperate*, the rudimentary manifestations of which can be seen when chimpanzees hunt, for instance, and then evolve into "even more sophisticated processes of *joint* intentionality, involving *joint* goals, *joint* attention, built for social coordination" (p. 34). He claims that

> new ecological pressures (e.g., the disappearance of individually obtainable foods and then increased population sizes and competition from other groups) acted directly on human social interaction and organization, leading to the evolution of more cooperative human lifeways (e.g., collaboration for foraging and then cultural organization for group coordination and defense). (p. 125)

Tomasello sees language as "the capstone of uniquely human cognition and thinking, not its foundation" (2014, p. 127), but while I agree with his emphasis on the amount and complexity of *prelinguistic* cultural and genetic evolution, without which there would be no language, when language arrived on the scene, this enabled *cumulative* cultural evolution (and, indeed, genetic evolution in response to it), a runaway process of ever swifter and more effective amassing of novel designs and discoveries. Language may not be the foundation, but I wouldn't call it the capstone; I would call it the *launching pad* of human cognition and thinking.

Laland, Odling-Smee, and Feldman (2000) introduced the concept of *niche construction*, the idea that organisms don't just respond to the selective environment they are born into; their activities can also revise the features of that environment quite swiftly, creating whole new selection pressures and relieving others. Then the niche their descendants operate in can be importantly different from the niche their ancestors coped with. Niche construction is not just an effect of the selective pressures of natural selection; it is also an important, even destabilizing, cause of new selective pressures, a crane of considerable lifting power in Design Space. There is no doubt that our species has engaged heavily in niche construction.

Steven Pinker (2003, 2010) calls our world the "cognitive niche," stressing that it is a *product* of human comprehension. Boyd, Richerson, and Henrich (2011) disagree with Pinker, proposing that it would better be called the "cultural niche," a platform of *competences* on which comprehension can grow. As we will see, the R&D that has constructed the niche we inhabit today is a changing blend of both Darwinian, bottom-up processes and top-down intelligent design. Our niche is certainly unlike that of any other species. It includes hardly any prey or predators (unless you're a fisherman or a surfer in shark-filled waters), where habitats are composed of almost nothing but artifacts and domesticated plants and animals, where social role, wealth, reputation, expertise, and style (of clothing, speaking, singing, dancing, playing) have largely supplanted stronger muscles, faster running, and keener eyesight as variable advantages that bestow genetic fitness.

So obvious is the role that language has played in enabling and enhancing this transformation that there is something of a tradition of corrective critique, insisting that at least primitive kinds of agriculture, fishing, clothing, religion, decoration, food preparation, and other standard items of culture could thrive and be transmitted *without* language. The ritual preparation of corpses for burial, for example, strongly suggests something akin to a belief in the afterlife, but it is hard to see how anything like a creed could be shared without verbal expression. Linguistic and nonlinguistic ways of transmitting hard-won information still coexist, after all, so, as usual, we can suppose that human interaction over the millennia led to the gradual adoption of ever more effective and systematic ways (memes), including ways of acquiring memes, such as the passage from brute imitation to apprenticeship (Sterelny 2012). Some of these meme transmissions required joint attention, some required (proto-)linguistic direction, and some required fully linguistic instructions, including mnemonic mantras and other devices, no doubt.

Consider this interesting exercise of imagination: ask how—in detail—any parent animal (a dog, a wolf, a dolphin, a hominin without language) could convey some of its hard-won experience to its young without language. Suppose a wolf, for instance, has learned the hard way that porcupines are to be avoided, not pounced on. How is she going to give the benefit of her experience to her cubs? Perhaps by leading them up to a handy porcupine and then making attention-grabbing *"stay away!"* signals of one sort or another to them. This would rely on joint attention and offspring trust/obedience (just what human parents do with their prelinguistic infants when warning them away from hot stoves and the like, with variable success). But in the absence of an example to occupy the joint attention of instructor and learner, she has no recourse. The capacity of language to direct attention to nonpresent things and circumstances is a huge enhancement.

Derek Bickerton's latest book (2014) takes this power of "displaced reference" to be the key innovation of language, but he puts a slightly different spin on the question we're all addressing, alluded to in his title: *More Than Nature Needs: Language, Mind, and Evolution.* "How did the human species acquire a mind that seems far more powerful than anything humans could have needed to survive?" (p. 1). According to Bickerton, "The cognitive gap between humans and nonhuman animals is evolution's Achilles heel" (p. 5), and it cannot be explained, in his opinion, by any account of the straightforward natural selection of increasingly powerful communicative behaviors leading eventually to language. For instance, social animals need to keep track of changes of rank, and who did what to whom, but such strategically valuable gossip could not have been transmitted

> until language had undergone a considerable degree of development. In other words, the earliest stages of language could not have expressed any gossip of the slightest interest, and since some other motive would have had to drive those stages, gossip as a grooming substitute cannot have constituted the selection pressure underlying the origin of language. (p. 63)

This is reminiscent of the familiar creationist arguments that until the eye, say, or the wing, or the bacterial flagellum is fully designed it is useless and hence will not be sustained over generations, a *you-can't-get-here-from-there* challenge to the evolutionist. But Bickerton is no creationist, and after displaying what he takes to be the flaws in others' attempts, he provides his solution to the "paradox of cognition" (p. 79):

> If we rule out normal selective processes and magical versions of evolution, what's left? The answer is just two things: words and their neural consequences. The invention—for it can only have been an invention albeit it not a conscious or even an intentional one—of symbolic units had crucial consequences for the brain. (p. 91)

I relish the irony of this passage, with its dichotomy between evolution and invention, and its postulation of an unconscious Promethean leap of creativity, because for more than a decade since we first met, Derek has taken great delight in challenging me to give him one example—*just one example*—of an otherwise inexplicable social or cultural phenomenon that memes could explain. He has now answered his own challenge with *two* examples: human language and cognition. He doesn't put his answer in terms of memes (which are absent from the index of his book), but consider a few vivid passages:

> How would the brain react to getting words in it? (p. 100)

> the brain's most probable reactions to its colonization with words. (p. 108)

> Uttering words is an autocatalytic process: the more you use them, the quicker you become, to a point where you can mentally assemble utterances before producing them. (p. 123)

Earlier in his book, there is a passage that shows exactly where he missed the boat:

> Bear in mind that (proto-)linguistic behavior had to pay off from the earliest utterances or its genetic infrastructure would never have reached fixation. So consider, given your favorite selection pressure, what those first utterances *could have been* and whether they could have been *useful enough* to be retained and added to. (p. 63)

Or *infectious enough*! He has ignored the possibility that once an uncomprehending copying habit was installed, driven by the sort of communicative efficacy *simple* ("animal") signaling could provide, selection pressures *on memes* could have taken charge regardless of the utility (to human fitness) of the memes being elaborated and spread.[79] As synanthropic species, they must have been unforget-table and attention grabbing, but not particularly useful, at least at the outset.

Building on Tomasello's work, Bickerton thinks the path from joint cooperation to full-fledged language must have led through *confrontational scavenging*: the climate changed, oblig-ing our ancestors to change their diet and become scavengers on the savannah, but they had to fend off predators and other scavengers, so they needed large groups, armed with sharp stone blades and probably spears. Foraging for kills to scavenge in large groups would have been inefficient—better to send out single or double scouts in many directions and have them report back to the base (like honey bees) whenever they found an exploitable kill. Like the bees, they would do best to provide information about where and how good the food source was. According to Bickerton, this was the birth of displaced reference, the semantic power (not found in mere alarm calls or mating calls) that could draw attention to unseen, unheard things. Bees, not bonobos, are the species to ponder.

There must have been some such rare sequence of thresholds or bottlenecks through which our ancestors groped. They certainly didn't think to themselves, intelligently, "Let's design a language, so we can coordinate our activities and lord it over everything in sight!" And it is almost as obvious that they didn't just get smarter and smarter and smarter than their cousin hominids, thanks to "better genes for intelligence," until they were smart enough to turn their caveman grunts and howls into grammatical language. Muta-tions without selection pressures to favor them evaporate in the next few generations. Somehow our ancestors got themselves into a circumstance where there was a rare opportunity for a big pay-off. What would have kept the other, similarly challenged hominids

---

79   Blackmore (1999) develops the conjecture that language could have begun as parasitic memes that eventually turned mutualist.

from rising to the occasion? Richerson and Boyd (2004) call this the adaptationist dilemma (p. 100), and note:

> We are getting confidently more uncertain about what was going on in the early Pleistocene, and knowing what you don't know is just as important as knowing what you do know. (p. 142)

The ultimate origin(s) of language and origin(s) of life are still unsolved problems, but in both cases there is no shortage of plausible just-so stories waiting to be refuted or turned into confirmed hypotheses, an embarrassment of riches to tempt future researchers.

Winding paths to human language

The further evolution of proto-linguistic phenomena, once a foothold has somehow been established, also provides plenty of scope for controversy and mixture:

1. First there was a proto-language of short utterances rather like vervet monkey alarm calls,[80] lacking productivity, and also lacking any distinction between imperatives and declaratives that could distinguish "run up into a tree!" from "leopard approaching!" (Bickerton 2009, 2014; see also Millikan 2004, on "pushmi-pullyu" representations). Evoked by critical event types in hominin life, these signals would be appropriate and recognized reactions to important affordances, and hence affordances themselves, elements in the hominin *Umwelt* with-

---

80 Are vervet calls innate or learned? The calls of isolated vervet populations are very similar, suggesting that the calls themselves are innately fixed, but adults show greater selectivity in their calls than juveniles, and their calls are attended to more often, so, as usual, there is a gradient between "instinct" and "learned behavior," not a dichotomy (Seyfarth, Cheney, and Marler 1980; Cheney and Seyfarth 1990).

out requiring any more semantic analysis on the part of the beneficiaries than is required for the alarm calls of other species.

2. Perhaps a gesture language rather like the signing languages of the Deaf came first, with vocalizations used for attention-grabbing and emphasis (Hewes 1973; Corballis 2003, 2009). Speaking *without* gesturing is a difficult feat for many people, and it might be that gesturing and vocalizing have traded places, with gestures now playing the embellishing role that was originally played by vocalizations. The vestigial hand movements so many of us find all but irresistible may in effect be fossil traces of the original languages.

3. Perhaps there was an auditory "peacock's tail" arms race, with male hominins vying to display their talent for musical vocalization, eventually including improvisations, like the competitive displays of nightingales and other songbirds. The syllables wouldn't *have* to have any meaning or function beyond their attractiveness, and productivity could be favored here, as it is in music generally, as a device for providing a mixture of (pleasing) novelty interspersed with (pleasing) familiarity. *Tralala, hey nonny nonny, derida derida, falalalalala, E-I-E-I-O.* Females would develop their own gift of meaningless gab as a byproduct of the critical abilities they would have needed to judge the competition. Like humans, many songbirds must learn their songs, which are definitely honest signals of superiority in the mating competition, and there are tantalizing neuroanatomical parallels between songbird brains and human brains (Fitch, Huber, and Bugnyar 2010), but as Hurford (2014) notes, competent language is developed before puberty, and "to attract sexual partners before one is able to reproduce would be a waste of effort and potentially harmful" (p. 172). This isn't decisive, and sexual selection (see, e.g., Miller 2000) may have played a role at some point in the elaboration of linguistic virtuosity.

A clear and comprehensive analysis of these and other options appears in James Hurford's *The Origins of Language: A Slim Guide*

(2014), and what particularly distinguishes his account is his recognition (without making it explicit) that these are not just *how come* questions, they are also *what for* questions.[81] For instance, language has two distinct compositional systems, "phonotactics" (governing which phonemes can follow which, independent of meaning—ruling out *fnak* and *sgopn* as English words, for instance) and "morphosyntax" (governing word order and the use of prefixes and suffixes to build meanings out of meanings). Why two compositional levels, one semantic and one not? The phonotactical compositional structure is largely dictated by the limitations of vocal control, hearing, and memory:

> Our tongues can only consistently get around a limited inventory of separate speech sounds, and our ears can only detect acoustic distinctions down to a certain level of subtlety.

So considerations of sheer physical economy and efficiency dictate the solution to the composition problem at the phonological level, but what drives the quest for compositionality in the first place?

> So if we can put these sounds into memorized sequences, and have enough memory capacity to store thousands of such sequences (i.e., words), that is an efficient solution for the task of expressing vast numbers of meanings, given a semantically compositional syntax. (p. 127)

The productivity of languages is "motivated by" the usefulness of being able to communicate many things about the world (p. 128).

---

81 Hurford is also fearless in adopting intentional language to characterize the (usually free-floating) rationales of the behaviors of nonhuman animals and young children. For example: "Now, the teeth-barer can see that when he bares his teeth, the other dog either submits or adopts a submissive pose . . . a signal tacitly understood by both. . . . We are not necessarily attributing any conscious calculation to the animals, although this 'anthropomorphic' way of describing the ritualization of the signal is convenient" (p. 41).

This is one of Hurford's ways of alluding to the free-floating ratio-nales that answer the *what for* questions about the evolution of lan-guage. The purpose or *raison d'être* of productivity is to magnify the expressive power of a communication system without having to mul-tiply the elements to an unwieldy number. But we should pause to consider Bickerton's observation that the earliest proto-language utterances could *not* communicate "vast numbers of meanings," so this new "task" would not have been an obvious step for proto-language users to attempt. (Compare it to the task of figuring out how to get visible, tempting fruit from branches just out of reach: you know the fruit is good, so the motivation to engage in strenuous trial-and-error attempts is readily available. A makeshift ladder is a likelier result than a makeshift grammar. Where is the tantalizing reward in the early days of language?) No inventor of language had the dream of creating such a useful communication system—any more than a cell actually *dreams* of becoming two cells—but the rationale, *in ret-rospect*, is clear enough. What process might uncover that rationale? Hurford is noncommittal in answer to the *how come* question, but on the topic of pronunciation he has a useful passage:

> Depending on age and personality, people end up talking like the people around them, often without conscious effort. The evolution of vowel systems is thus a case of "self-organization." A system evolves not through any deliberate planning, but through the accumulation over time of a myriad of little adjustments by individuals responding to immediate pres-sures. (p. 153)

This homing in on systematicity and productivity is apparently a process guided by two mutually reinforcing rationales. It was "in the interest" of audible memes, meaningful or not, to distinguish themselves from the competition but also to exploit whatever habits of tongue prevailed locally—when in Rome sound like the Romans sound or risk extinction—whereas it was "in the interests" of host/speaker/hearers to minimize the load on memory and articulation by keeping the repertoire of distinct sound-types fairly compact and efficient. No "conscious effort" is required because the immediate pressures are the selective pressures of differential replication.

A compelling instance of such a process is provided in a fascinating experiment by Claidière et al. (2014). Baboons, captive but free moving in their large enclosure, discover a place where they can get rewarded for correctly remembering which four out of sixteen squares in a grid had been illuminated a few seconds earlier. At the outset, when they are confronted with four randomly placed illuminated squares, they make a lot of mistakes, but gradually they improve their performance; they become quite good copiers. From then on, their responses, including any errors—like mutations— get transmitted to the next baboon as items to remember, and the patterns to remember gradually shift from the initially random arrangements to more memorable, connected, "tetrominos," salient groups of four squares forming lines or squares or L, T, or S shapes. See figure 12.1 of the color insert following p. 238.

This is Darwinian unconscious selection; the baboons are not *trying* to do anything but get the reward for pressing the right four squares, but over repetitions, the more readily perceived/remembered patterns survive while the others go extinct. The items in this instance have no semantic interpretation at all; they are memes well designed (by differential replication) to propagate in spite of providing no benefit *beyond a reward for copying*. As the authors say, their study shows

> that cultural transmission among non-human primates can result in the spontaneous emergence of efficient, structured, lineage-specific behaviours, therefore demonstrating that we share with our closest relatives many of the essential requirements for creating human-like culture. (p. 8)

In the early days of human language, a similar process could yield good building blocks of sound, memes that were ways of *articulating*, that could then be easily adopted as semantic building blocks—

memes that were ways of *communicating*: meaningful words, in short. A bounty of productively generated sounds looking for work (the better to avoid extinction) is a more productive workshop of "invention" than a passel of distinctions with no sounds yet to express them. It takes a particular sort of intelligent designer to coin an apt and useful neologism. We have plenty of such deliberate wordsmiths today, but in the early days of language, sounds already in circulation could have been more or less unconsciously adopted to serve on particular occasions, the coinciding in experience of a familiar sound and a salient thing (two affordances) being wedded on the spot to form a new word, whose meaning was obvious in context.

That populates the lexicon with phonology and semantics, but where does grammar come in? Isolated, conventionally fixed articulations are, like alarm calls, limited in semantic variety: *hello, ouch, yikes, aaah, scram*. When and why did the noun-verb distinction emerge from the proto-linguistic signals or calls that do not distinguish between declarative and imperative import? Every language, Hurford argues, has—and needs—a topic/comment distinction (roughly, what you are talking about, and what you are saying about it). But some languages have barely discernible distinctions between nouns and verbs. Moreover, some languages that do have the distinction honor the SOV (subject, then object, then verb) pattern and others are SVO (like English: *Tom eats steak*, not *Tom steak eats)*. (Welsh is a VSO language.) Some languages make heavy use of subordinate clauses, and some don't. Hurford distinguishes what is obligatory (and says why—*what for*) from what is optional (and may have only a historical, *how come* explanation) and puts the elements in order of appearance in languages. Function words, prepositions like *of* and *for* and *off* in English, often have descended from verbs or nouns, and articles are often the number one (*un* in French, *ein* in German) or derived from the number one. Content words are almost never descended from function words. This and other historical clues derived from centuries of scholarship inform Hurford's hypotheses about the gradual growth of grammars.

One of the most telling of these hypotheses concerns the variation in complexity (of both grammar and pronunciation) in the world's languages. In the small hunter-gatherer groups in which our ancestors lived until very recently,

> group identity was a force for social cohesion, in competition with other groups. . . . Any exogamy [marrying outside the group] was likely to be with near neighbours, who spoke a similar language. Children were raised with little contact with outsiders. There was little or no motivation to communicate with outsiders. Hence the languages of such small groups were free to evolve in their own idiosyncratic ways, uninfluenced by other languages. There is a robust negative statistical correlation between the morphological complexity of a language and the size of the population that speaks it. (Hurford, p. 147)

In other words, a small, isolated community is rather like an island, in which evolution can generate a bounty of novelties that couldn't survive in a larger competitive world. The memes that go to fixation may or may not provide their hosts with important benefits; they could just as well be commensals or parasites that manage to thrive in this refuge, until forced into competition by invading memes.

"Contact between adults speaking different languages tends to produce varieties of language in which morphological complexity is stripped out" (p. 148). The obvious way of seeing the "motivation" here is that speakers unconsciously "gravitate" to simpler utterances, in response to the incomprehension of their interlocutors. This "gravitation" might in some instances really be something like erosion, a simple economizing response to the physical demands of utterance, an effect of laziness or thrift, coming up with a shortcut that others copy. But it also might be accomplished not by any such slide down a gradient under the influence of economy, but by mutation and selection of ways of making oneself understood, as homed in on by following subtle cues of facial expression and other reactions from interlocutors, with little or no deliberate or methodical guidance by the speakers. Such gradual "gravitational" changes could just as well be intensifications or elaborations as energy-saving, time-saving simplifications. As usual, though, we can see Darwin's bridge of "unconscious selection" leading seamlessly to "methodical selection" and eventually to "intelligent design," which itself can be seen to come in degrees, with relatively unimaginative trial and error (like the repetitive yelling of naïve tourists trying to get the natives to understand "plain English") giving way to insightful innovations scaffolded by gestures and pantomime, and quickly settling on nonce "conventions" whose meaning is readily seen in context (you shake your head and blurt out "smaller" to the fishmonger who knows no English, and from then on, the more diminutive species on sale are known to both of you as *smollar*).

The first words were no doubt assigned to the "things we had concepts for"—which means simply: we were well equipped by either our innate genetic endowment or the experience we had had to date to discriminate these affordances, attend to them, track them, and then deal appropriately with them under normal circumstances. Just what manages to anchor these competences in our brains is still a largely unknown matter, and there may be many different ways of embodying (embraining?) the mastery of an affordance. Hurford sees that we can proceed in the absence of a good neuroscientific theory of this:

> The relationship between words and things, i.e., meaning, is indirect, mediated by concepts in the heads of language users. So we have three kinds of entity: linguistic entities such as words and sentences, mental entities such as concepts, and worldly objects and relations, such as dogs and clouds, and eating and being higher than. . . . In terms of the brain, we know that concepts are somehow stored there, but we have little idea of exactly how. . . . If you are uneasy about this, think of the nineteenth-century search for the source of the Nile. People knew the Nile must have a source, as all rivers do, and they knew it was somewhere in the middle of Africa. Eventually it was located. The phrase *the source of the Nile* was not a

meaningless phrase just because no one had yet pinpointed its referent. (p. 60)[82]

Once we have among these affordances a host of ways of uttering, labeling, signaling, and saying, the opportunity for something a bit like symbiosis arises: two affordances unite to form something new, a *concept* in the specifically human sense of a word with an understood meaning. How do we get these different kinds of "things"—words and objects—in our manifest images to play together? When we are infants, we discover all these words and all these objects. Children devote a lot of their exploration time to juxtaposing things—blocks, dolls, sticks, bits of food and bits of trash, and every other manipulable item they can reach.

> What's this? I can look it all over, put it in my mouth, stick it up my nose, smash it with my palm, bang on it, squish it, drop it, throw it, grab it, put it on my head. Meanwhile I can babble and coo and get my tongue and ears familiar with those pronounceable "things" as well. Pretty soon I can wonder what *these things* are *called*, and what *this sound* means.[83]

Out of a rather chaotic jumble of opportunities, regularities can emerge, with only intermittent attention and hardly any intention. When things become familiar enough, they can be *appropriated*: my block, my dolly, my food, and my words—not, at first, consciously thought of *as mine* but just handled as possessions. With discrimination and recognition comes the prospect of reflection: judgments that these two things are the same and those two are different can lead to the recognition of the higher-order pattern of *sameness* and

---

82 Among explorers making valuable forays into this *terra incognita*, the best starting from what we know about language and thinking are Jackendoff (2002, 2007, 2007b, 2012) and Millikan (1984, 1993, 2000, 2000b, 2002, 2004, 2005, and forthcoming).

83 Is this *conscious* wondering? Not necessarily; it can be mere epistemic hunger, of the free-floating variety that "motivates" explorations in all animals.

difference, which then become two additional "things" in the manifest image of the child. Like the prebiotic cycles that provided the iterative processes from which life and evolution itself emerged, these iterated manipulations provide an engine of recombination from which the densely populated manifest image of a maturing human child can be constructed. We will look more closely at how the manifest image becomes *manifest to the child*, that is, part of conscious experience, in chapter 14.

Brains are well designed for picking up affordances of all kinds and refining the skills for responding to them appropriately. Once a brain starts being populated with pronounceable memes, they present as opportunities for mastery, and the pattern-finding powers of the brain get to work finding relations between them and the other available affordances. As noted in chapter 9 (p. 190), children learn on average seven words a day for the first six years of their lives, and most of these words are not deliberately taught to them by pointing ("Look, Johnny, a *hammer*. Look Lucy, *seagulls!*"), let alone by definition ("A *convertible* is a car with a roof that folds down."). Children acquire the meanings of most of these words gradually, usually not even noticing the onset of comprehension, by a process that is not much like deliberate hypothesis formulation and testing, except in its capacity to secure results reliably: unconscious, involuntary statistical analysis of the multifarious stimuli they encounter.[84]

Can grammatical and morphological "rules" be acquired by bottom-up processes, competent and uncomprehending? Yes, since nobody learns the grammar of their *first* language "top-down," by learning explicit generalizations ("There are three genders in German: masculine, feminine, and neuter.") or explicit rules ("Nouns and adjectives in French must agree in gender and number."). Jack-

---

84 The feasibility of such a process has been known since the invention of "latent semantic analysis" by Landauer and Dumais (1998), a forerunner of the "deep learning" algorithms being applied these days, by IBM's Watson, Google Translate, and a host of other impressive applications (see chapter 15).

endoff (1994) vividly dramatizes the issue with his "paradox of language acquisition":

> An entire community of highly trained professionals [theoretical linguists], bringing to bear years of conscious attention and sharing of information, has been unable to duplicate the feat that every normal child accomplishes by the age of ten or so, unconsciously and unaided. (p. 26)

But there are two different ways a bottom-up process could acquire this know-how: either some kind of deep learning, unconscious pattern-finding process or by genetic inheritance. Actually, it is better to say that there are two extremes between which controversy reigns over where on the spectrum the truth lies. At one extreme the work is done by a pattern-finding competence that is completely general and has nothing specific in it about language, and at the other extreme, there is an almost-complete innate system (universal grammar) that just needs to have its "parameters" set for one language or another by experience (rather like customizing your word processor to suit your typing style, except that this would be unwittingly done by the infant language-learner). It may be irresistible in the heat of polemical battle to pretend your opponent occupies one or the other of the extremes, but in fact intermediate positions are perfectly possible, and more defensible. The *mainly learning* end of the spectrum has been recently occupied by machine learning modelers and linguists who are struck—as we have often been struck in these pages—by *ubiquitous gradualness*, as much a feature of grammatical categories as of species and subspecies. Consider the differences between

> idioms like *one fell swoop* and *in cahoots* that are quite impervious to internal analysis, and
> *that doesn't cut any ice* and *kick the bucket*, whose meanings cannot readily be derived by analyzing their parts (and hence also have to be learned as autonomous lexical items), and

pass muster and *close quarters*, which are analyzable if you know a
little about military practices but familiar enough to be learnable and usable without that knowledge, and

prominent role, mixed message, and *beyond repair,* which are "pre-
fabs" (Bybee 2006), "conventionalized but predictable [in
meaning] in other ways," and

where the truth lies and *bottom-up process,* which can be understood in context by anybody who knows the meanings of the
components.

These variations *wear their histories on their sleeves* (a nice middle-
ground case), since we can often reconstruct the *grammaticalization*
process that takes *frequently replicated combinations* and gradually
hardens them into units that can then replicate on their own as
combinatorial units. But even if "in principle" all grammatical regularities could evolve in the linguistic community and be installed
in the individual without any help from genes (the *mainly learning*
pole), there is a strong case to be made for an innate contribution.

The most influential argument for a dedicated, innate Language
Acquisition Device (or LAD) (Chomsky 1965, p. 25) is the "pov-
erty of the stimulus" argument, which claims that a human infant
simply doesn't hear enough grammatical language (and ungram-
matical language corrected—trial and *error*) in the first few years
of life to provide the data required to construct a "theory" of its
grammar. Birds that have never seen a nest being built can straight
off build a usable, species-typical nest, thanks to their innate nest-
building competence. Infants who have never been instructed in
the fine points of adverbs and prepositions could in the same way be
equipped for language. Since the grammatical competence has to
come from somewhere, runs the argument, it must be at least partly
innate, an internal set of principles or rules or constraints that
allow—force, in fact—the child to narrow down her search through
the Vast space of possibilities, a directed, not random, process of
trial and error. Such a crane (see chapter 4) would indeed account
for the remarkable ease with which children home in on the gram-

mar of their native tongue: they are choosing from a relatively small set of learnable languages restricted by the built-in constraints of the LAD. But Chomsky made the claim suspect in many quarters through his adamant resistance to any attempt to account for the design of the LAD by natural selection! Chomsky's view made the LAD seem more like a skyhook than a crane, an inexplicable leap in Design Space that miraculously happened, a gift from God, not the result of arduous R&D by evolution over many generations.

Perhaps somewhat in response to this criticism, Chomsky (1995, 2000, 2000b) dramatically revised his theory and has since defended the minimalist program, which discards all the innate mechanisms and constraints of his earlier vision and proposes that all the work they were designed to do could be accomplished by just one logical operator, called *Merge*. (See Hauser, Chomsky, and Fitch 2002 for a famous defense of this position in *Science*, and Pinker and Jackendoff 2005, for a detailed rebuttal with useful references to the torrent of critical literature provoked by the essay.) Merge is an all-purpose combining operator, which "recursively joins two elements (words or phrases) into a binary tree bearing the label of one of them" (Pinker and Jackendoff 2005, p. 219). According to Chomsky, then, the one cognitive talent required for language, the competence that human beings have but other animals lack, is the logical operation known as *recursion*, and it could be installed without the lengthy R&D of natural selection, in a sort of one-step cosmic accident, a found object—lucky us—not an evolved tool.

Recursive functions, in mathematics and computer science, are functions that can "take themselves as arguments," which means only that after you have applied the function once, yielding a new entity from an old, you can apply the same function to that new entity and so forth, ad infinitum. This twist allows the generation of multiple nestings, like Russian dolls. True recursion is a powerful mathematical concept, but there are cheap look-alikes that will do a lot of the same work without being properly recursive. The standard example of recursion in natural language is the embedding of subordinate clauses: *This is the cat that killed the rat that ate the cheese that sat on the shelf that graced the house that Jack built.* Obviously you can in principle make an endless sentence by piling on the clauses ad infinitum, and one of the hallmarks of natural language is its (in principle) infinity. There is no longest grammatical sentence of English. But there are other ways of getting this infinity. *Tom ate a pea and another pea and another pea and . . .* will also generate an infinitely long (and infinitely boring) sentence, but that is just iteration, not recursion. And even when recursion is involved in natural language, there is a pretty clear limit of how many recursive embeddings competent speakers can keep track of. If you can do seven, you're a black belt sentence-parser, but I wonder if you realize that I intend you to be persuaded by my informal demonstration that even you have limits that I can only guess to be in the neighborhood of seven, in this sentence which has only six. Moreover there is at least one language, Piraha, in the Amazon jungle, that reportedly has no such embeddings at all (Everett 2004), and there are also relatively simple functions that look recursive until you try to apply them to themselves more than once (or twice or . . . *n* times for any finite number *n*). In Microsoft Word, for instance, there are the typographical operations of superscript and subscript, as illustrated by

base^{power}

and

human$_{female}$

But try to add another superscript to base^{power}—it *should* work, but it doesn't! In mathematics, you can raise powers to powers to powers forever, but you can't get Microsoft Word to display these (there are other text-editing systems, such as TeX, that can). Now, are we sure that human languages make use of true recursion, or might some or all of them be more like Microsoft Word? Might our interpretation of grammars as recursive be rather an elegant mathematical idealization of the actual "moving parts" of a grammar?

Pinker and Jackendoff present a strong case, with much empirical evidence, against the ideas that Chomsky has adduced in support of his minimalist program and also the ideas he claims follow from it.[85] They show how the claim that Merge can do all the work done by the earlier systems is either false or vacuous, since the Chomskyans then reintroduce most of the design features officially jettisoned by minimalism as constructions implemented via Merge. Ironically, then, if we ignore Chomsky's antagonism to the hypothesis that natural selection is responsible for LAD, his proposed elementary Merge operation looks like a plausible candidate for an early adaptation of grammar, from which all later grammar memes have descended.

What's more, we could plausibly conjecture that Merge itself was no fortuitous giant step, no saltation in design, but a gradual development out of more concrete versions of Merge that can be seen in the manipulations of children (and adults) today: put block on block on block; use your big hammer stone to make a smaller hammer stone to make a still smaller hammer stone; put berry in pile, put berry pile in bigger pile, put bigger pile in still bigger pile; put pile in cup in bowl in bag, and so forth. But are any of these processes *real* recursion? That is a misguided question, like the question: Are the hominins involved real *Homo sapiens*? We know that gradual transitions are the rule in evolution, and a gradual emer-

---

85   See also Christiansen and Chater, "Language as Shaped by the Brain" (a *Behavioral and Brain Sciences* Target article, 2008) and the extensive commentaries thereupon for a vigorous debate on whether genetic evolution or cultural evolution has played a greater role in creating our linguistic competence. Christiansen and Chater defend a position largely consonant with mine, but they misconstrue the memetic approach in some regards (see Blackmore 2008, for the details), and overstate the case against genetic evolution playing a major role. By now it should be clear that my verdict is that hardware (genetically transmitted brain structure) follows software (culturally transmitted brain structure), but that there is no way, at present, to tease out the current proportions. Echoing Gibson, we may say that the information is in the sound, but there has to be equipment in the head for picking it up, and we don't yet know how much of the information is embedded/presupposed in that equipment.

gence of (something like) real recursion—real enough for natural language—would be a fine stepping-stone, if we could identify it.

And notice that *if* something like Merge eventually proves to be a hard-wired operation in human brains, as Chomsky proposes, it wouldn't then be a skyhook. That is, it wouldn't be the result of a chance mutation that, by cosmic coincidence, happened to give our ancestors an amazing new talent. The idea that a random mutation can transform a species in one fell swoop is not a remotely credible just so story; it has more in common with comic book fantasies like the Incredible Hulk and all the other action heroes whose encounters with freak accidents grant them superpowers.

Over the years, much of the argumentation in theoretical linguistics at its most abstract has involved hard-and-fast "necessary and sufficient conditions" or "criteria" or "difference-makers"— between noun and verb, topic and comment, sentence and clause, and, in particular, between Language A and Language B. In short, linguists have often been tempted to posit *essences*. But when—if ever—do any two speakers speak (exactly) the same language? We could say that each actual speaker has an idiolect, a dialect with a single native user, and while your idiolect of English and mine may be practically indistinguishable (which explains our ability to communicate so efficiently), when we disagree on whether a particular sentence is grammatical or whether a particular kind is a subset of another kind, in the end there is no authority to appeal to aside from majority rule. And when should we treat local majorities as overruling global majorities? One linguist/philosopher of my acquaintance once bravely insisted that neither Joseph Conrad nor Vladimir Nabokov spoke English, *strictly speaking*. Only *native* English speakers speak English! But which native English speakers? Londoners or Brooklynites or Valley Girls or Kiwis from New Zealand? The quandaries facing the linguistic taxonomist are remarkably similar to the quandaries facing the pre-Darwinian definer of genera, species, and varieties. If we follow the Darwinians in adopting *population thinking*, these problems are exposed as artifacts of misplaced essentialism. Populations of what? Memes.

In 1975, Chomsky made the case for the LAD by noting: "A normal child acquires this knowledge [of grammar] . . . without specific training. He can then effortlessly make use of an intricate structure of specific rules and guiding principles to convey his thoughts" (p. 4). If we reposition Chomsky's Merge, or something like it, as an early candidate for a transitional innovation on the way to modern languages, then we can reconcile early and late Chomsky by saying the "intricate structure of specific rules and guiding principles" are not so much explicit rules as deeply embedded patterns in *ways of speaking* that consist of a series of improvements wrought by evolution, both cultural and genetic, in response to the success of proto-languages. As we have seen again and again in these chapters, we, like other animals, are the unwitting beneficiaries of brilliantly designed systems for accomplishing ends that pay for the R&D required, and this is yet another case of evolved competence requiring little or no comprehension.

The evolutionary origin of language is an unsolved, but not insoluble, problem, and both experimental and theoretical work has made progress in formulating testable hypotheses about gradual, incremental evolutionary processes, both cultural and genetic, that could transform the more primitive talents of our ancestors into the verbal dexterity and prolixity of modern language users. The arrival of language then set the stage for yet another great moment in evolutionary history: the origin of comprehension.

In the next chapter we will see how, as linguistic competence grew, it not only accelerated cultural evolution; it permitted the process itself—the process of cultural evolution—to evolve into something less Darwinian, less bottom-up, paving the way for top-down comprehension, one of the most recent fruits on the Tree of Life, and the inauguration of the age of intelligent design. The creativity of individual human beings can be seen to echo, at high speed and in concentrated forms, the R&D processes that created it.

13

The Evolution of Cultural Evolution

You can't do much carpentry with your bare hands, and you
can't do much thinking with your bare brain.
　—Bo Dahlbom

Darwinian beginnings

Recall our preview in chapter 7 (p. 148).

Generations of naturalists have established that animal
parents can impart skills and preferences to their offspring
without benefit of verbal instruction, and these "animal traditions"
(Avital and Jablonka 2000) are a sort of memetic evolution, but animals' memes do not, in general, open up opportunities for further
memes the way words do. There is none of the snowballing accumulation of the sort that language permits, and as noted in chapter 12,
there are ecologically relevant, indeed life-threatening, facts about
perceptually absent circumstances (e.g., what to do if you encounter a bear) that simply cannot be conveyed without language. Displaced reference is, as Bickerton and others argue, a giant step in
Design Space.

It is time at last to flesh out the central claim of part II of this book:

Human culture started out profoundly Darwinian, with uncomprehending competences generating various valuable structures

in roughly the way termites build their castles. Over the next few hundred thousand years, cultural exploration of Design Space gradually de-Darwinized, as it developed cranes that could be used to build further cranes that lifted still more cranes into operation, becoming a process composed of ever more comprehension.

How far could *H. sapiens* get with only ape-level comprehension? We may figure this out some day, if we get enough evidence to fix rough dates on such milestones as cooperative hunting and scavenging, control of fire, building habitation, making various tools and weapons, but we know that everything changed once language got in place. In terms of the diagram, human culture started in the lower left corner, and gradually spread out to include pockets of growing comprehension (up the vertical *y* axis), more top-down control (along the horizontal *x* axis) and more efficient directed search (on the front-to-back *z* axis). I am claiming that these three dimensions naturally get traversed in a diagonal direction from

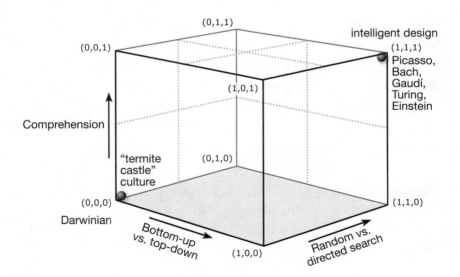

FIGURE 13.1: Darwinian space.

pure Darwinian toward the (ultimately unreachable) summit of intelligent design, because the phenomena arrayed along each dimension are ordered by their use of more and more semantic information and hence design *improvements*, which require more R&D, more gathering and exploiting and refining, building, as always, on the foundations established by earlier (more Darwinian) R&D. Some steps are larger than others, but we don't need what Freeman Dyson called a gift from God to have arrived where we are today. We will look closely at the controversies surrounding the current state of cultural evolution after we have reviewed the Darwinian, and semi-Darwinian, and hemi-semi-demi-Darwinian phenomena that led to it.

We set the stage in chapter 12: Memes, like viruses, are symbionts dependent on the reproductive machinery of their hosts, which they exploit for their own ends, so for there to be a population explosion of memes, there has to have been a preexisting (or concurrently evolving) instinct to imitate or copy, which would "pay for itself" by providing some (genetic) fitness benefit to our ancestors. The ancestors of our nearest nonhuman relatives apparently didn't enjoy conditions conducive to this, or didn't enjoy them for long enough, to join us on the stepping-stones. Chimpanzees and bonobos, for instance, don't exhibit the interest, the focused attention, the imitative talent required to kindle the cumulative cultural wildfire that marks us off from the other hominids.[86] Future research may tip the balance in favor of one of the rival hypotheses about how this happened, but we have seen enough, I think, to rest assured that it didn't take a miracle to put us on the path.

---

86   It is worth reminding ourselves of the possibility that today's chimpanzees and bonobos may have *lost* cognitive competence or curiosity during the six million years that separate them from our common ancestor. Cave fish whose ancestors could see are now blind. Closer to home, when the land bridge between Australia and Tasmania subsided about 10,000 years ago, an isolated, and then shrinking, population lost much of the technology they had had for millennia, including bows and arrows, boats, and maybe even the ability to kindle fire. See Diamond (1978), Henrich (2004, 2015), and Ridley (2010) for the fascinating details.

Once under way, we became the apes with (meme-)infected brains. Like viruses and other symbionts by the trillions that have taken up residence in our bodies, the invaders evolved to be more effective reproducers, holding their own against the competition within bodies and winning the dispersal competitions, spreading to new hosts. They must have included enough mutualists and commensals among the parasites not to kill off their hosts, though it is entirely possible that waves of meme infection did just that before one wave finally happened to be benign enough to secure a long-term foothold. (We can imagine a group of would-be ancestors afflicted by a dance craze so potent that it led them to abandon hunting and gathering, leading to mass starvation, or an unfortunate lineage whose initiation into manhood was a circumcision ritual that cut off more and more, in a misfiring variation of the peacock's tail.) Similarly, in the early days life may have evolved many times only to fizzle out, again and again, until one origin got enough of the details right (right enough) to survive indefinitely.[87]

The earliest memes, whether they were pronounceable proto-words or silent behavioral habits of other sorts, were synanthropic, not yet domesticated, and they had to be particularly "infectious" until the suite of genes and memes for enhancing the copying routines were in place. Probably only a minority of them had any functional capacities at all beyond their reproductive power. Bad habits, but *catchy* bad habits, would have been a price worth paying for a few really good habits, and once some of the good habits were in place, there would be time for both cultural and genetic R&D to go back and clean up some of the excesses, in much the way our genetic micromachinery has evolved measures for dealing with the troublemakers in the genome, such as genomic parasites, trans-

---

87 This sentence is, in fact, something of a tautology: life started and died, started and died, started and died, until eventually it didn't die, because it got just enough things right to keep it from dying. Not very informative, but not vacuous, since it is a coherent and even confirmable (in principle) alternative to any skyhook theory of miraculous origins of life.

posable elements, and segregation distorters (see ch. 12, p. 255). Once verbal communication became not just a Good Trick but an obligatory talent for our species, there would be steady selective pressure in favor of organic modifications that enhanced or streamlined the process of language acquisition. Chief among the important innovations is altriciality (prolonged infancy), which extended the time offspring are dependent on parents for protection, nourishment, and—not at all inadvertently—education. This huge increase in "face time" was further enhanced by gaze monitoring, which in turn enabled shared attention, and, as Tomasello notes, shared *intention*. (The only other mammal that regularly engages in gaze monitoring is the domestic dog, and it monitors the gaze of its human master, not other dogs; see, for example, Emery 2000, and Kaminski 2009.) Compare the "whites of the eyes" of human beings to the dark sclera surrounding the pupils of other apes; this is probably a recent adaptation to enhance gaze monitoring, a fine example of cultural/genetic coevolution, a genetic response to a novel behavior designed to enhance meme transmission (for an overview, see Frischen et al. 2007).

What was it like for these pioneer meme hosts? Was it like anything? Their heads were being filled with tools, toys and junk, all "made of information," occupying brain tissue and exploiting brain energy, but which of these invaders, if any, made themselves known in some way to their hosts? The germs and viruses that occupy our bodies fly beneath our radar except when they provoke large-scale somatic changes: pains, palpitations, vomiting, sneezing, dizziness, and the like. As we saw in chapter 4, the ontology implicit in an elevator's design accounts for its competence in controlling its activities without the elevator having to comprehend what it is doing, without its having to be *conscious* in any but the most minimal sense of its ontology (it just has to be *sensitive* to it, to discriminate its elements and react differently to the differences). This is the ontology the engineers have hit upon for the elevator they designed. Our ancestors' ontologies could similarly have included words and other memes: they could have used these words competently, and benefited from having words, for instance, without yet *realizing* that they were using *words*, without words being *manifest* to them *as words* in their manifest image. In a minimal sense they would *notice* words—they would be perceptually sensitive to them and respond differentially, just like the elevator— but they wouldn't have to *notice* their *noticing*. Words could pass between them, perceptually, entering and inhabiting their bodies rather like vitamins or their gut flora: valuable symbionts whose value didn't depend on being recognized or even appreciated by their hosts.

Still, the natural home of memes is in our manifest image, not our scientific image (where the vitamins and gut flora are found). Memes can travel through the air only by generating human-perceivable effects, so they are, in general, *available* for noticing, even while many enter inconspicuously and unannounced.[88] Memes, then, unlike viruses and microbes, are affordances we are equipped from the outset to notice, to recognize, to remember, to respond to appropriately. They are items in good standing in *our* ontology even while they are typically "invisible" to other animals, not part of their ontology at all. Here we should recall the distinction Sellars made between the manifest image and the (more primitive) original image (see ch. 4, fn. 15, p. 61). When we notice our memes, and start to own them, and reflect on them, we have moved from the original image to the manifest image, the world we live in *and know that we live in*.

The free-floating rationales of human communication

Competence without comprehension is as ubiquitous in human life as in animals, bacteria, and elevators, but we tend to overlook this possibility and assign appreciation of the rationales of successful

88   The role of "covert entry" of information plays an important role in humor. See Hurley, Dennett, and Adams 2011.

human action to the clever actors involved. This is not surprising. After all, we tend to attribute more understanding to wolves and birds and honey bees than we can support with evidence: the use of the intentional stance to interpret behavior across all species carries with it a tacit presupposition of rationality, and whose rationality can it be if not the behavers'? The very idea of free-floating rationales is a strange inversion of reasoning, as we can see by following a family of controversies over many years provoked by some celebrated work in the philosophy of language.

One of the landmarks of twentieth century philosophy of language is H. P. Grice's account (1957, 1968, 1969, 1989) of the necessary conditions for communication, or what he called "non-natural meaning." Grice's central claim was a three-part definition of what it is for a person to mean something by doing something. As clarified by Strawson (1964) and others, to mean something by doing *x*, *S* must *intend*

(1) *x* to produce a certain response *r* in a certain audience *A*,
(2) *A* to recognize *S*'s intention (1), and
(3) *A*'s recognition of (1) to function as at least part of *A*'s reason for
    A's response *r*.

This elegant recursive stack captures a lot. *A* may learn a good deal from something *S* does without *S* *meaning something by doing it.* For instance (1) if *S* screams in pain or falls over in a faint, *A* may be informed that *S* is in some agony, but it may not have been *S*'s *intention* to so inform *A*; it may be involuntary pain behavior. If (2) *S does* have the intention to (mis-)inform *A* but does not intend *A* to realize this (think of a malingering soldier trying to get sent to the field hospital, or the injury-feigning bird), it also doesn't count as communication. Finally, (3) if it isn't *A*'s re*cognition* of this communicative intention that grounds *A*'s belie*f,* this is not a case of communication, however informative. For instance, my wife leaves the light on in the kitchen so I will be sure to see the dirty dishes that I should clean up; she wants me to hop to it but isn't necessarily *sending me a message*; in fact, she'd just as soon I *not* recognize her intention to draw my attention to my duties.

Grice's analysis rings true for most who have encountered it, and it has inspired a flood of further theoretical work, much of it dealing with an apparently endless supply of ingenious counterexamples and problems (for an insightful and detailed overview, see Azzouni 2013). It draws attention to subtle differences in how human beings can interact, and in the process highlights subtleties that seem to be absent in animal communication systems. Does the anecdote about the vervet monkey who issues a false eagle alarm just when his tribe is losing a confrontation with a rival tribe, thereby provoking a cessation of hostilities that regains ground provide a case of (*intentional*) deception, or is it a case of dumb luck? Stotting gazelles don't need the chasing lions to distinguish deliberate showing off from natural exuberance; the lions will leave them alone in any case if they stot. The inviting conclusion to draw—and many (including me) accepted the invitation—is that Grice's nested conditions show that human communicative behavior is of a different order from simpler, mere animal behavior: involving at least fourth-order intentionality:

S (1) *intends* A to (2) *recognize* that S (3) *intends* A to (4) *believe* that *p*.

Not only philosophers have been inspired by Grice. Anthropologist/psychologists Dan Sperber and Deirdre Wilson (1986) used Grice as a launching point for a rival theory of meaning, but they didn't demur from the general appreciation of Grice's insights:

From a psychological point of view, the description of communication in terms of intentions and inferences also makes good sense. The attribution of intentions to others is a characteristic feature of human cognition and interaction. (pp. 23–24)

Ruth Millikan (1984), in contrast, was at first firmly dismissive:

If I believed that Jack the Ripper was under my bed, I would not crawl into bed and instantly fall asleep. In crawling so into bed, I clearly do not believe "Jack is under my bed." But it does not follow that I believe "Jack is not under my bed." Indeed, I may never have heard of Jack. Likewise, from the fact that if I had reason to believe that a speaker did not intend that I comply with an imperative then likely I would not comply, it does not follow that in normal cases of compliance with imperatives I believe that the speaker intends compliance. (p. 61)

On the next page she softens this:

First there is the problem just what it *is* to have a certain belief or intention. Then there is the problem what it is to *use* a belief in doing something, as opposed, for example, to merely having it *while* doing something. And we must also ask whether there is any need to interpret Gricean intentions and beliefs as things actually used while speaking and understanding, or whether they may have another status entirely yet still do the work the Gricean theorists want them to do of distinguishing a kind of "meaning" that all non-natural signs have.

Millikan, who stands out as a distinguished non-Gricean, eventually concludes:

Only Gricean beliefs that were actually used in the process of speaking and understanding could be of any interest to the theory of meaning. (p. 66)

But this has not deterred others from adopting the Gricean program in one way or another. Recently, my colleague Jody Azzouni (2013) has confronted the Gricean analysis head on, and it is worth noting the pains he must take to soften his readers up for his assault:

My primary aim in this book isn't to refute Gricean and neo-Gricean approaches once and for all. Griceanism is too widespread and its numerous proponents are far too varied in their individual approaches. And besides [he adds in a footnote], who needs to make *that many* enemies all at once? (p. 4)

For students of the literature, Azzouni's patient and imaginative dissection is a must read,[89] but I highlight just one of the amusing intuition pumps he concocts to shock us into a non-Gricean perspective:

Imagine two people on an island who speak two unrelated languages. He holds up an object and starts to mime an action while she tries to figure out what he's trying to say. It probably doesn't get any more Gricean than that. She recognizes that he has communicative intentions; he recognizes that she recognizes that he has communicative intentions (and so on, iteratively maybe). Based on this, on the context, and on some other background mutual knowledge (they've been doing this sort of thing for a while, perhaps, and they've built up a repertoire of meanings), she tries to determine what he's trying to communicate and, derivatively, what the object he's holding up means.

It doesn't get more Gricean than that, but (let's be honest) it doesn't get much more frustratingly atypical either. Real Gricean communication is a real pain (as anyone who's ever been forced to engage in it will *angrily* tell you). (p. 348)

Azzouni's example shows how everyday communication is hugely *unlike* Gricean communication, but then we have a puzzle: How could Grice have persuaded so many people that he was onto something original and important? Did he provoke a mass intellectual hallucination? Yes, in a way. Azzouni's next sentence points in the right direction: "Ordinary language may *originate* evolutionarily speaking in events somewhat like real Gricean communication

89   Azzouni's book offers many perspectives congenial to this book (e.g., his discussion of words as tools, pp. 83ff), and a few, such as his dismissal of the type-token distinction, that can be nicely brought into registration with the discussions here, with the help of a few of his other innovations, such as his divorce of quantification and ontological commitment (see his various methodological interludes).

events, but a great deal has changed since then, and, in particular, a great deal has changed in our brains" (p. 348).

What Grice did—without realizing it—was to reverse engineer human communication, adducing the free-floating rationales that would naturally be uncovered by eons of cultural and genetic evolution once the basic Good Trick of using words as tools had been established. Just as when we explain the phenomena of the stotting gazelle and the murderous cuckoo chick, the *façon de parler* that naturally springs to our minds as the appropriate terms for explanation is the intentional stance, treating the agents as rational agents whose reasons are being exposed, without noting that these reasons did *not* have to be understood, noted, or even represented by the agents. Grice's point was, or should have been, that human communicators (once they catch onto the full power of the tools they have at their disposal) have the *competence to exploit* these features—and the competence to avoid being exploited by others exploiting those features.

Remember: I have been suggesting that the acquisition of a language—and of memes more generally—is very much like the installation of a predesigned software app of considerable power, like Adobe Photoshop, a tool for professionals with many layers that most amateur users never encounter. With human communication, there is much variation, and most uses of the system are rudimentary, routine, guided by habits that are themselves beneath the ken of most observers (and self-observers). But the tools admit of some very sophisticated applications. Some people are natural manipulators, impression creators, masters of indirection and almost subliminal blandishment, and others are bluff, direct, naïve, unguarded in their speech—novice users of the tools, you might say—but neither kind of people *have* to comprehend the reasons why their everyday communication tools have all the options that they do.

Our systems of (verbal) communication have been brilliantly designed, by a process of natural cultural selection that has endowed them with a host of features that new users only gradually acquire and may never deeply comprehend. They may develop acute sensitivities to risks and dangers and opportunities in communication that they themselves cannot analyze. They may be able to "smell" a lie or an instance of false humility, for instance, without even seeing the speaker's face or hearing the tone of voice. Something about the choice of words "didn't ring true" in some way that is impenetrable to their efforts of analysis. Other speakers of a language may have the "gift of gab" or more specialized gifts—the power to comfort, to persuade, to seduce, to amuse, to inspire. This all "comes naturally" to some and is largely beyond the talents of others. Some, especially those on the autism spectrum who are especially "high function-ing," like Temple Grandin, manage to devise, with much effort and ingenuity, a genuine TOM, a *theory* of mind, to help them inter-pret the kaleidoscopic social world that most of us can "perceive directly." We are also gradually acquiring theoretical understand-ing, and genuine comprehension, of the reasons for the features of our communication systems. Those who excel at this research often sell their services as intelligent designers of communication—pub-lic speaking coaches, marketing consultants, advertisers—but as we can confirm in other areas of human endeavor—jazz comes partic-ularly to mind—*theory* is often no match for *ear*, even when it is good theory. If you have ever observed a brilliant scientist with Asperger's syndrome craft speech acts and other social ploys with impressive speed while unable to erase the telltale signs of contrivance, you can appreciate that if Grice's analysis were a performance theory,[90] it would be applicable to a small minority of speakers.

90   Sperber and Wilson see that there is a role for a competence model to play. Of their own account, they say:

> Our claim is that all human beings automatically aim at the most efficient information processing possible. This is so whether they are conscious of it or not; in fact, the very diverse and shifting conscious interests of individu-als result from the pursuit of this permanent aim in changing conditions. (p. 49)

Azzouni points out that Sperber is positing an optimality model, which "faces a problem of testability. For in general, any cognitive process (that has been pres-sured by evolutionary factors) is going to *resemble* an optimality strategy with respect to the costs of the processing versus the value of what's yielded by that processing" (p. 109).

Projects of reverse engineering human activity via the intentional stance have been invented quite often by philosophers, though not in that guise. G. E. M. Anscombe, in her obscure masterpiece *Intention* (1957), talking about Aristotle's account of the practical syllogism, says

> if Aristotle's account were supposed to describe actual mental processes, it would in general be quite absurd. The interest of the account is that it describes an order which is there whenever actions are done with intentions. (p. 80)

"An order which is there"? Where? Not in the brain, not in the environment, not *represented* anywhere. I submit she was crediting Aristotle with the discovery of the free-floating rationales that "govern" our most rational practical behavior *without always being represented*.

So even when words themselves have been appropriated as domesticated tools, and hence a part of the manifest image, it is possible that the free-floating rationale of the design of some of our practices is unimagined by us. Grice can be seen to have worked it out, seen the "order which is there" when people engage in nonnatural meaning, and simply presented it as an account of their intentional states at the time. That's how natural our practice of overendowing people with reasons is: it permits us to fall for mass hallucinations.

Using our tools to think

It may be natural to imagine that when the toddler first says "doggie!" as the dog walks by, she understands that she is labeling the thing she sees with a spoken word, but it is not obligatory. We must be particularly careful here not to confuse competence and comprehension. In fact, it is probably *practically* useful (however theoretically mistaken) for adults to presume much more understanding in their children's early utterances than the children actually have, since by thus "scaffolding" them into the conversation, they help their children grow into their language (McGeer 2004). Thus proto-labelings pave the way for labelings, and proto-requests pave the way for requests. It may take hundreds of hours of "conversation" before the child really gets the hang of it, and even longer before children understand what they are doing when they are having a conversation.

Recall (from chapters 9 and 10) that a meme is a *way*, a way of behaving or doing, internal or external, that can be transmitted from host to host by being copied. In other words, as we shall see later in this chapter, it is like a software app or applet, in fact *very* like an app: a relatively simple informational structure designed (by natural selection, in the early days) that can add a talent, a bit of know-how, useful or useless. Such apps are usually visible or audible to others when you run them—when the word gets heard or uttered, when the action gets observed or executed—and also visible or audible to yourself, in the course of normal self-monitoring; you hear yourself speak, you see or feel yourself gesture, act, move, and so on. But it is one thing to hear yourself speaking and quite another to notice that you're saying something. The child in primary school makes a telling slip of the tongue and calls his teacher "Mommy," to the amusement of the class and his own red-faced embarrassment—when he notices! Even adults often speak without realizing they are speaking; they are monitoring their verbal output, correcting mispronounced words, perhaps, but still quite oblivious to the import of their actions—almost like talking in their sleep. We know they are monitoring themselves because they sometimes—too late—do a double-take: "Did I just say that? Did I really utter those words!?"

I am proposing that we should imagine our pioneer language-user ancestors jabbering away as volubly as you like, getting plenty of benefits from these new habits even though they occasionally scared off their prey. Sometimes they barely managed to communicate, and sometimes they communicated with eloquence (but unwitting, not studied eloquence), and they only gradually came to recognize, retrospectively, what they had been doing. I submit that just as it never dawns on antelopes what they are doing when they are stotting, it didn't have to dawn on these ancestors of ours what they were doing with words until they switched from unconscious selection, as Darwin so aptly called it, to methodical selection, of fully domesticated memes. Children today make such a swift transition from babbling to speaking and understanding that it is all too easy to see it as sudden phase shift (like Helen Keller's [1908] iconic trip to the water pump, an unforgettable anecdote although of uncheckable accuracy). But in the early days of language it was probably a much slower, less efficient process, not yet benefiting from genetic adjustments that streamlined acquisition by turning up the curiosity level for speech sounds, for instance, so that more attention—brain effort—was devoted to their analysis. I speculate that there was a gradual, incremental process of growing competence leading to self-monitoring, leading to reflection, leading to the emergence of new *things to think about*: words and other memes in our manifest image.

It is this series of small steps, I am claiming, that both elevated words and other memes into *our* ontology, our *manifest* image, opening huge vistas to our inbuilt curiosity, and permitted the beginnings of "top-down" exploration of Design Space.[91] I presented a simple version of this innovation in *Consciousness Explained* in 1991, where I imagined ancestors who, having acquired the habit of asking simple questions when they were stumped about something, discovered that sometimes they discovered the answers they were seeking even when no one *else* was there to hear them (pp. 193ff). They found themselves answering their own questions. They invented talking to themselves, which had some immediate—and

91   As I was completing this book, I received a copy of Peter Carruthers's *The Centered Mind* (2015), which a quick skim shows to be loaded with good ideas often congenial to and even anticipatory to my own about how this human talent for self-questioning gives us unique cognitive powers. A proper assessment of his claims is beyond the scope of this book but is clearly called for. In the Appendix, I list other pertinent new books that I encountered too late to incorporate into my thinking.

immediately appreciated—benefits. How can talking to yourself help? Why isn't this like paying yourself for making yourself lunch? The air of paradox vanishes as soon as we recognize that we may "know things" in one part of our brain that cannot be accessed by other parts of the brain when needed. The practice of talking to yourself creates new channels of communication that may, on occasion, tease the hidden knowledge into the open.

"It seemed like a good idea at the time" is a stock comic phrase in the aftermath of big mistakes, but far from being a badge of stupidity, this rueful phrase, if uttered truthfully, is a clear mark of intelligence; a thinker who can accurately recall what she actually thought and how she felt about it is more than halfway to taking a major step toward debugging her own thought processes so she won't fall in that trap again. The more subtle the error, the more valuable the habit of self-monitoring that is enabled by anyone who talks to themselves on a regular basis. (The next time you try to solve a puzzle, consider indulging in a vocalized soliloquy; it's a good way to notice gaps in your thinking.)

It *may* be possible for animals without language to "wrack their brains" for dormant clues, though there are scant signs of any such covert behavior, but in any case our practice of explicit self-questioning has the further huge advantage of making our broodings more readily memorable, so that we can review them with hindsight. And once you have a habit of going into question-posing mode, all your R&D becomes much more top-down, using more directed search, and relying less on random variation and retention. Randomness (or what counts as randomness in evolution, the decoupling of the generation of candidates to test from the criteria of success) is not eliminated; even the most sophisticated investigations often are expedited by deliberately "random" bouts of trial and error. But the search space can be squeezed by using information already acquired in other contexts to rule out large regions as unlikely or irrelevant—but only if the thinker can be reminded of it in a timely fashion. Talking to yourself, asking yourself questions, or even just the inner rehearsal of relevant words ("key words"), is an efficient way of probing the networks of association attached to each word, reminding you of overlooked possibilities that are likely to be relevant to your current perplexity.

Turing's invention of the digital computer can serve as our parade case of top-down intelligent design, and from what we can glean from the records of those heroic years, his path was no beeline to glory, but a meandering exploration of possibilities, sidetracks, false starts, adjusted goals—and plenty of serendipitous help from encounters with other thinkers thinking about other problems. The ideal of a conscious human mind is one in which all its knowledge is equi-accessible, always available when needed, never distorted, a pandemonium of experts ready to lend a timely hand. The Bayesian brain of an animal or child is well equipped to do *some* of this bottom-up canvassing for clues without benefit of more advanced thinking tools, but an adult mind can—on rare occasions—exert significant discipline on the crew, prioritizing, squelching wasteful competition, and organizing the search parties.

In Hurford's book on the evolution of language he says at one point: "Non-human animals, at a certain evolved level, have rich mental lives, a kind of natural intelligence that allows them to negotiate their world well" (2014, p. 101). The second part of the sentence is undeniable, but whether one has to have a "rich mental life" in the sense of a conscious life like ours in order to get around well in the world is not yet established. I have argued that the brain can be a Bayesian expectation-generator without the affordances thereby tracked and exploited being owned, in effect. Whether nonhuman animals can think about thinking, or in any way go "meta" on their own mental states, is an empirical question that is not yet answered, and a "rich mental life" with no capacity to escalate reflexively is really not all that rich. How valuable is a compelling and accurate perception of, say, odors, in a mind that cannot do much comparing and recalling and marveling and musing and noticing the results of these secondary and tertiary reactions?[92] Note: I am *not* saying that nonhuman animals and human infants are not conscious (I have been postponing all discussion of consciousness). If consciousness itself is best seen as admitting of degrees (as I have argued and will defend again here in due course), then we can conceive of varieties of consciousness that are "rich" in some ways and not in others.

Wealth is in fact a good concept to use to leverage this perspective. Consider the lucky chap who owns the land on which a pirate's cache of gold ingots is buried. In one sense he is rich, but if he doesn't know about it, or can't find it, and hence can't put his wealth to any use, his wealth is Pickwickian at best.[93] Similarly, before the infant can track, notice, reflect on, compare, conjure up, recall, and anticipate the various differences in her world that are now available (in some sense) to her, we might better speak of potential riches, and if an animal can never occupy this reflexive perspective, the riches are not even potential. (Showing your parrot lots of Nova documentaries will guarantee a vivid and detailed parade of information passing into its eyes and ears, but don't expect the parrot to get even a smattering of science education from the experience.)

I am still postponing a frontal attack on the puzzles of consciousness, so please note that I am not claiming (or denying) that tracking, noticing, reflecting on, and so on are necessarily conscious. I am claiming that these are very important response competences in the absence of which consciousness (whatever it is) would offer little or nothing to the growing competence, and hence comprehension, of our ancestors. In chapter 5, I argued that comprehension comes

92 Recent experiments demonstrate some rudimentary capacity for meta-representation in birds and mammals. In particular, animals given an opportunity to choose between a high-risk high-reward option and a safer bet exhibit their preference for the latter in instances where they have been given grounds for mistrusting their hunches (Crystal and Foote 2009; and other essays in Cook and Weisman 2009).

93 Recall Plato's wonderful image (in the *Theaetetus*) of one's knowledge as an aviary filled with birds; they are your birds, but do they come when you call?

in degrees and is in fact a product of competence, not an independent source of competence. Now we can see how this product is created. Basic affordances, genetically transmitted endowments of Nature enhanced by Bayesian learning to capitalize on patterns in the environment, provide a wealth of competence for all locomotors from insects and worms to elephants and dolphins. Some of the more impressive of these competences are naturally taken as signs of comprehension—of sorts. The talents of a tool-using crow or a tourist-wise grizzly bear in Yellowstone National Park demonstrate levels of adaptiveness to unique situations that invite the verdict of understanding. One might even say, that's what understanding *is* in these behavioral domains: not communicable theoretical knowledge but deeply practical know-how.

If that know-how is the limit of nonhuman comprehension (so far as we can tell), then what does human comprehension add to it? Not just the ability to communicate the know-how (or factual knowledge) to conspecifics but the ability to treat whatever topic is under consideration as itself a thing to be examined, analyzed, inventoried, thanks to our capacity to represent it explicitly via words, diagrams, and other tools of self-stimulation. You can't do much thinking with your bare brain, but armed with these simple tools, an explosion of thoughtful exploration becomes available. As so often before, a competence designed by natural selection—now both genetic and memetic—creates a gift, an uncomprehending competence (in this case, the mastery of a language), which in turn provides ever expanding levels of further competence. These are meta-competences, in which we use our thinking tools to think about not just food, shelter, doors, containers, dangers, and the other affordances of daily life but also about *thinking about* food and shelter, and about *thinking about thinking about* food and shelter, as this sentence is demonstrating.

Philosophy has been the chief academic home of meta-representation for several thousand years. It can be alternatively gripping and amusing to watch Plato and Socrates, and later Aristotle, struggling so cleverly with the quite novel task of trying to understand *every-*

thing, including the very state of understanding itself. The reflective steps—examining the meaning of meaning, the understanding of understanding, using words to analyze words—eventually led to the recognition that, as Doug Hofstadter once put it, "Anything you can do I can do meta." It doesn't always yield insight, and sometimes threatens to lose the participants in a hall of mirrors with no clear anchoring in the real world, but such are the excesses of a meme (a meta-meme) of tremendous power.

The age of intelligent design

This recursion, piling up stacks of meta-representations, triggered the MacCready Explosion that has blossomed during the last ten or twenty thousand years, and which is still picking up speed. I suggested above that memes were very like software, and now I want to extend and sharpen that claim.

Consider Java applets. You probably encounter them every day, almost certainly if you spend any time on the Internet, but they are largely invisible, by design. Java is the invention (by James Gosling and others in 1991) that is most responsible for giving the Internet its versatility; it lets you download little programs—Java applets—from websites that permit you to solve crossword puzzles, play Sudoku, explore maps, enlarge photos, participate in adventure games with other players halfway around the world, and do a lot of "serious" computing, too. A website designer can write programs in the Java programming language without having to know whether the computers of the users who go to that website will be Macs or PCs (or Linux machines), because a Java applet always runs on a JVM (a Java Virtual Machine) specifically designed to run on either a Mac or a PC or a Linux machine. The appropriate JVM is automatically downloaded to your computer and installed in a few seconds, and then the Java applet runs on the JVM, like magic. (You may have noticed Java updates being downloaded to your computer—or you may not have noticed! Ideally, you can forget about which JVM is installed on your computer and expect any website you go to will either have Java applets that already run on your JVM or will unobtrusively install the appropriate JVM update then and there.)

The JVM is the interface that compiles the Java code into code that will run on whatever hardware you are using. The Java slogan is WORA: Write Once, Run Anywhere, which means that design problems need to be solved just once. The excellence of this arrangement is easy to see when we compare it to the EVM (English Virtual Machine) installed in your brain's wetware. That is, you're an English speaker, or you couldn't be reading this book. When I wrote this chapter I had many problems of exposition and clarification to solve, but I didn't have to know any details at all about the neuro-anatomy of the different brains that might end up reading it; I wrote it once (well, I wrote many drafts, but published only one), and now it can run "anywhere"—on any brain that has an EVM. Some brains don't; the book will have to be translated into other languages if its effects are to be experienced by all. Otherwise, would-be readers will just have to learn English.

Think how easy this makes it for teachers, guides, informants, trainers (and also con artists, propagandists, and religious prose-lytizers) to design a pitch that will have a good chance of installing still further memes in your brain, informing you, arming you with new tools, influencing you. I simply write down my invention, and when you read what I write, you download a new app to your necktop, as simple as that. My book *Intuition Pumps and Other Tools for Thinking* (2013) consists of 72 chapters, each presenting at least one thinking tool for you to download, along with advice on how to make and critique such thinking tools. Of course that book presupposes a fair amount of prior general knowledge on many topics, so you have to have installed apps for those thinking tools as well. To pick a page at random (p. 71), among the words I use without explanation are the following: *daisy, starfish, anteater, olives, blob, tongue, inherited, childhood, category*. If you don't happen to have any of these familiar tools in your kit, the page will be rough going. More difficult terms (*ontology, Umwelt, manifest image, flicker-fusion rate, mass term*, and *sortal*) are explained as I go along; new apps depending on other new apps depending on old apps.

As Jackendoff (2002) has observed, many experiments with animals as subjects require that they first be trained to perform a very particular task, for a small food reward in most cases. If you want a cat or a monkey or a rat to press a lever whenever they see a square rather than a triangle, or when they hear *beep* rather than *boop*, it may take hundreds or even thousands of training trials to get the animal to be a reliable and attentive subject. When cognitive scientists want human subjects to perform the same task, a simple briefing in a language the subjects understand (including a promise of a payment to be received at the end of the session), along with a few practice trials, turns most human volunteers into flawless task executors for the duration of the experiment. In effect, people can download and execute a virtual machine with no need for trial and error or associative learning, taking on hundreds of roles on demand, swiftly and reliably.

Thus the neural plasticity and habits of attention required to turn us into (relatively uncomprehending) imitators provides the basis for acquiring language, which in turn grounds the cognitive virtuosity needed for hearing, *understanding*, and following instructions. Comprehension of language thus yields a much more general comprehension of behavior and of the whole world. But again, we mustn't exaggerate the effectiveness of this transfer that language makes possible. Some skills are harder to transmit than others, and some call for substantial rehearsal, and mnemonic tricks, as noted in chapter 6. The fanciful names and superimposed shapes for the constellations are among the surviving relics of these pioneering efforts at transfer of information, but we needn't assume that some clever astronomer devised the whole scheme as a self-conscious improvement in pedagogy; probably the associations were the result of gradually accreted traditions of storytelling, with their utility only slowly dawning on the participants, while second-rate, less memorable associations went extinct—failing to reproduce amidst the competition—until only the mnemonic champions remained.

The parallel between Java applets and memes (or between digital computers and brains) is not exact, of course, as we have already seen, but let me underline a few points. It is true that a computer without an operating system (Windows, Mountain Lion, Linux, etc.) is all but disabled, and it is true that you can't do much thinking with your bare brain for much the same reasons. But brains, unlike digital computers, don't have centralized, hierarchical control, with timing pulses, traffic cops to determine priorities, and so forth. So I can't *just* download my app to your brain and let it run; control must be accomplished by negotiation, diplomacy, and even, on occasion, pleas, threats, or other emotional nudges. In general I have to secure your attention, your cooperation, even—to some degree—your trust, because you are and *ought to be* vigilant against possible manipulation by other agents. Computers were initially designed to be obedient executors of every task offered to them—no questions asked—but with the growth of the Internet, the growing presence of "malware" designed to hijack computers for nefarious purposes has obliged the designers of operating systems to include layer upon layer of protection, so that the default state today when any new software arrives is DON'T RUN THIS CODE without an express permission granted by the computer user or administrator. We human users are now the weakest link in the security, and ever more seductive "phishing" lures arrive almost daily, intelligently designed to win our trust.

With the help of figure 13.1, we can picture human culture accumulating at an ever swifter pace, since more directed search and more top-down problem-setting leads to more efficient problem-solving. Among the enabling innovations were such brilliant "inventions" as writing, arithmetic, money, clocks, and calendars, each contributing a novel and fecund system of representation that provided our manifest image with ever more portable, detachable, manipulable, recognizable, memorable *things to do things with*, to exploit in our growing mastery of other things. These were, so far as anyone can tell, Darwinian "inventions," that is, inventions without inventors or foresighted authors, more like bird wings than helicopter blades.

The free-floating rationales of the features and structures of these inventions have been gradually captured, represented, and celebrated by later beneficiaries, retrospective reverse engineers who could explain to the world the particular utility of phonemic representation of words, zero as a number, hard-to-counterfeit coins, representing time with a line or a circle or a volume, using a fixed short cycle of names for days. All of these culturally transmitted artifacts, abstract or concrete, are unmistakeably well-designed tools for thinking, but they were not the brainchildren of particular individual intelligent designers.[94]

This is also true of many—probably most—of the tools for other purposes that we have added to our toolkits over the millennia. In *The Secret of Our Success* (2015), anthropologist Joseph Henrich explores in considerable detail excellent artifacts, such as kayaks, and argues that it is very unlikely that any one designer/fabricator could have been responsible for more than a few features, a few improvements over the artifacts that were his models. The lore that builds up around these artifacts is the product of retrospective theorizing, not original invention, in most cases. It probably is composed mostly of byproducts of attempts to pass on the skills to apprentices, second-hand commentaries that may improve the understanding of the whole process—or may be "false consciousness," persuasive but mistaken "theories" of the topic.

The adoption and use of new memes to solve problems typically generates greater comprehension in those who are doing the solving, but not always. Sometimes problem "solvers" stumble on solutions without noticing or while *mis*understanding what they have done; their triumphs spread, and often their reputations hitch a

---

94 There is room for debate and speculation about these cases, and some historical grounds for assigning leading roles to some individuals, not so much for *invention* as for *refining and spreading the meme*, either by fiat (in the case of rulers establishing the local currency or calendar) or by influential exploitation (in the case of writing systems or mathematical notation, for instance). To take a recent example, Al Gore didn't "invent the Internet" and didn't claim to have done so, but he does deserve credit for his foresighted support and popularization of it.

ride. If you have ever made a "brilliant" chess move and only appreciated its power retrospectively, and then kept that secret to yourself, you know the phenomenon. There is a general generous tendency to credit innovators with more prior comprehension than they actually deserve, and this helps perpetuate the myth of the Godlike powers of our famous geniuses, and by extension, of all of us. Are we not the *rational animals* celebrated ever since Aristotle? The same illusions of comprehension that tend to infect our appreciation of the clever ploys of animals (and plants, and bacteria, as we have seen) distort our assessments of our fellow humans—and of ourselves.

One of the rules of thumb of a well-designed scam is that it should exploit our self-regard as clever detectors of fraud, giving us opportunities to "prove" the power of our skepticism before putting our judgment to the test. We are often shocked to discover how vulnerable we can be to scams and hoaxes, and in the very recognition of our victimhood on those occasions we often commit the sin a second time, attributing more cleverness to the con artists who fooled us than they deserve, for after all, their tricks are usually ancient and were field-tested and well polished by generations of earlier hoaxers, before being passed on. A classic instance is the old shell game, which dates back at least to ancient Greece. You are shown three walnut shells or inverted cups, with a pea shown under one of them; the shells are then rapidly moved on the table: now where's the pea? Place your bets. You may have encountered a small group of people crowding around a shell manipulator, following the action and placing their bets on where the pea is. Appropriately cautious, you join the crowd, your hand carefully protecting the wallet in your pocket, and watch for a while. Some folks win, most folks lose, but in those cases, you—clever you—can actually detect the misdirection, the sleight of hand, of the operator. After suitably scouting the whole scene, you get out your wallet and watch like a hawk for the surreptitious move you've practiced tracking. You make your bet—and *lose!* How can that be? Simple: there are two sleight-of-hand moves: the one you've been allowed to detect and track, and the one the manipulator hasn't done yet, and it is, in fact, practically invisible to all but practiced magicians and con artists. Aside from you, the crowd is composed of accomplices—shills—who have put on a good show of being a little stupid, a little inattentive, too trusting, so that you can "explain" their losses and prepare to do otherwise when you enter the game. There is some genuine skill in the manipulator, who knows how to make the key move, and knows how long to withhold it while waiting for you to put down your money, but the trick is new only to you.

In more august social settings, the same idealized aura of comprehension tints our perception, across agriculture, commerce, politics, music and art, religion, humor—in short, across all human civilization. The farmers with the largest harvests must know what they are doing, likewise the most profitable stockbrokers, the most popular musicians, the most reelected politicians. While recognizing that dumb luck plays a hard-to-detect role, we follow the general rule of seeking the advice of the (apparently) successful. The models of classical economics, with their assumption of rational agents using the available information to optimize their market decisions, exhibit the undeniable attractions of the rationality assumption. By oversimplifying and idealizing actual transactions in the world, they turn messy human activity into a somewhat predictable, somewhat explicable set of phenomena, and they work quite well for many purposes. We may not be perfectly rational, but we are not *so* irrational. The explanation of the law of supply and demand, for instance, is transparent: since, in general, buyers will try to minimize expenses and sellers will try to maximize net receipts, the price for any commodity will rise and fall with its availability and desirability. It isn't a law of physics; it's a dictate of rational behavior, but you would be foolish indeed to bet against the law of supply and demand. (If you have inside knowledge of something relevant that the market doesn't share, you may seem to cheat the law of supply and demand, but you are actually exploiting it: any edge you gain will be dictated by the broad tendency of your victims to behave as the law prescribes.)

When we look at traditional accounts of cultural evolution, in the neutral sense of cultural change over time, the dominant theme is the economic model—as if all human cultural evolution took place near the ceiling of comprehension, and all involved objects *valued for good reason*. Cultural items, on this model, consist of *goods*, which are rationally acquired, maintained, repaired, and then bequeathed to the next generation, along with the know-how to preserve and enjoy them. We protect what we value, we transmit it to those whose interests we wish to enhance, and we invest in the infrastructure required to keep the goods in working order. This model doesn't work for cultural junk—the deplorable but infectious habits that spread in spite of our disvaluing them[95]—or for stealthy cultural variants that reproduce without our taking notice of them (the new meaning of "begs the question," for example), but it works quite well for the parts of human culture deemed treasures by at least some members of society. Grand opera is kept alive by patrons, who also endow music *conservatories*, and offer prizes for accomplished reproductions of the legacy items. Museums and libraries, universities and historic buildings, famous battlegrounds and monuments all require costly maintenance, and those who value these work hard to ensure that funds are secured to preserve them. Many people devote their lives to preserving traditional crafts and activities: weaving, embroidery, woodworking, blacksmithing and Morris dancing, the Viennese waltz, dressage, and the Japanese tea ceremony. Religions devote a very large portion of the donations they collect from their members to preserving, heating, and furnishing their edifices, and paying their staffs, often with scant funds left over for helping and feeding the poor.

When we peruse the heights of contemporary human culture, all the excellent artifacts, abstract and concrete, that maintain our

---

95   Asked by a student for an example of infectious cultural junk that is hard to eradicate, I replied "Well, it's like, when, like, you use a phrase which, like, isn't really, like, doing any serious work, but, like, you go on, like, using it." To which he replied, "I, like, understand the point, but I wanted, like, an example."

health and security while freeing us from mindless toil, and filling our newfound free time with all manner of art, music, entertainment, and adventure, we find a bounty of examples of intelligently designed entities. We know they were intelligently designed because we have witnessed and recorded the processes many times and have conversed with the designers, hearing their explanations and justifications, learning their techniques and goals, their aesthetic norms, and in many cases checking their proofs and documentations.

We are indeed living in the age of intelligent design, and it goes back several millennia—as far back as we have documentation. The builders of the pyramids knew what they were doing and had articulated goals and plans, which they understood and executed with precision, organizing thousands of human workers in a process not at all like termite castle construction: it relied on top-down control and an impressive level of comprehension. That is not to say that the pyramid builders didn't rely on a massive amount of know-how—memes—that had been refined and optimized by relatively mindless differential replication over earlier millennia. Every generation has a legacy of accumulated knowledge most of which has, by now, been practically confirmed in thousands of applications, including memorable failures by rebels who decided to buck tradition and try other methods and standards. While much of this hard-won knowledge has also been expressed, explained, analyzed, and rationalized in treatises on every imaginable topic, we shouldn't make the mistake of thinking that the authors of those treatises were often the inventors or designers of the principles and practices they teach. In general, from Aristotle's day to the present, the explanations and justifications of our storehouse of general knowledge are a kind of Whig history, written by the victors, triumphantly explaining the discoveries and passing over the costly mistakes and misguided searches.

If we want a clear idea of intelligent design, we need to resist this retrospective mythmaking and give some attention to the history of the failures, and to all the second-rate design we can identify around us. For one thing, it heightens our appreciation of the best work. For instance, my understanding of how magnificent great art is has been amplified by hours spent in second-rank art museums in Europe, where one can see hundreds of minor works in the styles and with the same aspirations as Raphael and Rembrandt and Rubens and Rodin, and see immediately how lifeless and unbalanced most of them are compared with the masterpieces. The world's music libraries contain the scores of thousands of classical symphonies that will probably never be played again, for good reason. (There are perhaps a few unheralded masterpieces among the lackluster compositions, but few musicians are eager to engage in the mind-deadening task of prospecting for them.) This huge backlog of nearly invisible cultural accumulation plays a role not unlike that of all the organisms in all the species who died childless but whose competition tested the mettle of those whose descendants live on today. Haydn and Mozart and Beethoven had to grow their talents in a world that included lots of slightly less talented, or maybe just unlucky, composers.[96]

In the course of all this deliberate husbandry of domesticated memes, there is a lot of conscious, conscientious decision-making, and the economic model does about as well accounting for the patterns of decisions in these areas as it does in more commercial arenas. The law of supply and demand governs the future trajectories of both dressage academies and lollipop makers. But even in the most bureaucratic and rationalized of institutions, there are patterns of change—of evolution—that resist capture by the economic model, that appear as mere noise or happenstance. Some of this obstreperous variation is no doubt sheer noise, as good as random, but much of it is driven not by the rational pressures of the marketplace, but by arms races of Darwinian memetic evolution: there is an incessant competition among memes for replication, and accounting for the

96   Writers have resurrected some of the also-rans in memorable works: Robert Browning's poem about Andrea del Sarto (a famously "flawless" painter who lived in the shadow of Raphael and Michelangelo) and Peter Shaffer's play *Amadeus*, a fictional version of the tribulations of Antonio Salieri, eclipsed by the young Mozart.

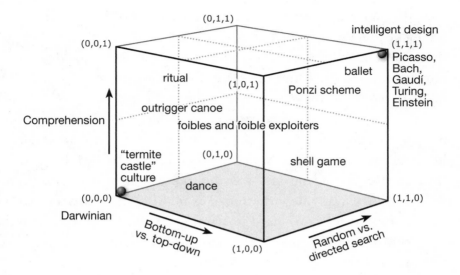

FIGURE 13.2: Darwinian space of cultural evolution with intermediate phenomena.

biases that fuel the competition requires descending from the ceiling of pure rationality (see figure 13.2) and looking at the middle ground, where semi-comprehending agents engage in semi-well-designed projects that generate a host of *foibles*, fresh new targets to attract *foible-exploiters*, equally semi-comprehending agents who sense, with varying degrees of accuracy, that their interests will be well served by adopting memes, adapting memes, revising memes designed (by a mixture, now, of natural selection and intelligent design) to take advantage of the weaknesses they seem to detect in their quarry. These arms races are the latest wave in the creative arms races that have driven genetic evolution for three billion years, and they differ mainly in involving a significant degree of comprehension to lubricate and accelerate the innovations and reactions. This comprehension is overwhelmingly the product of language and other media of communication.

Adding these varieties of information exchange to the scene changes the selective environment in revolutionary ways. Standard Darwinian evolution by natural selection, lacking all foresight, rolls through populations that also lack foresight. The survivors are lucky to happen to be better equipped for the new conditions than the rest of their generation, and there is little or no advance warning of new risks and opportunities. If an individual wildebeest hits on a novel evasive move that saves it from predation, there is no swift way for all the other wildebeests in the herd to adopt it, and "word of it" certainly can't be spread to other herds across the savannahs. With us, there are well-developed traditions of gathering and passing along new advice as soon as it becomes available (whether it is good or bad advice). All the fables and folk tales and nostrums, all the worldly wisdom you acquired "at mother's knee," is supplemented by gossip and rumor, and further enhanced thanks to writing, printing, and all the high-tech media of today. This speeds up and intensifies the arms races, and complicates them with subplots and side bets as people furnish their imaginations with all the information and use it to dream up variations.

It is harder today to find "marks" for the Nigerian prince scam than it was twenty years ago, because it is so well-known that it can be a reliable topic for humor (Hurley, Dennett, and Adams 2011). Why, you may wonder, does the scam survive, and why do the scammers persist in using the same old dubious tale of a Nigerian prince who wants to transfer a fortune and needs your help? Because the scammers don't want to waste their time and effort on smart people. The cost of sending out millions of initial e-mail lures is trivial compared to the cost (in time and effort) of reeling in a sucker who has bitten on it. You can't afford to attract too many "bites." Most potential victims who start down that path will soon get suspicious and easily discover the truth and then back out without making any down payments, and some may even enjoy playing the scammers for as long as possible. To minimize the cost of reeling in fish that will shake off the hook before being landed, the initial lure is *designed* to be so preposterous that only a very dull, sheltered, credulous person would seriously engage it. In short, there is a skeptic filter built into the scam, so that effort can be concentrated on the most vulnerable victims.

Political advisors and advertisers, trend analyzers, and speculators aggressively observe and perturb the memosphere, prospecting for new moves, new opportunities, new cracks in the armor of skepticism and caution that all but the most naïve people use to protect themselves. Everybody wants to "go viral" with their pet messages, while finding new ways of ignoring the surfeit of attention-grabbers that assault them. This fluidity of information transmission in human culture, and its use in combatting, discrediting, discarding, but also revising, improving, adapting and spreading, new memes pushes Darwinian meme evolution into the background.

Have I just admitted that memetics cannot be a valuable theoretical tool in modeling today's (and tomorrow's) cultural evolution? Consider again the claim by Steven Pinker, quoted in chapter 11:

> The striking features of cultural products, namely their ingenuity, beauty, and truth (analogous to organisms' complex adaptive design), come from the mental computations that "direct"—that is, invent—the "mutations," and that "acquire"— that is, understand—the "characteristics." (1997, p. 209)

This is close to the truth about some cultural products, but, as we have seen, by insisting that invention and understanding are the sole phenomena at work, Pinker restricts our attention to cultural treasures only, and exaggerates the role of intelligent design in creating these. Moreover, Pinker underestimates the distortion that is built into our linguistic practices: a demand for reasons that outstrips our ability to feed it. When we acquire our mother tongues, we are initiated into the space of reasons, not free-floating rationales anymore, but *anchored* rationales, *represented* reasons we own up to and endorse, our answers to all the "why" questions. The transition is awkward, because we don't always have good answers, but that doesn't stop us from playing the game.

Why do you build your boats this way?

Because that's the way we've always built them!

But why is that a good reason?

Because it just is!

The normativity of reason-giving imposes itself even when we are at a loss for an answer. There is an obligation to have reasons that you can give for your behavior. This is the intentional stance in action, always *presuming* rationality: There *must be* a good reason or we wouldn't be doing it this way! Faced with this demand, the temptation, often succumbed to, is to pick a likely reason out of thin air and advance it, with a knowing air, thereby giving birth to a handy bit of ideology that might even be right. If no plausible reasons occur to us, the wisdom of our ancestors can be wrung in when necessary:

> They knew the reasons, and we are grateful that they taught us, even though they didn't tell us why.

It's a short step from here to

> God works in mysterious ways.[97]

Infants, before they have language, must be trained much as you would train a dog. *Don't touch! No! Come! Stop that! That's good. Now do it again!* Parents don't presume that the infants will understand why; they're just babies. But as they begin to speak, we begin to give them reasons: *Don't touch—hot! Don't touch—dirty! Eat!—its good for you!* Obedience, even blind obedience, is useful as a basis; there will

---

97   The work on reason-giving and normativity descended from Sellars at Pittsburgh, via Brandom, McDowell, and Haugeland, has never stressed, to my knowledge, that these all-important human practices are systematic generators of false ideology whenever the demand for reasons exceeds the available supply.

be time for explaining and arguing later. "Because I said so!" is an important stage. And then, as we grow up, we are introduced to the norms of human society, which include preeminently the presumption of rationality in each other, especially in adults—but not, of course, in our domesticated animals. A cow pony exhibiting insight into the tasks she executes with her rider delights and surprises us. How clever! Sheep dogs are thrilling examples, and they are thrilling because they are exceptional. We don't hold animals up to the standards we hold for each other.

This is an ineliminable feature of language use; the presumption of comprehension is the default, so much so that when it is violated, we can become disoriented. In a foreign land you need to ask directions and optimistically try your native tongue, just in case you're speaking to somebody who happens to know it. The native smiles and listens, nodding and bright eyed, so you carry on confidently, and when you stop, your interlocutor reveals he hasn't understood a word. Bump! What a let down!

We lead our daily lives bathed in the presumption of understanding, in the strangers we encounter as well as our family and friends. We expect people to expect us to have reasons, reasons we can express, for whatever it is we are trying to do, and we effortlessly make up reasons—often without realizing it—when a query catches us without a pre-thought answer. It is against this background that memetics looks so wrongheaded, so subversive, as Pinker says. We are the reasoners! We are the intelligent designers! We are the explainers of everything. It is an affront to suggest that some of our brainstorms are just that: cerebral disturbances caused by invading memes duking it out for dominance. We are in charge!

And we *are* in charge, to a gratifying degree. That is the triumph of the memes invasion: it has turned our brains into minds—*our* minds—capable of accepting and rejecting the ideas we encounter, discarding or developing them *for reasons we can usually express*, thanks to the apps installed in our necktops.

But we don't always *have* reasons, and that is one thing that distinguishes us from the imaginable but impossible rational agents of GOFAI, the ultimate dream of intelligent design. In such knowledge systems, reasons are represented "all the way down," in the sense that all thinking is seen as theorem proving. When a GOFAI tells you something, you can, in principle, always demand a reason, and it should be forthcoming, since whatever the GOFAI told you had to have been generated by rational inference from its database of axioms. It can show you the proof, and you can criticize it step by step if you wish.

The advent of "deep learning" and Bayesian methods has been viewed with mixed emotions by many in cognitive science. Why? The fact that these new cognitive fabrics work so well is astonishing and delightful, and their applications are going to sweep the world, but . . . although they will give us great answers to hard questions like never before, *they won't be able to tell us why.* They will be oracles who cannot engage in the game of reason-giving. (More on this in the last chapter.)

Pinker, Wilde, Edison, and Frankenstein

> The first duty in life is to be as artificial as possible.
> —Oscar Wilde

In a typically lucid commentary on my Harvard Mind-Brain-Behavior Lectures of 2009, Steven Pinker expanded his critique of memes:

> Design without a designer is crucial for biological evolution—but it is perverse for cultural evolution: there really is a designer—the human brain—and there's nothing mystical or mysterious about saying that. (April 23, 2009, online at https://www.youtube.com/watch?v=3H8i5x-jcew)

Pinker and I agree, of course, about biological (genetic) evolution being design without a designer, and we agree that the human brain's powers as a designer depend on its having been designed (until very recently) without a designer. What I have been arguing, however, is that much of this R&D is, and must be, *memetic*, not *genetic* evolution. Long before we had designer brains, we had brains that were acquiring design-without-a-designer in the form of invading memes.

But there is more explanatory work for memetics to do even once we are well into the age of intelligent design. In his commentary, Pinker introduces two good examples: the origin of the word *acne*, which descended from an uncorrected misprint of the Greek word *acme*, meaning pinnacle, which he contrasted with the coining of the term "palimony" introduced by "a stroke of creation or insight or wit"—a portmanteau combination of *pal* and *alimony*. But we don't know that this is true. The term was first used in public by the divorce attorney Marvin Mitchelson in 1977 when his client, Michelle Triola Marvin, filed an unsuccessful suit against the actor Lee Marvin, to whom she had not been legally wed (Wikipedia, s.v. "palimony"), but we don't know whether Mitchelson's creation was a stroke of wit or the result of a tedious generation of variations—"alimony, ballimony (hmm), calimony, dallymony (maybe) . . . galimony (ho hum), . . . palimony—Hey, I can use that!" I know two inveterate punsters who have confessed to me that this is what they do, obsessively, with almost every phrase they hear, discarding 99% and more of what they swiftly consider. Not *very* intelligent design, with its laborious and promiscuous semidirected search—or is it?

A curious feature of our appreciation of wit or genius is that we prefer not to know how it is done. We think of Oscar Wilde as a great wit. It would no doubt diminish his reputation considerably if we learned that he lay awake most nights for hours, obsessively musing "Now what would I reply if somebody asked me . . . and what might my pithy opinion be of . . . ?" Suppose we learned that he patiently worked out, and polished, several dozen bons mots per night, ingeniously indexing them in his memory, all freshly canned and, if the occasion should arise, ready to unleash without skipping a beat—for brevity is indeed the soul of wit. They would still be *his* witticisms, and their funniness would depend (as the Polish comedian said) on timing. Timing is important for almost any intelligent act—which is why it is intelligent to anticipate and presolve problems wherever possible. Wilde's brute force witticism-production scheme might disappoint us, but it draws attention to the fact that all intelligent responses depend on costly R&D, and it doesn't make much difference how the work is distributed in time so long as the effects are timely.

Thomas Alva Edison's famous line about genius being "1% inspiration and 99% perspiration" provokes in many people the unkind thought that this may be true of Edison, and a few other dogged but dimwitted inventors, but *real* genius reverses the percentages—just as Picasso boasted! This attitude clouds our insight into the relationship between creativity and consciousness—one of the distortions of Cartesian gravity, which we are beginning to feel ever more strongly, as we approach consciousness. We *want* our minds to be "inspired" and "uncanny," and we tend to view attempts to dissect these, our most precious gifts, as impertinent assaults (and philistine to boot). "Anybody who is foolish enough to think our minds are all just so much neural machinery must have a pathetically impoverished acquaintance with the magic of our minds!"

Lee Siegel (1991), in his remarkable book on Indian street magic, includes a passage I have often quoted:

> "I'm writing a book on magic," I explain, and I'm asked, "Real magic?" By real magic people mean miracles, thaumaturgical acts, and supernatural powers. "No," I answer: "Conjuring tricks, not real magic." Real magic, in other words, refers to the magic that is not real, while the magic that is real, that can actually be done, is not real magic. (p. 425)

Even many neuroscientists, committed naturalists, tend to flinch when they start closing in on how the neural machinery works its wonders, and a tempting bit of modesty is to deny that they are even trying to tackle the "Hard Problem" (Chalmers 1995), leaving the "real magic" of consciousness as a cosmic mystery for future centuries to track down, with the help of currently unimaginable revolutions in physics. Those who think this way ought to contemplate a trade secret divulged by one of the best stage magicians, Jamy Ian Swiss (2007): "No one would ever think that we would ever work this hard to fool you. That's a secret, and a method of magic." If you don't even *try to imagine* how it might be done with a mountain of drudgework, you'll be permanently mystified. I mentioned in chapter 2 that magicians have no shame; neither do brains.

Since there is no such thing as "real magic," whatever trajectory a genius mind manages to pursue must ultimately be accounted for in the cascade of cranes that have been designed and erected over the last few billion years. One way or another, the mind was lifted (by good design) into a region of Design Space from which its further forays were admirably swift and effective. Whether the "inspiration" is 1% or 99%, it must be a feature of a vehicle for exploring Design Space that has its own nonsupernatural R&D history, typically an unfathomable combination of genes, education, life experience, mentoring, and who knows what else—diet, the chance overhearing of a strain of music, emotional imbalances temporary or permanent, and, yes, sometimes "insanity"—mental illness or pathology that has fortuitous benefits (for the art world if not the artist).

When we try to figure out "how much" of the invention we can attribute to the creativity of the individual artist, we get at least a rough idea by imagining the particular region of Design Space traversed to arrive at the final product. A thought experiment focuses the issue:

> Suppose Dr. Frankenstein designs and constructs a monster, Spakesheare, that thereupon sits up and writes out a play *Spamlet*. Who is the author of *Spamlet*?

First, let's take note of what I claim to be irrelevant in this thought experiment. I haven't said whether Spakesheare is a robot, constructed out of metal and silicon chips, or constructed out of human tissues—or cells or proteins or amino acids or carbon atoms. As long as the design works and the construction was carried out by Dr. Frankenstein, it makes no difference what the materials are. It might well turn out that the only way to build a robot small enough and fast enough and energy-efficient enough to sit on a stool and type out a play is to construct it from artificial cells filled with beautifully crafted motor proteins and other carbon-based nanorobots. That is an interesting technical and scientific question but not of concern here. For exactly the same reason, if Spakesheare is a metal-and-silicon robot, it may be allowed to be larger than a galaxy, if that's what it takes to get the requisite complication into its program—and we'll just have to repeal the speed limit for light for the sake of our thought experiment. Since such technical constraints are commonly declared to be off-limits in these thought experiments, so be it. If Dr. Frankenstein chooses to make his AI robot out of proteins and the like, that's his business. If his robot is cross-fertile with normal human beings and hence capable of creating what is arguably a new species by giving birth to a child, that is fascinating, but what we will be concerned with is Spakesheare's purported brainchild *Spamlet*. Back to our question:

Who is the author of *Spamlet*?

In order to grasp on this question, we need to look inside and see what happens in Spakesheare. At one extreme, we find inside a file (if Spakesheare is a robot with a computer memory), or a memorized version of *Spamlet*, all loaded and ready to run. In such an extreme case, Dr. Frankenstein is surely the author of *Spamlet*, using his intermediate creation, Spakesheare, as a mere storage-and-delivery device, a particularly fancy word processor. All the R&D work was done earlier and copied to Spakesheare by one means or another.

We can visualize this more clearly by imagining a subspace of Design Space, which I call the Library of Babel, after Jorge Luis Borges's classic short story by that name (1962). Borges invites us to imagine a warehouse filled with books that appears to its inhabitants to be infinite. They eventually decide that it is not, but it might as well be, for it seems that on its shelves—in no order, alas—are all the possible books. Now we are ready to return to that needle in a haystack, *Spamlet*, and consider how the trajectory to this particular place in the Library of Babel was traversed in actual history. If we find that the whole journey was completed by the time Spakesheare's memory was constructed and filled with information, we know that Spakesheare played no role in the search. Working backward, if we find that Spakesheare's only role was running the stored text through a spell-checker before using it to guide its typing motions, we will be unimpressed by claims of Spakeshearian authorship. This is a measurable, but Vanishingly small, part of the total R&D. There is a sizable galaxy of near-twin texts of *Spamlet*—roughly a hundred million different minor mutants have but a single uncorrected typo in them, and if we expand our horizon to include one typo per page, we have begun to enter the land of Vast numbers of variations on the theme. Working back a little further, once we graduate from typos to thinkos, we have begun to enter the land of serious authorship, as contrasted with mere copy editing. The *relative* triviality of copy editing, and yet its unignorable importance in shaping the final product, gets well represented in terms of our metaphor of Design Space, where every little bit of lifting counts for something, and sometimes a little bit of lifting moves you to a whole new trajectory. As usual, we may quote Ludwig Mies van der Rohe at this juncture: "God is in the details."

Now let's turn the knobs on our thought experiment, as Douglas Hofstadter has recommended (1981) and look at the other extreme, in which Dr. Frankenstein leaves most of the work to Spakesheare. The most realistic scenario would be that Spakesheare has been equipped by Dr. Frankenstein with a virtual past, a lifetime stock of pseudo-memories of experiences on which to draw while responding to its Frankenstein-installed obsessive desire to write a play. Among those pseudo-memories, we may suppose, are many evenings at the theater, or reading books, but also some unrequited loves, some shocking close calls, some shameful betrayals, and the like. Now what happens? Perhaps some scrap of a "human interest" story on the news will be the catalyst that spurs Spakesheare into a frenzy of generate and test, ransacking its memory for useful tidbits and themes, transforming—transposing, morphing—what it finds, jiggling the pieces into temporary, hopeful structures that compete for completion, most of them dismantled by the corrosive processes of criticism that expose useful bits now and then, and so forth. All this multileveled search would be somewhat guided by multilevel, internally generated evaluations, including evaluation of the evaluation . . . of the evaluation functions as a response to evaluation of . . . the products of the ongoing searches. Now if the amazing Dr. Frankenstein had actually anticipated all this activity down to its finest grain at the most turbulent and chaotic level and had hand designed Spakesheare's virtual past, and all its search machinery, to yield just this product, *Spamlet*, then Dr. Frankenstein would be, once again, the author of *Spamlet*, but also, in a word, God. Such Vast foreknowledge would be simply miraculous. Restoring a smidgen of realism to our fantasy, we can set the knobs at a rather less extreme position and assume that Dr. Frankenstein was unable to foresee all this in detail, but rather he delegated to Spakesheare most of the hard work of completing the trajectory in Design Space to a literary work to be determined by later R&D occurring within Spakesheare itself. We have now arrived, by this simple turn of the knob, in the neighborhood of reality, for we already have actual examples of impressive artificial authors that Vastly outstrip the foresight of their own creators. Nobody has yet created an artificial playwright worth serious attention, but an artificial chess player—IBM's Deep Blue—and an artificial composer—David Cope's EMI—have both achieved results that are, *in some respects*, equal to the best that human creative genius can muster.

We can adapt the thought experiment of Spakesheare's trajectory through Design Space to analyze other realms of human creativity and discovery. The refinement of the text of *Spamlet* can be precisely measured in some dimensions because all the possible variants reside in Borges's Library of Babel, and their differences—typos and thinkos and boring digressionos—can be identified and evaluated. We can identify culpable plagiarism and laudable inspiration—though there is no bright line separating these realms. Now that we have 3D printers to play with, we have a host of formats for representing any three-dimensional object (a clothespin, an electric can-opener, a Rodin sculpture, a Stradivarius violin) as a digital file that can be compared to its variants. When Picasso sees a bicycle seat and handlebars as just what he needs to make a bull's head, he is not really plagiarizing, but he is helping himself to some carefully designed and optimized shape-and-materials for his own purposes, an artistic practice honored with its own term: *objets trouvés*.

Thus all concrete inventions and appropriations—tools, instruments, art objects, machines—can be "placed" uniquely in a multidimensional space, the Library of Objects, if you like, and their discovery/invention lineages mapped, rather like the lineages in the Tree of Life, but with a lot of anastomoses. How similar must two things be, in which dimensions, to "count"—as a single design worth protecting by patent or copyright? Don't expect there to be easy objective answers to such questions. Abstract inventions particularly defy being put "in alphabetical order" in any space of any particular dimensionality, a refreshing note for all antireductionists to savor, but the *conveying* of abstract inventions from mind to mind still relies on some combination of words and images or other recordable, transmissible representations. Turing's brilliant fable/image of a machine chugging along on a paper tape, reading and erasing (or not) a mark and then moving on, is not an optional feature of his invention. This gives us another perspective on the law's requirement that patents be of demonstrated utility, and that a musical innovation must have some "fixed expression" to be copyrightable (see p. 133). The multidimensional Design Space in which the Tree of Life has grown has now generated a daughter Design Space, for our brainchildren, and it has more dimensions and more possibilities than its parent, and we are (so far) the only species that can explore it.

Steven Pinker is right that "the human brain really is a designer," but this should be seen not as an alternative to the memetic approach, but as a continuation of the memetic approach into the age of gradually de-Darwinizing semi-intelligent design. Our traditional vision of genius portrays it as *completely* unlike natural selection in its creative powers, and it is no accident that genius is often seen as divine, supernatural, Godlike. After all, we created God in our own (manifest) image, a natural overextension of the intentional stance into the cosmos. Pinker, a stalwart and imaginative Darwinian and naturalist, has done what he can to launder the mystery out of our conception of the mind (1997) and has been roundly cudgeled for his intrepid Darwinian forays into the humanities, but here Cartesian gravity deflects him slightly from the course that will take us all the way from bacteria to Bach.

Bach as a landmark of intelligent design

> It is in perfect accordance with the scheme of nature, as worked out by natural selection, that matter excreted to free the system from superfluous or injurious substances should be utilized for [other] highly useful purposes.
> —Charles Darwin, 1862

> Judging a poem is like judging a pudding or a machine. One demands that it work. It is only because an artifact works that we infer the intention of an artificer.
> —W. Wimsatt and M. Beardsley, 1954

Turing, Picasso, Einstein, Shakespeare, all are excellent examples of intelligent designers, but I want to look more closely at Johann Sebastian Bach (1685–1750), who combines the virtues of intelligent designers in a particularly informative way. On the one hand, he was born into a musical family—his father and uncles were all professional musicians of note, and of his twenty (!) children, four who survived to adulthood were composers or musicians of note—so there is a strong likelihood that he had an unusually strong dose of "genes for music." (We mustn't ignore the role of culture, however, in explaining how music "runs in families"; being raised in a musical family, whatever your genes are, means you will be surrounded by music, encouraged to participate in the family's musical life, and be imbued with the importance of music.) On the other hand, he supplemented his "God-given" musical competence with extensive study; he was an expert on counterpoint and harmony, and a broadly informed student of earlier composers, whose works he analyzed in depth. He was also something of an eighteenth-century technocrat, an organist, and also an expert on the design, repair, and maintenance of those magnificent mechanical marvels, honored ancestors of today's electronic synthesizers.

Compare Bach to another truly great composer, Irving Berlin, who couldn't read music, could play the piano in only one key (F#, so he could play almost everything on the black keys!), and had to rely on a "musical secretary" to write down and harmonize his great melodies. He apparently exerted very strict control over the harmonizations written down by his assistant but did it all "by ear" with no knowledge of harmony theory. Once he became rich and famous (very early in his career) he had a special piano built, with a sliding keyboard, so he could keep playing in F# while rolling the keyboard into position to transpose into other keys.

Could such an unschooled composer be brilliant? Absolutely. Among his ardent admirers as a composer were Cole Porter and George Gershwin, both highly trained musicians. Gershwin, no slouch himself, called Berlin "the greatest songwriter that has ever lived" (Wyatt and Johnson 2004, p. 117). Berlin is not by any means the only great composer who couldn't read music. The Beatles are probably the most notable example, with at least several dozen songs that will go on to inspire covers and adaptations and homages across the musical spectrum from hip-hop to string quartets.

Leonard Bernstein, the Harvard educated, classically trained composer and symphony conductor, published an essay in 1955 with the amusing title "Why don't you run upstairs and write a nice Gershwin song?" in which he recounted his failed attempt, in collaboration with another sophisticated musical friend, to write a hit Tin Pan Alley song. Nobody was more appreciative of the gulf between technical expertise and success (both financial and musical) than Bernstein, who ended his essay with a wistful confession: "It's just that it would be nice to hear someone accidentally whistle something of mine, somewhere, just once." Two years later, his wish came true, with *West Side Story*, but Bernstein, for all his genius, was no overnight sensation, and neither, for that matter, was Bach. In spite of Bach's prodigious output as a composer, at the time of his death he was known mainly as a great organist, and it was roughly fifty years after his death when other composers, critics, and musicians began to secure his fame and drive his memes to fixation in the history of Western music.

The failure of Bach's compositions to spread like wildfire was not due to a lack of interest on Bach's part in securing their ready replication. He was, in fact, a pioneer in the fine art of borrowing success from tradition. In Leipzig he was *Kapellmeister* (Music Director) of the Thomas Church and other churches in Leipzig for over twenty-five years until his death in 1750. One of his duties was to compose and perform cantatas, elaborate songs for choir and soloists, accompanied by a chamber orchestra or sometimes just an organ. He wrote hundreds of these, one for each Sunday in the church calendar, each tailored to fit the liturgy and ceremony of the occasion. He is reported to have composed five full year-cycles (*Jahrgänge*) of which three still exist—so roughly a hundred of his cantatas are apparently lost forever.

Beginning in his second year, he adopted the policy of basing each cantata on a *chorale*, a preexisting Lutheran hymn that would be familiar to most of the congregation. These melodies had proven themselves over the decades or centuries, demonstrating by their survival that they were well adapted to the ears and minds of their hosts. Thus Bach gave himself a fine head start in the project of coming up with musical memes that would have staying power, if not go viral. If you were faced with the task of composing fifty hummable theme songs a year, you might follow his lead by helping yourself to "Amazing Grace," "I've Been Workin' on the Railroad," "Jingle Bells" (composed by James Lord Pierpont), "Oh Susanna!" and "Way Down upon the Swanee River" (composed by Stephen Foster), and "You Are My Sunshine" (composed by Jimmie Davis). Bach took the chorale melody as the skeleton and then built a breathtakingly beautiful musical body to bring it new life. He was the Dr. Frankenstein of musical reincarnation.

No doubt some music lovers, reading these juxtaposed paragraphs on Bach, Bernstein, Berlin, and the Beatles will be tempted to declare their judgment confirmed: those who dare dabble in memes are appalling philistines, unable to tell quality from quantity. In fact, I hope I have succeeded in provoking that sentiment, precisely in order to unpack it—and send it packing. There need be no conflict at all between the objective, scientific study of culture, high and low, on the one hand, and the articulation of aesthetic judgments on the other. Memeticists are not denied entry to the arena of critical judgment just because they insist on trying to measure, predict, and explain the differential reproduction of memes.

To balance the equation, those in the arts and humanities who tend to regard best sellers, graphic novels, and platinum albums as ipso facto artistic trash need to jettison that prejudice and learn to discern the considerable interplay between quality and quantity across the spectrum of the arts. I might add that scientists who disdain all attempts to explain science to the general public need to take on board much the same advice. I have on several occasions polled large audiences *of scientists* to confirm that they were inspired to become scientists by reading the works of Stephen Hawking, E. O. Wilson, Richard Dawkins, Douglas Hofstadter, Steven Pinker, and other excellent communicators of science. It is a shame, really, that the arts and humanities have not managed to generate many great counterpart explainer/celebrators from their ranks. Leonard Bernstein was one, and Sir Kenneth Clark was another, but that was half a century ago. Who since then has even tried to do work of sufficient *quality* to reach an audience of large *quantity*? Winton Marsalis and Stephen Greenblatt come to mind. Perhaps there would be less hand-wringing about the forlorn plight of the arts and humanities if more intelligent design in those fields had been devoted to spreading the word more widely.

It is an objective cultural fact worth discovering that today more people can hum "White Christmas" than "Jesu Joy of Man's Desiring." Pointing in the other direction is the fact that Brahms was not just wildly popular during his lifetime, but also the first famous recording artist (a colleague of Edison made a cylinder recording of him playing a portion of his first Hungarian dance in 1889). Quantity isn't to be equated with quality, but success in propagation is, in the end, as necessary for memes (however excellent) as it is for organisms. Most organisms leave no issue, and most published books have readerships in the dozens, not thousands, before going out of print for good. Even the greatest works of genius must still pass the test of differential replication. Today, Herman Melville's *Moby-Dick* (1851) is rightly regarded as one of the greatest novels in the English language, yet it went out of print in his lifetime and was all but forgotten until 1919, when the centenary of Melville's birth occasioned some favorable retrospective criticism. This was followed by two film versions (departing radically from the novel), and the Chicago edition of 1930, with the unforgettable Rockwell Kent woodcut illustrations. This led finally to the Modern Library Giant edition of 1943, which included the brilliant Kent illustrations. Thereafter the "immortality" of *Moby-Dick* was assured.

Perhaps, as some critics have proposed, the truly greatest works of art always travel well, through time and space, and somehow find a way of striking a chord in people's hearts, no matter how removed they are from the circumstances of the creation of the work. Or perhaps not. The processes by which "neglected masterpieces" get resurrected often include striking juxtapositions of features that have little to do with the excellences thoughtfully designed by their authors and more to do with "striking a chord" in the serendipitous encounters that ensue. So even if our brains are, at their best, intelligent designers, explaining the dominant patterns in culture requires divorcing the circumstances of creation from the vicissitudes of survival.

In chapter 2 a question was posed and postponed: Why have there been so few famous female geniuses? Is it genes or memes or a mixture of both? Our present vantage point suggests that the answer will lie more in features of culture than in cortex—but not by supporting the discredited mantra from the 1960s: boys and girls are "biologically" the same; all differences are due to socialization and other cultural pressures. That is politically correct nonsense. Male and female brains are not exactly alike. How could they be, given the differences in paternal and maternal biological roles? There are dozens of reliably detectable differences of neuroanatomy, hormonal balance, and other physiological signs, and their genetic sources are not in doubt. Moreover, these physical differences issue in differences in cognitive and emotional competence and style that are statistically significant (see, e.g., Baron-Cohen 2003). But there is plenty of variation *between* women, and *between* men, so some women excel at tasks that mostly men excel at, and vice versa.

Besides, we're asking what explains the *fame* of more male than female intelligent designers, and that property is, as we have just noted, quite loosely and unreliably associated with the quality of the designer's mind. What makes a song or a joke go viral and what makes a person's name and reputation go viral are about equally inscrutable, with plenty of worthy candidates of all kinds for natural selection to pick, quite blindly, for amplification. That doesn't mean that it is sheer luck that determines who gets famous—the imbalance between men and women virtually guarantees that this cannot be the case. (Could have a gene for luck that women lack? Not a chance.) So there must be some imbalance in the average starting lines, with men getting more chances for stardom than women in the famous-genius crap shoot. For past centuries the explanation for this is obvious: for millennia, very few women ever got a chance to develop their talents. Even in recent years, moreover, there are lots of anecdotes about female scientists, for instance, who deserved most of the credit for the discoveries and proofs that bestowed fame on their male colleagues. And now that women are finally beginning to make large inroads in male-dominated fields, science is shifting to ever larger teams of researchers, where the opportunities to stand out are curtailed by the administrative demands of the projects. The same is true in other domains of intelligent design. Where are the Babbages and Edisons and Watts of today? Working with teams at Google and Apple and Amazon, legendary among their colleagues, perhaps, but living in relative anonymity in the wider world. As we shall see in chapter 15, the heroic age of intelligent design is beginning to wane just when its would-be heroines are finally getting to prove their powers.

The evolution of the selective environment
for human culture

At the dawn of human culture, ancestors hosted beneficial memes with no more understanding than they had for their genetically transmitted instincts. They didn't need understanding to acquire the new competences, and they didn't gain much understanding from having the new competences. The significant difference that culture made in the early days was just that these culture-borne *ways* (of behaving) could be reshaped during a single generation of hosts and transmitted between nonrelatives and acquired throughout life, not just at birth. As memes accumulated and became more and more effective at inhabiting their hosts (becoming more helpful or less of a hindrance, or else subjugating their hosts for their own benefit), the manifest image became populated with more and more affordances, more and more opportunities to track, more and more things to do things with, more and more things—words—to use as tools to help keep track of things, and so forth. Some memes were tools, some were toys, some were distractions, some were crippling parasites. They all depended on cultural replication to survive.

This competitive environment created evolutionary arms races with the same inevitable burgeoning of technology and counter-technology found in human warfare: lies and threats and bluffs and probes and tests—like the *shibboleth* test, for instance, which could be used as a sort of linguistic passport. (Can you pronounce *shibboleth* correctly? If you can't, you're not one of us.) Systems of best practices—bartering, promising, mutual informing, warning before striking, for instance—could prove themselves or at any rate go to fixation locally, creating traditions that were known by all, and known to be known by all, making them part of the more-or-less fixed behavioral environment in which plans could be made, discussed, accepted, or rejected. How much of this could be accomplished without full-fledged grammatical language is anyone's guess, at present, but without getting all the elements in proper order of historical appearance we can still see how a gradual growing sophistication of behavior could arise by small innovations, adjustments, and refinements, building excellently designed cultural habits and institutions without much help from intelligent design. "First learning, then adaptations for learning" (Sterelny 2012, p. 25). Once the value of swift learning was revealed by the relative success of the swifter learners, ways of accelerating the process could evolve, both culturally and genetically. One of the most valuable innovations was the practice of putting marks in the environment to take a load off personal memory, one of the first forays of "the extended mind" (Clark and Chalmers 1998). Marks then evolved into number systems and written languages, which enhanced the power of discursive teaching, and within a few millennia we have Socrates and Plato and Aristotle talking about talking, thinking about thinking, imagining republics, theorizing about tragedy and comedy. The age of intelligent design is in full swing. Merely Skinnerian and Popperian creatures couldn't keep up with Gregorian creatures, their minds overflowing with new tools for making ever swifter and more accurate assessments of the complex environment confronting them. Brute force trial and error would no longer suffice; you had to comprehend to compete.

Now we live in a world where memes without clear authors still go viral, such as fads and fashions, pronunciation shifts and buzzwords, but they have to compete with memetically engineered inventions, created with foresight and purpose by professional memesmiths who play a significant role in contemporary society as authors, artists, composers, reporters, commentators, advertisers, teachers, critics, historians, speechwriters. Pinker is right that the successful brain-children of these human beings are intelligently designed, without mystery, but they swim in an ocean of semi-intelligently designed, hemi-semi-demi-intelligently designed, and evolutionarily designed competitors, all still dependent on getting into new human brains to continue their lineages. This may change. Cultural evolution has been de-Darwinized by its own products, but its Darwinian ancestry is still very much in evidence, and synanthropic, unauthored memes, like the bacteria that outnumber and outweigh us, still surround us every day.

Part III

TURNING OUR MINDS INSIDE OUT

14

Consciousness as an Evolved User-Illusion[98]

Keeping an open mind about minds

At last we are ready to put the pieces together and examine human consciousness as a system of virtual machines that evolved, genetically and memetically, to play very specialized roles in the "cognitive niche" our ancestors have constructed over the millennia. We are ready to confront Cartesian gravity head on and tackle some big questions:

1. How do human brains achieve "global" comprehension using "local" competences without invoking an intelligent designer?
2. Do our minds differ from the minds of other animals, and if so, how and why?

98   The Danish science journalist Tor Nørretranders published the Danish edition of his book *The User Illusion: Cutting Consciousness Down to Size*, in 1991, the same year my *Consciousness Explained* appeared, with my account of consciousness as a user-illusion. Neither of us was in a position to cite the other, obviously. The English translation of Nørretranders's book appeared in 1999. As I noted in my book (p. 311, fn. 9) among those who suggested somewhat similar forerunners of the idea were Kosslyn (1980), Minsky (1985), and Edelman (1989).

3. How did our manifest image become manifest to us?

4. Why do we experience things the way we do?

A brief review: Evolution has endowed all living things with the wherewithal to respond appropriately to their particular affordances, detecting and shunning the bad, detecting and obtaining the good, using the locally useful and ignoring everything else. This yields competence without comprehension, at every level from the molecular on up. Since there *can be* competence without comprehension, and since comprehension ("real" comprehension) is expensive, Nature makes heavy use of the Need to Know principle, and designs highly successful, adept, even cunning creatures who have no idea what they are doing or why. Reasons abound, but they are mostly free-floating rationales, undreamt of by those who benefit from them. As reverse engineers we can work out the ontology of affordances in the *Umwelt* of trees, fleas, and grizzly bears while remaining entirely agnostic about whether "it is like anything" to be them.

There can be reasons why trees do things without their *having* those reasons ("in mind"). Is there anything a flea can do for reasons that demonstrate that, unlike the tree, it *has* the reasons, it somehow "appreciates" the reasons that govern its actions? It might be that it is not "like anything" to be a flea, any more than it is like something to be an automatic elevator. And while we're at it, what makes us so sure it is like something to be a grizzly bear? It certainly *seems* to us that there must be something it is like to be a grizzly bear—just watch them, and listen to them! It seems more obvious that it is like something to be a grizzly bear than like the flea hiding in the grizzly bear's fur—doesn't it? But perhaps our imagination is playing tricks on us here. We know that it is like something to be *us* for the simple reason that we talk about it every day, in avowals, complaints, descriptions, poems, novels, philosophy books, and even peer-reviewed scientific papers. This is a central feature of our manifest image, *and that objective fact would be evident to any "Martian" scientists who studied us long enough to learn our languages.* Our introspective divulgences are behaviors that are just as observable and measurable as our acts of eating, running, fighting, and loving. Is there anything we can do—aside from talking about it—that some other animals can also do that would clinch the case for their having consciousness more or less like ours? If the Martian scientists framed the question of whether other Earthlings—dolphins, chimpanzees, dogs, parrots, fish—were like the talking Earthlings, what would they point to, what would impress them, and why? That is not just a legitimate scientific question; it is an obligatory question, but one that thinkers commonly excuse themselves from addressing. They beg off, saying something along the lines of:

> I have no idea where, on the complexity scale of life, to draw the line—are worms, fish, reptiles, birds conscious? We may never know, but still we do know that we human beings are not alone in being conscious. That is obvious.

That is not acceptable, for two reasons. First, the idea that there is and must be a line drawn even if we don't know where to draw it is profoundly pre-Darwinian; there might be all manner of variations from poet to possum to peacock to perch to protozoon, and no "essence" of consciousness to discover. The fact that Nagel's (1974) famous formulation "what is it like?" is now treated as a crutch, a nudging gesture that lacks content but is presumed to point to the sought for (and cosmic) distinction, should be seen as an embarrassment, not an acceptable bit of temporizing—or even, by some philosophers, as a fundamental element of theory. Second, it ties our hands by appeal to an intuition—it is nothing more—that might be mistaken in some way. Temporary agnosticism about consciousness is fine—I have just advocated it—but not agnosticism saddled with the proviso that *of course other animals are conscious even if we can't say what that means.* That is, at best, the expression of confidence in the eventual triumph of the manifest image over the scientific image, in the face of a long history of defeats. It used to be obvious that the sun circled the earth, after all. People who won't let themselves even *think* about whether grizzly bears are "conscious like us" (whatever that means) are succumbing to ideology, not common sense. Their motives may well be honorable—they are eager to expand the circle of the beings owed moral consideration because they can *suffer*—but until we can identify the important features, the features that matter, and explain why, this gesture is worse than empty. Worse, because it indefinitely postpones addressing the hard and important questions about exactly what suffering is, for instance, and whether insects or fish or daisies can suffer. We do have to "draw the line," for moral reasons, and most people (but not Jains, for instance) are comfortable with policies that call for peremptorily killing mosquitoes and deer ticks and tsetse flies, and *Plasmodium falciparum* (the microorganism that causes malaria—can a protozoan suffer?). Most people agree that rats can be exterminated but not their rodent cousins squirrels, who have been called rats with good PR by an insightful comedian. But we should keep our science neutral, just in case it surprises us with some important exceptions to folk wisdom.

And now, do you feel the pinch of Cartesian gravity? "Not like anything to be a grizzly bear? Are you kidding?" No, I am not kidding; I am putting the burden of demonstration on those who want to argue that some special phase shift occurs somewhere along the (apparent) continuum of growing adroitness that would put trees and fleas (or fleas and grizzly bears—take your choice) on different sides of a cosmic divide. There may be such a boundary, but unless "it's being like something" permits the organisms on one side to *do something important* (it might well be to *suffer*—but we'd have to find some objective way of detecting that) that they couldn't do if they were on the other side, it will be a boundary with nothing but folk tradition behind it. I am not denying the existence of such a boundary; I am postponing the issue, exploring how far we can get without postulating such a boundary, which is the way any scientific investigation should proceed. If you find yourself unable to tolerate that even-handedness, you are overcompensating for the effects of Cartesian gravity and disabling yourself from participation in the investigation. (It is worth remembering that Descartes solved the problem by fiat: only humans were conscious; animals were mindless automata.) If we shrink from his verdict, we still have to draw the *moral* line somewhere in the meantime, and let's err on the safe side, but unless we suspend *scientific* judgment until we have a better idea of what we're judging, we have no grounds for confirming, or moving, that line when we learn more. In the United Kingdom, the law since 1986 rules that the octopus (but only *Octopus vulgaris*— not any other cephalopods) is an "honorary vertebrate" entitled to legal protection. You may legally throw a live lobster or worm or moth into boiling water, for instance, but not an octopus; it has the same protection enjoyed by mammals and birds and reptiles. Should that law be expanded or retracted, or did the legislators get it right the first time? If we want a defensible answer to that question, we need to identify—and then bracket—our gut intuitions. We mustn't let our moral intuitions distort our empirical investigation from the outset.

The feats of clueless agents should not be underestimated. The termites' castle, the cuckoo chick's ovicide, and many other behavioral marvels are accomplished with only the sort of behavioral comprehension that amounts to practical know-how, unarticulated and unconsidered. When we human observers/explainers/predictors confront this well-designed excellence, we automatically set about figuring out the reasons why plants and animals do what they do, reverse engineering them with the aid of the intentional stance. And, as we have seen, when we do this it is common and natural to impute more understanding to an organism than it actually has, on the reasonable grounds that the behavior is manifestly clever, and whose cleverness is it, if not the organism's? Ironically, if we were creationists, we could comfortably attribute all the understanding to God and wouldn't feel so compelled to endow the organisms with it. They could all be God's marionettes. It was Darwin's discovery and exposure of the mindless process of natural selection, with its power to generate free-floating rationales, that freed our imaginations to continue reverse engineering all Nature's marvels without feeling an obligation to identify a mind that harbors the reasons we uncover.

How do human brains achieve "global" comprehension using "local" competences?

Language was given to men so that they could conceal their thoughts.
    —Charles-Maurice de Talleyrand

Language, like consciousness, only arises from the need, the necessity, of intercourse with others.
    —Karl Marx

Consciousness generally has only been developed under the pressure of the necessity for communication.
    —Friedrich Nietzsche

There is no General Leslie Groves to organize and command the termites in a termite colony, and there is no General Leslie Groves to organize and command the even more clueless neurons in a human brain. How can human comprehension be composed of the activities of uncomprehending neurons? In addition to all the free-floating rationales that explain our many structures, habits, and other features, there are the anchored reasons we represent to ourselves and others. These reasons are themselves *things* for us, denizens of our manifest image alongside the trees and clouds and doors and cups and voices and words and promises that make up *our* ontology. We can *do things* with these reasons—challenge, reframe, abandon, endorse, disavow them—and these often covert behaviors would not be in our repertoires if we hadn't downloaded all the apps of language into our necktops. In short, we can think about these reasons, good and bad, and this permits them to influence our overt behaviors in ways unknown in other organisms.

The piping plover's distraction display or broken-wing dance gives the fox a reason to alter its course and approach her, but not by getting it to trust her. She may modulate her thrashing to hold the fox's attention, but the control of this modulation does not require her to have more than a rudimentary "appreciation" of the fox's mental state. The fox, meanwhile, need have no more comprehension of just why it embarks on its quest instead of continuing to reconnoiter the area. We, likewise, can perform many quite adroit and *retrospectively* justifiable actions with only a vague conception of what we are up to, a conception often swiftly sharpened in hindsight by the self-attribution of reasons. It's this last step that is ours alone.

Our habits of self-justification (self-appreciation, self-exoneration, self-consolation, self-glorification, etc.) are ways of behaving (ways of *thinking*) that we acquire in the course of filling our heads with culture-borne memes, including, importantly, the habits of self-reproach and self-criticism. Thus we learn to plan ahead, to use the practice of reason-venturing and reason-criticizing to presolve some of life's problems, by talking them over with others and with ourselves. And not just talking them over—imagining them, trying out variations in our minds, and looking for flaws. We are not just Popperian but Gregorian creatures (see chapter 5), using thinking tools to design our own future acts. No other animal does that.

Our ability to do this kind of thinking is not accomplished by any dedicated brain structure not found in other animals. There is no "explainer-nucleus" for instance. Our thinking is enabled by the installation of a virtual machine made of virtual machines made of virtual machines. The goal of delineating and explaining this stack of competences via bottom-up neuroscience alone (without the help of *cognitive* neuroscience) is as remote as the goal of delineating and explaining the collection of apps on your smartphone by a bottom-up deciphering of its hardware circuit design and the bit-strings in memory without taking a peak at the user interface. The user interface of an app exists in order to make the competence accessible to users—people—who can't know, and don't need to know, the intricate details of how it works. The user-illusions of all the apps stored in our brains exist *for the same reason*: they make our competences (somewhat) accessible to users—*other* people—who can't know, and don't need to know, the intricate details. And then we get to use them ourselves, under roughly the same conditions, as guests in our own brains.

There might be some other evolutionary path—genetic, not cultural—to a somewhat similar user-illusion in other animals, but I have not been able to conceive of one in convincing detail, and according to the arguments advanced by the ethologist and roboticist David McFarland (1989), "Communication is the only behavior that requires an organism to self-monitor its own control system." Organisms can very effectively control themselves by a collection of competing but "myopic" task controllers, each activated by a condition (hunger or some other need, sensed opportunity, built-in priority ranking, and so on). When a controller's condition outweighs the conditions of the currently active task controller, it interrupts it and takes charge temporarily. (The "pandemonium model" by Oliver Selfridge [1959] is the ancestor of many later models.) Goals are represented only tacitly, in the feedback loops that guide each task controller, but without any global or higher level representation. Evolution will tend to optimize the interrupt dynamics of these modules, and nobody's the wiser. That is, there doesn't have to be anybody home to be wiser!

Communication, McFarland claims, is the behavioral innovation which changes all that. Communication requires a central clearing house of sorts in order to buffer the organism from revealing too much about its current state to competitive organisms. As Dawkins and Krebs (1978) showed, in order to understand the evolution of communication we need to see it as grounded in manipulation rather than as purely cooperative behavior. An organism that has no poker face, that "communicates state" directly to all hearers, is a sitting duck, and will soon be extinct (von Neumann and Morgenstern 1944). What must evolve to prevent this exposure is a private, proprietary communication-control buffer that creates opportunities for *guided* deception—and, coincidentally, opportunities for self-deception (Trivers 1985)—by creating, for the first time in the evolution of nervous systems, explicit and more globally accessible representations of its current state, representations that are detachable from the tasks they represent, so that deceptive behaviors can be formulated and controlled without interfering with the control of other behaviors.

It is important to realize that by communication, McFarland does not mean specifically *linguistic* communication (which is ours alone), but *strategic* communication, which opens up the crucial space between one's actual goals and intentions and the goals and intentions one attempts to communicate to an audience. There is no doubt that many species are genetically equipped with relatively simple communication behaviors (Hauser 1996), such as stotting, alarm calls, and territorial marking and defense. Stereotypical deception, such as bluffing in an aggressive encounter, is common, but a more productive and versatile talent for deception requires McFarland's private workspace. For a century and more philosophers have stressed the "privacy" of our inner thoughts, but seldom have they bothered to ask why this is such a good design feature. (An occupational blindness of many philosophers: taking the manifest image as simply *given* and never asking what it might have been given to us *for.*)

How did our manifest image become manifest to us?

Here is yet another strange inversion: this practice of sharing information in communicative actions with others, giving and demanding reasons, is what creates our personal user-illusions. All organisms, from single cells to elephants, have a rudimentary "sense of self." The amoeba adroitly keeps the bad stuff out and lets the good stuff in, protecting its vital boundaries. The lobster "knows enough" not to rip off and eat its own appendages. The free-floating rationales of the behavior of all *organisms* are *organized* around self-protection. In our case, the behaviors include a host of covert thinking behaviors we pick up in the course of enculturation, a process requiring lots of overt interaction with conspecifics. Practice makes perfect, and the sharpening and extending of these talents depends on a heightened level of mutual accessibility. The frolicking of puppies and bear cubs hones their abilities to perceive and anticipate each other's moves and to perceive and modulate their own actions and reactions, a fine preparation for the more serious activities of adulthood. We humans need to develop a similar rapport with each other as we learn to communicate, and this requires perceiving *ourselves* in the execution of these behaviors. This is what gives us a less rudimentary, more "selfy" sense of self. We need to keep track of not only which limbs are ours and what we're doing with them but also which thoughts are ours and whether we should share them with others. We can give this strange idea an almost paradoxical spin: it is like something to be you *because* you have been enabled to tell us—or refrain from telling us—what it's like to be you!

When we evolved into an *us*, a communicating community of organisms that can compare notes, we became the beneficiaries of a system of user-illusions that rendered *versions* of our cognitive processes—otherwise as imperceptible as our metabolic processes—accessible to *us* for purposes of communication. McFarland is far from being the first to express the idea that explaining ourselves to others is the novel activity that generates the R&D that creates the architecture of human consciousness, as the epigraphs at the beginning of this section show. The idea purports to provide the basis for the long-sought explanation of the evolution of distinctively human consciousness. If it is mistaken, then at least it provides a model for what a successful account would have to accomplish. A number of thinkers have recently been homing in on related and congenial ideas: among them Douglas Hofstadter's "active symbols" (1979, 1982b, 1985 [esp. pp. 646ff], 2007), and three books in 2013, by psychologist Matthew Lieberman, neuroscientist Michael Graziano, and philosopher of cognitive science Radu Bogdan.

The evolution of memes provides the conditions for the evolution of a user interface that renders the memes "visible" to the "self" which (or who) communicates with others, the self as a *center of narrative gravity* (Dennett 1991), the author of both words and deeds. If joint attention to a shared topic is required (see the discussion of Tomasello in chapter 12), there have to be things—affordances— that both the first and the second person can attend to, and *this is what makes our manifest image manifest to us*. If we didn't have to be able to talk to each other about our current thoughts and projects, and our memories of how things were, and so forth, our brains wouldn't waste the time, energy, and gray matter on an edited digest of current activities, which is what our stream of consciousness is. The self who has limited access to what is happening in its brain is well designed to entertain new memes, spread old memes, and compare notes with others. And what is this self? Not a dedicated portion of neural circuitry but rather like the end-user of an operating system. As Daniel Wegner put it, in his groundbreaking book *The Illusion of Conscious Will* (2002), "We can't possibly know (let alone keep track of) the tremendous number of mechanical influences on our behavior because we inhabit an extraordinarily complicated machine" (p. 27). Isn't it remarkable how easily we can follow Wegner into this apparently dualistic vision of ourselves as distinct *occupants* of our bodies! These machines "we inhabit" simplify things for our benefit: "The experience of will, then, is the way our minds portray their operations to us, not their actual operation" (p. 96).

Curiously, then, our *first-person* point of view of our own minds is not so different from our *second-person* point of view of others' minds: we don't see, or hear, or feel, the complicated neural machinery churning away in our brains but have to settle for an interpreted, digested version, a user-illusion that is so familiar to us that we take it not just for reality but also for the most indubitable and intimately known reality of all. That's what it is like to be us. We learn about others from hearing or reading what they say to us, and that's how we learn about ourselves as well. This is not a new idea, but keeps being rediscovered apparently. The great neurologist John Hughlings Jackson once said, "We speak, not only to tell others what we think, but to tell ourselves what we think" (1915). I, and many others, have misquoted the novelist and critic E. M. Forster as saying, "How do I know what I think until I see what I say?" Although Forster does have a version of this line in his book of criticism *Aspects of the Novel* (1927), he means it sarcastically and alludes to an earlier anecdote from which he draws it. This viral mutation of the Forster meme has spread widely, according to R. J. Heeks (2013), who shows that the quote in context is meant to disparage the writing method of André Gide:

> Another distinguished critic has agreed with Gide—that old lady in the anecdote who has accused her nieces of being illogical. For some time she could not be brought to understand what logic was, and when she grasped its true nature she was not so much angry as contemptuous. "Logic! Good gracious! What rubbish!" she exclaimed. "How can I tell you what I think till I see what I say?" Her nieces, educated young women, thought that she was passée; she was really more up-to-date than they were. (Forster 1927, p. 71)

I am happy to set the record straight—or straighter, since I can find no trace of the anecdote about the lady and her nieces—but want to suggest that Forster walked by an important, if counterintuitive, possibility without noticing it. Our access to our own thinking, and especially to the causation and dynamics of its subpersonal parts, is really no better than our access to our digestive processes; we have to rely on the rather narrow and heavily edited channel that responds to our incessant curiosity with user-friendly deliverances, only one step closer to the real me than the access to the real me that is enjoyed by my family and friends. Once again, consciousness is not just talking to yourself; it includes all the varieties of self-stimulation and reflection we have acquired and honed throughout our waking lives. These are not just things that happen in our brains; they are behaviors that we engage in (Humphrey 2000, 2006, 2011), some "instinctively" (thanks to genetic evolution) and the rest acquired (thanks to cultural evolution and transmission, and individual self-exploration).

Why do we experience things the way we do?

If, as Wegner puts it, "our minds *portray* [my emphasis] their operations to us," if (as I have just said) your individual consciousness is rather like the user-illusion on your computer screen, doesn't this imply that there *is* a Cartesian Theater after all, where this portrayal happens, where the show goes on, rather like the show you perceive on the desktop? No, but explaining what to put in place of the Cartesian Theater will take some stretching of the imagination.

We can list the properties of the tokens on the computer desktop: blue rectangular "files"; a black, arrow-shaped cursor; a yellow-highlighted word in black Times New Roman 12-point font; and so forth. What are the corresponding properties of these internal, re-identifiable private tokens in our brains? We don't know—yet. In chapter 9 we considered the way that bare meanings, with no words yet attached, could occupy our attention in consciousness, especially in the tip-of-the-tongue phenomenon. These are genuine tokens, tokens of memes or of sensation-types we are born with, or of other remembered affordances that can be recognized and re-identified even if they have no names (yet). Close your eyes and imagine a blue capital A. Done? You just created a token in your brain, but we can be sure that it isn't blue, any more than the tokens of "o" that occur in a word-processing file are round. The tokenings occur in the activity of neural circuits, and they have an important role to play in directing attention, arousing associated tokens, and modulating many cognitive activities. They contribute to directing such fundamental actions as saccadic eye movements and to initiating such higher level actions as awakening dozens of apps—memes—that are, as always, bent on getting new tokens of themselves—offspring—into the arena. Look:

tigr strp

The visual experience I just provided probably awakened the words "tiger" and "stripe" in your mind, and probably—did you notice?—these tokens had a specifically "auditory" cast, the long *i* in both words being somewhat highlighted in contrast to the almost unpronounceable visual stimuli that awakened them. These words then populated your neural workspace with *represen-*

tations of black and orange stripes that were not themselves black or orange, of course. (Were you actually aware of orange and black tiger stripes in your visual imagination? Maybe not, because the activation wasn't quite strong enough in your case, but you can be quite sure that the subpersonal [and subconscious] tokens were activated, since they would "prime" your answers to other questions in an experimental setting.)

All this subpersonal, neural-level activity is where the actual causal interactions happen that provide your cognitive powers, but all "you" have access to is the results. You can't tell by introspection how "tigr" got succeeded by "tiger" which then got succeeded by a "mental image" of a tiger, "focusing" on its stripes. When you attempt to tell us about what is happening in your experience, you ineluctably slide into a metaphorical idiom simply because you have no deeper, truer, more accurate knowledge of what was going *inside* you. You cushion your ignorance with a false—but deeply tempting—model: you simply reproduce, with some hand waving and apologies, your everyday model of how you know about what is going on *outside* you.

Here's how it happens. Let's start by reminding ourselves of something familiar and well understood: we send a reporter to observe some part of the external world—a nearby house, let's say—and report back to us by cell phone. Our phone rings, and when we answer, he tells us that there are four windows on the front side of the house, and when we ask him how he knows; he responds, "Because I'm looking right at them; I see them plain as day!" We typically don't think of then asking him how the fact that he sees them plain as day explains how he knows this fact. Seeing is believing, or something like that. We tacitly take the unknown pathways between his open eyes and speaking lips to be secure, just like the requisite activity in the pathways in the cell towers between his phone and ours. We're not curious on the occasion about how telephones work; we take them for granted. We also don't scratch our heads in bafflement over how he can just open his eyes and then answer questions with high reliability about what is positioned in front of him in the light, because we all can do it (those of us who are not blind). How does it work? We don't know and are not usually curious about it.

When we do get curious, and ask him to describe, not the outside world, but his *subjective experience* of the outside world, his *inside* world, we put him on the spot, asking him to perform a rather unnatural act, and the results—unless he is a practiced *introspector* of one school or another—tend to be disappointing: "I dunno. I look out and see the house. That is, I think I see a house; there's a house shaped thing that seems to be about fifty yards away, and it has four window-looking things on it . . . and if I close my eyes and reopen them, it's still there and . . ."

The relative accessibility and familiarity of the outer part of the process of telling people what we can see—we know our eyes have to be open, and focused, and we have to attend, and there has to be light—conceals from us the utter blank (from the perspective of introspection or simple self-examination) of the rest of the process. We have no more privileged access to that part of the process than we do to the complicated processes that maintain the connectivity between our reporter's cell phone and ours.

> *How do you know there's a tree next to the house?*
> Well, there it is, and I can see that it looks just like a tree!
> *How do you know it looks like a tree?*
> Well, I just do!
> *Do you compare what it looks like with many other things in the world before settling upon the idea that it's a tree?*
> Not consciously.
> *Is it labeled "tree"?*
> No, I don't need to "see" a label; besides, if there were a label, I'd have to read it, and know that it was the label for the thing it was on. I just know it's a tree.

Imagine now that you could just spread your toes and thereby come to have breathtakingly accurate convictions about what was currently happening in Chicago. And imagine not being curious about how that was possible.

> *How do you do it?*
> Not a clue, but it works, doesn't it? If I close my toes tight, I can no longer do it, and when I open them up again, whatever strikes my curiosity about current events in Chicago is instantly available to me. I just know.
> *What is it like?*
> Well, it's sort of like seeing and hearing, as if I were watching a remote feed television, but yet it's not quite like that. I just find all my Chicago-curiosity satisfied effortlessly.

Explanation has to stop somewhere, and at the personal level it stops here, with brute abilities couched in the familiar mentalistic language of knowing and seeing, noticing and recognizing, and the like. The problem with the first-person point of view is that it is anchored in the manifest image, not the scientific image, and cannot avail itself of the resources of the scientific image. The standard presumption is that when we challenge a reporter, "I know because I can see it" is an acceptably complete reply, but when we import the same presumption into the case where a subject is reporting on mental imagery or memory (or the imagined opened-toed Chicago clairvoyance), for instance, we create an artifact. What our questions *directly* create, or provoke, are answers, as in the dialogues above. What they indirectly create are ideologies based on those answers. You can ask yourself what your subjective experience is, and see what you want to say. Then you can decide to endorse your own declaration, to believe it, and then pursue the implications of that creed. You can do this by talking aloud to yourself, talking silently to yourself, or "just thinking" to yourself about what you are currently experiencing.

That is the extent of your access to your own experience, and it does not differ much from the access another person can have to those experiences—*your* experiences—if you decide to go public with your account. Your convictions are no doubt reliable but not infallible. Another person could help you test them and perhaps get you to adjust them in the face of further experiences. This is the way to study consciousness scientifically, and I have given it an ungainly but accurate name: *heterophenomenology*, the phenomenology of the *other* person's experience, as contrasted with *autophenomenology*, the phenomenology of one's own experience. There is a long-standing tradition to the effect that somehow autophenomenology is a more intimate, more authentic, more direct way of getting at the objects of experience, that adopting the "first-person point of view" is the key strategic move in any promising study of consciousness, but that is itself a delusion. Heterophenomenology is more accurate, more reliable, less vulnerable to illusion than autophenomenology, once you control for lying and other forms of noncooperation with the investigation, and you can get a better catalogue of *your own* experience by subjecting yourself to all the experimental circumstances in which consciousness is studied. You can be *shown* features of your own experience of which you had no inkling, both unimagined absences and weaknesses and surprising abilities you didn't know you have.

Collaborating with other investigators on the study of your own consciousness (adopting, if you like, the "second-person point of view") is the way to take consciousness, as a phenomenon, as seriously as it can be taken. Insisting, in resistance to this, that you know more about your own consciousness just because it's yours, is lapsing into dogma. By shielding your precious experience from probing, you perpetuate myths that have outlived their utility.

We ask a subject to tell us how many windows there were in his bedroom in the house he grew up in, and he closes his eyes for a moment and replies "two." We ask: How do you know? "Because I just 'looked' . . . and I 'saw' them!" He didn't literally look, of course. His eyes were closed (or were staring unfocused into the middle distance). The "eyes part" of the seeing process wasn't engaged, but a lot of the rest of the vision process was—the part that we normally don't question. It's sort of like seeing and sort of not like seeing, but just how this works is not really accessible to folk-psychological exploration, to introspection, or self-manipulation. When we confront this familiar vacuum, there is an almost irresistible temptation to postulate a surrogate world—a mental image—to stand in for the part of the real world that a reporter observes. And we can be sure of the existence of such a surrogate world in one strained sense: there has to be *something* in there—something in the neural activity—that reliably and robustly maintains lots of information on the topic in question since we can readily confirm the fact that information can be extracted from "it" almost as reliably as from real-world observation of a thing out there. The "recollected image" of the house has a certain richness and accuracy that can be checked, and its limits gauged. These limits give us important clues about how the information is actually embodied in the brain, and we mustn't just jump to the conclusion that it is embodied, as it seems to be, in an image that can be consulted.[99]

From this perspective, our utter inability to say what we're doing when we do what we call framing mental images is not so surprising. Aside from the peripheral parts about what we're doing with our eyes, we are just as unable to say what we're doing in a case of seeing the external world. We just look and learn, and that's all we know. Consider the subpersonal processes of normal vision, and note that at some point they have to account for all the things we can do, thanks to having our eyes open: we can pick blueberries, hit baseballs, recognize landmarks, navigate through novel terrain, and read, for instance. Also, thanks to these processes, our internal cortical states suffice to guide our speaking-subsystem in the framing of descriptive speech acts. We are making steady progress on this subpersonal story, even if large parts of it remain quite baffling today. We can be confident that there will be a subpersonal story that goes all the way from the eyeballs to oral reports (among other things), and in that story there will *not* be a second

99   This is where the experimental and theoretical work on mental imagery by Roger Shepard, Stephen Kosslyn, Zenon Pylyshyn, and many others comes into play.

presentation process with an ego (a self, a boss, an inner witness) observing an inner screen and then composing a report. As I never tire of saying, *all* the work done by the imagined homunculus in the Cartesian Theater has to be broken up and distributed around (in space and time) to lesser agencies in the brain.

Well then, let's try to break up the self into some plausible parts. What diminutions, what truncations, of the observing reporter might do the trick? Perhaps "agents" that were full of convictions but clueless about how they came by them—rather like oracles, perhaps, beset with judgments but with nothing to tell us (or themselves, of course) about how they arrived at that state of conviction. Ray Jackendoff addressed this issue some years ago (1996) and offered a useful prospect, recently elaborated on by Bryce Huebner and me, introducing the concept of a subpersonal *blurt* (2009):

> The key insight is that a module "dumbly, obsessively converts thoughts into linguistic form and vice versa" (Jackendoff 1996). Schematically, a conceptualized thought triggers the production of a linguistic representation that approximates the content of that thought, yielding a reflexive *blurt*. Such linguistic *blurts* are proto-speech acts, issuing subpersonally, not yet from or by the person, and they are either sent to exogenous broadcast systems (where they become the raw material for personal speech acts), or are endogenously broadcast to language comprehension systems which feed directly to the mind-reading system. Here, *blurts* are tested to see whether they should be uttered overtly, as the mind-reading system accesses the content of the *blurt* and reflexively generates a belief that approximates the content of that *blurt*. Systems dedicated to belief fixation are then recruited, beliefs are updated, and the *blurt* is accepted or rejected, and the process repeats. (Huebner and Dennett 2009, p. 149)

This is all very impressionistic and in need of further details, but a similar idea of "narrative" making has still more recently been developed by Gustav Markkula (2015), who argues persuasively that the human activity of (roughly) asking and telling ourselves what it is like to be us, creates artifacts of imagination that we take to be the "qualia" so beloved by philosophers of consciousness who yearn to reinstate dualism as a serious theory of the mind.

Hume's strange inversion of reasoning

But still, comes the objection, why does it have to be like anything to see, hear, and smell? Why does there *seem* to be an inner theater with a multimedia show going on whenever we're awake? Even if we grant that there must be a subpersonal story, in the scientific image, that can satisfactorily explain all the behaviors and emotional responses, the decisions and verbal reports I make, it must leave *me* out of the story! And putting me and my qualia back in the world, not leaving them out, is a task still to be done. The best response I know to this challenge is what I call Hume's strange inversion of reasoning, because he articulated a prescient account of one case—our experience of causation itself—long before Darwin and Turing discovered their inversions. There are complications and controversies aplenty about Hume's theory of causation, and some of his account, influential as it has been for several centuries, is largely abandoned today, but one central idea shines through and provides an important insight about the relation between the manifest and scientific images, and about the nature of our conscious experience in general, not just our experience of causation.

We seem to *see* and *hear* and *feel* causation every day, Hume notes, as when we see a brick shatter a window or hear a bell ring on being struck, but all we ever directly experience, Hume insists, is sequence: A *followed by* B, not A *causing* B. If Hume were wrong about this, animated cartoons would be impossible: when representing Bugs Bunny chomping on a carrot, the animators would have to add not just a sound track, with a loud *crunch* synchronized with the bite, but some kind of *cause track* that would *show us directly* somehow that Bugs's teeth closing actually causes, and doesn't just immediately precede, the disappearance of half the carrot and the crunching noise we hear. There is no such need, of course. Mere succession of one frame of film by the next is all it takes to create the *impression* of causation. But, Hume notes, the impression of causation we experience comes from inside, not outside; it is itself an effect of a habit of expectation that has been engrained in us over many waking hours. (Hume insisted that these expectation habits were all learned, acquired during normal infancy, but contemporary research strongly suggests that we are born with a sort of automatic causal sense, like a reflex, that is ready to "see" causation whenever our senses are confronted by the right kind of sequence of stimuli.) Seeing A, we are *wired* to expect B, and then when B happens—this is Hume's master stroke—we *misattribute* our perceptual reaction to some external cause that we are somehow directly experiencing. (We think we *see* Bugs Bunny's cartoon teeth *cause* the lopping off of the carrot.) In fact, we are succumbing to a benign user-illusion, misinterpreting our fulfilled expectation of an ensuing B as somehow coming from the outer world. This is, as Hume says, a special case of the mind's "great propensity to spread itself on external objects" (1739, I:xiv). The "customary transition" in our minds is the source of our sense of causation, a quality of "perceptions, not of objects," and, as he notes, "the contrary notion is so riveted in the mind" that it is hard to dislodge. It survives to this day in the typically unexamined assumption that all perceptual representations must be flowing inbound from outside.

Here are a few other folk convictions that need Hume's strange inversion: sweetness is an "intrinsic" property of sugar and honey, which causes us to like them; observed intrinsic sexiness is what causes our lust; it was the funniness out there in the joke that caused us to laugh (Hurley, Dennett, Adams 2011). Oversimplifying somewhat, in these instances the causes and effects in the manifest image have become inverted in the scientific image. You can't find intrinsic sweetness by studying the molecular structure of glucose; look instead to the details in the brains of sweetness seekers. It is how our brains respond that causes "us" (in the manifest image) to "project" an illusory property into the (manifest) world. There are structural, chemical properties of glucose—mimicked in saccharine and other artificial sweeteners—that cause the sweetness response in our nervous systems, but "the intrinsic, subjective sweetness I enjoy" is not an internal recreation or model of these chemical properties, nor is it a very special property in our non-physical minds that we use to decorate the perceptible things out there in the world. It is no property at all; it is a benign illusion. Our brains have tricked us into having the conviction, making the judgment, that there seems to be an intrinsically wonderful but otherwise undescribable property in some edible things: sweetness. We can recognize it, recall it, dream about it, but we can't describe it; it is ineffable and unanalyzable.

There is no more familiar and appealing verb than "project" to describe this effect, but of course everybody knows it is only metaphorical; colors aren't literally projected (as if from a slide projector) out onto the front surfaces of (colorless) objects, any more than the idea of causation is somehow beamed out onto the point of impact between billiard balls. If we use the shorthand term "projection" to try to talk, metaphorically, about the mismatch between manifest and scientific image here, what is the true long story? What is literally going on in the scientific image? A large part of the answer emerges, I propose, from the predictive coding perspective we explored briefly in chapter 8 (How do brains pick up affordances?).

Here is where Bayesian expectations can play an iterated role: our ontology (in the elevator sense) does a close-to-optimal job of cataloguing the things in the world that matter to the behavior our brains have to control. Hierarchical Bayesian predictions accomplish this, generating affordances galore: we expect solid objects to have backs that will come into view as we walk around them, doors to open, stairs to afford climbing, cups to hold liquid, and so forth. But among the things in our *Umwelt* that matter to our well-being are *ourselves*! We ought to have good Bayesian expectations about what we will do next, what we will think next, and what we will *expect* next! And we do. Here's an example:

Think of the cuteness of babies. It is not, of course, an "intrinsic" property of babies, though it seems to be. What you "project" onto the baby is in fact your manifold of "felt" dispositions to cuddle, protect, nurture, kiss, coo over, . . . that little cutie-pie. It's not just that when your cuteness detector (based on facial proportions, etc.) fires, you have urges to nurture and protect; you *expect* to have those very urges, and that manifold of expectations just *is* the "projection" onto the baby of the property of cuteness. When we expect to see a baby in the crib, we also expect to "find it cute"—that is, we *expect* to *expect* to feel the urge to cuddle it and so forth. When our expectations are fulfilled, the absence of prediction-error signals is interpreted by our brains as confirmation that, indeed, the thing in the world we are interacting with really has the properties we expected it to have. Cuteness as a property *passes the Bayesian test for being an objective structural part of the world we live in*, and that is all that needs to happen. *Any further "projection" process would be redundant.* What is special about properties like sweetness and cuteness is that their perception depends on particularities of the nervous systems that have evolved to make much of them. They have a biased or privileged role in the modulation of our control systems—we care about them, in short.

Here we must be very careful not to confuse two independent claims. The properties of sweetness and cuteness depend on features of our nervous systems and hence are in that limited sense subjective, but that must not be taken to mean that sweetness, say, is an *intrinsic* (subjective) property of conscious experience! Hume's strange inversion is wonderful but incomplete: when he spoke of the mind's "great propensity to spread itself on external objects," this should be seen not as a stopping point but as a stepping-stone to a further inversion. Hume's image brilliantly conjures up the curious vision of the mind painting the external world with the proprietary ("intrinsic") hues properly worn by the mind's internal items— impressions and ideas, in his vocabulary. But there is no such paint (which is why I once dubbed it "figment"). We need to push Hume's inversion a little harder and show that the icons of the user-illusion of our minds, unlike the user-illusion of our computers, don't need to be rendered on a screen.

A red stripe as an intentional object

One more example should at least clarify my point, if not succeed in persuading everybody—as Hume says, the contrary notion is so riveted in our minds. Look steadily at the white cross in the top panel of figure 14.1 of the color insert (following p. 238) for about ten seconds, and then switch your gaze to the white cross in the bottom panel. What do you see?

"I see an American flag, red white, and blue."
"Do you see a red stripe at the top on the right?" (Do the experiment again.)
"Yes, of course. There is a fuzzy, faint red stripe to the right of the field of blue with the stars."

But think: there are no red stripes on the page, on your retina, or in your brain. In fact, there is no red stripe anywhere. It just seems to you that there is a red stripe. Your brain "projects" a nonexistent red stripe onto the world. (It is important that the illusory stripe doesn't appear to you to be in your head; it appears to be on the page, as if projected there by a video projector in the middle of your forehead.) The phenomenon in you that is responsible for this is *not* a red stripe. It is a representation of a red stripe in some neural system of representation that we haven't yet precisely located and don't yet know how to decode, but we can be quite sure it is neither red nor a stripe. You don't know exactly what causes you to seem to see a red stripe out in the world, so you are tempted to lapse into Humean misattribution: you misinterpret your sense (judgment, conviction, belief, inclination) that you're seeing a red stripe as arising from a subjective property (a *quale*, in the jargon of philosophy) that is the *source* of your judgment, when in fact, that is just about backward. It is your *ability to describe* "the red stripe," your judgment, your willingness to make the assertions you just made, and your emotional reactions (if any) to "the red stripe" that is the source of your conviction that there *is* a subjective red stripe.

This is an instance of a kind of mistake that has been much examined in other quarters of philosophy: mistaking the *intentional object* of a belief for its *cause.* Normally, when you're not being tricked by your senses, or by tricksters, when you believe in something (in the existence of something with certain features in your vicinity, for instance), it is because just such a thing with just those features has caused you to believe in it, by stimulating your sense organs. You believe in the apple in your right hand *because* that very apple has *caused* you to believe in its existence, reflecting light into your eyes, and exerting a downward force on your palm. In this sort of normal case, we can say, carefully setting aside quibbles, that the apple, the intentional object of your belief is also the (primary, or salient) cause of your belief. But there are well-known abnormal cases: mirages, optical illusions, hallucinations, and complementary color afterimages—and pranks. Suppose a group of us decide to play a mean trick on Otto: we concoct a phony person, Dan Quale, and proceed to plant e-mails, text messages, and birthday cards to Otto from Dan Quale along with footprints, phone calls, and very close—but not too close—encounters that Otto is maneuvered into having with the elusive (in fact fictitious) Quale. Pretty soon Otto believes Dan Quale is a real person, with a fairly detailed recent biographical trail, a voice, a stature, and a lot more. That Dan Quale is the intentional object of a manifold of beliefs Otto has. The beliefs are all *about* Dan Quale, even though Dan Quale does not exist. Lots of other people exist, and lots of footprints and e-mails and all the rest, but not Dan Quale. Otto's beliefs about Dan Quale have a panoply of semi-organized causes, none of which is a person named Dan Quale. But Otto doesn't know this. He is really quite sure that Dan Quale exists—he's seen him, talked to him on the phone, has letters from him, and so forth. He wants to meet Dan Quale. Whom does he want to meet? Not any of the pranksters; he knows them all and has no particular desire to see any of them again. Dan Quale doesn't exist but is the intentional object of Otto's quest—just like Ponce de Leon searching for the Fountain of Youth. Ponce *had an idea in his mind* of the Fountain of Youth (we might say, loosely speaking), but that mental state was not the object of his quest. He already had it! He wasn't seeking an idea; he was seeking a fountain. And Otto isn't seeking his mental states about Dan Quale. He's seeking a man, driven in that search by his mental states.

Now apply the same analysis to the red stripe. If you didn't know about complementary color afterimages, you might well be naïvely confident that there really was a red stripe, visible to others from the same vantage point, in the "external" world. If *that* is what you believe, then the intentional object of your belief does not exist, and among its causes are a green striped flag picture and a lot of neural events in your visual cortex, none of which are red or even appear to be red. You are not that naïve and know very well that no such external red stripe exists, and this may mislead you to the conjecture (or, in many cases, the adamant conviction) that you couldn't be wrong—*there is* a "subjective" red stripe in your mind. You *see* it! Well, you sorta see it. In support of this theoretical postulation, you may ask: How *could* I be having an experience of a horizontal red stripe unless something somewhere is red and horizontal? The somewhat rude answer to this rhetorical question is, "Easy. If you can't conceive of this, try harder."

This moment marks the birth of qualia, an artifact of bad theorizing. The intentional object of your beliefs is not in doubt: you believe with all your heart and soul—not that there is a red stripe *out there* but—that there is a red stripe *in here* (something with the qualia of a red stripe): after all, you can "look at it," "concentrate on it," "recall it," "enjoy it," "compare it with other such things in memory." Qualia are supposed to be the somehow internal, subjective properties that we are acquainted with more directly, when we are slightly less directly acquainted with their normal external causes—real red stripes, and so on in the world. When you make this move, you are positing an *internal cause* that has the same properties as the *intentional objects* that normally cause your perceptual beliefs—except that these are private, subjective versions, somehow, of the public, objective properties of redness and so forth. But once you realize that the intentional objects of mistaken beliefs simply don't exist, anywhere, you have no need in your theory or conjecture for weird internal something-or-others with mysterious properties. Dan Quale, the intentional object of Otto's Dan-Quale-beliefs, isn't made of ectoplasm or fictoplasm or anything. Neither is Santa Claus or Sherlock Holmes. So when you seem to see a red stripe when there is no red stripe in the world as its source, there need be no *other thing* (made of red figment) that is the "real seeming" you take yourself to be experiencing.

What is there in its place? What *does* explain your conviction that there is a red stripe? The presence in your brain of *something*—yes, of course there has to be something in your brain that is responsible—but it is something in the medium of neural spike train activity, not some other medium: a remarkably salient, information-rich, subpersonal state, a token of a red stripe representation that is no more red nor striped than those neural word tokens described in Chapter 9 were loud or soft (or red or black). That is the cause of your belief in the red stripe, but it is not the intentional object of your belief (because it isn't red or striped).

But even if this is a *possible* explanation of all my subjective states, how do we know there isn't a qualia medium somehow in our minds, if not in our brains? How do we know that the "naïve" theory is mistaken? Suppose that instead of answering the rhetorical question rudely, we surrender to it, for the moment, and see what follows. Let's suppose then that there *is* a subjective property of some kind that "explains" your current introspective convictions and abilities. Let's suppose, that is, that when you experience what seems to be a horizontal red stripe, there really is, somewhere, a horizontal-shaped red quale (whatever that is) and it is somehow the cause or source of your conviction that you are experiencing a horizontal red stripe, and that this *rendering* in some unknown *medium* is caused or triggered by the confirmation (the absence of disconfirmation) of all the expectations generated by the normal operation of your visual system. Just to make the supposition as clear as possible, here is a somewhat expanded version of the purported explanation of the red afterimage effect:

> Fixating on the real green stripes in front of you for a few seconds fatigues the relevant neural circuits in the complementary color system, which then generate a false signal (red, not green), which does not get disconfirmed so long as the fatigue lasts, so somewhere fairly high in the process betwixt retina and, um . . . the philosophical conviction center, a red stripe-shaped quale is rendered, and it is the appreciation of this quale that grounds, fuels, informs, causes, underwrites the philosophical conviction that right now you are enjoying a stripe-shaped red quale.

This spells out the idea behind the rhetorical question: We need *something* like this—don't we?—to *explain* the undeniable fact that it sure seems to you there's a red stripe right now. You're not just saying this (the way a robot might, if programed to be a model of complementary color afterimages); you believe it with all your heart and soul.

Fine. So now we have qualia installed in our sketchy model of the process. What next? Something would have to *have access* to the rendering in that medium (otherwise, the rendered qualia would be wasted, unwitnessed, and unappreciated, like a beautiful painting locked in an empty room). Call whatever it is that has this access the inner observer. Now what do you suppose an appropriate reaction to this rendering by this inner observer would be? What else but the judgment that there sure seems to be a red stripe out there, part of an apparent American flag? But that conclusion had already been arrived at in the course of the nondisconfirmed expectations. A red stripe in a particular location in visual space had already been identified by the system; that conclusion was the information that informed the inner rendering (the way a bitmap informs the rendering of colors on your computer screen). The postulation of qualia is just doubling up the cognitive work to be done. There is no more work (or play) for consciousness to do.

This is the importance of always asking what I have called the Hard Question (1991, p. 255): *And then what happens?* Many theorists of consciousness stop with half a theory. If you want a whole theory of consciousness, this is the question you must ask and answer after you have delivered some item "to consciousness" (whatever you take arrival in consciousness to amount to). If instead you just stop there and declare victory, you've burdened the Subject or Self with the task of reacting, of doing something with the delivery, and you've left that task unanalyzed. If the answer you give to the Hard Question ominously echoes the answer you gave to the "easy" questions about how the pre-qualia part of the process works, you can conclude that you're running around in a circle. Stop. Reconsider.

Doggedly pursuing the idea that qualia are both the causes and the intentional objects (the *existing* intentional objects) of introspective beliefs leads to further artifactual fantasies, the most extravagant of which is the idea that unlike our knowledge of all other kinds of causation, our knowledge of mental causation is infallible and direct: we can't be wrong when we declare that our subjective beliefs about the elements of our conscious experience are caused by those very elements. We have "privileged access" to the *causes* or *sources* of our introspective convictions. No logical room for any tricksters intervening here! We *couldn't be* victimized by any illusions here! You might be a zombie, unwittingly taking yourself to have real consciousness with real qualia, but I *know* that I am not a zombie! No, you don't. The only support for that conviction is the vehemence of the conviction itself, and as soon as you allow the theoretical possibility that there *could* be zombies, you have to give up your papal authority about your own nonzombiehood. I cannot prove this, yet, but I can encourage would-be consciousness theorists to recognize the chasm created by this move and recognize that they can't have it both ways.

What is Cartesian gravity and
why does it persist?

René Descartes wasn't the first great thinker to try to give an account of the human mind, but his vision, as presented in his *Discourse on Method* (1637) and *Meditations* (1641) was so vivid and compelling that it has strongly influenced all subsequent thought on the topic. His pioneering investigations of brain anatomy were intrepid and imaginative, but his tools and methods were unable to plumb more than a tiny fraction of the complexities he exposed, and the only available metaphors—wires and pulleys and fluids rushing through hoses—were too crude to furnish his imagination with a measure of the possibilities for a materialistic model of the brain as the mind. So he can hardly be faulted for jumping to the conclusion that the mind he knew so much about "from the inside" must be some *other* thing, a thinking thing (*res cogitans*) that was not material at all. He got off on the wrong foot, then, by taking the "first-person point of view" as his direct, and even infallible, epistemic access to consciousness, a step which anchored him in a user-illusion that systematically distorted the investigation from the outset. But what else could he do? Looking at brain tissue was preposterously uninformative compared with reflecting on his thoughts, the sensations and perceptions he enjoyed or abhorred, the plans he concocted, and the emotions that altered his moods.

Ever since, philosophers, psychologists, and other scientists have relied heavily on introspection as at least a bountiful source of hints (and problems), while postponing asking the question of how this marvelous treasure trove was possible. After all, it was "self-evident"; our conscious minds are filled with "ideas" and "sensations" and "emotions" of which we have "knowledge by *acquaintance*" that— most thinkers agreed—surpassed in intimacy and incorrigibility every other kind of knowledge. The primacy of "first-person experience" has been implicit in the practices of most investigators if not always a declared axiom. Sometimes it has even been upheld as fundamental methodological wisdom: John Searle (1980) lays it down categorically: "Remember, in these discussions, always insist on the first person point of view. The first step in the operationalist sleight of hand occurs when we try to figure out how we would know what it would be like for others" (p. 451).[100] Indeed for many philosophers, the central problem has been not how to provide a scientific account of conscious experience, but how to penetrate the "veil of perception" and get from "in here" to the "external world," and Descartes's *Meditations* was the inaugural exploration of that way of thinking.

The price you pay for following Searle's advice is that you get all your *phenomena*, the events and things that have to be explained by your theory, through a channel designed not for scientific investigation but for handy, quick-and-dirty use in the rough and tumble of time-pressured life. You can learn a lot about how the brain does it—you can learn quite a lot about computers by always insisting on the desktop point of view, after all—but only if you remind yourself that your channel is systematically oversimplified and metaphorical, not literal. That means you must resist the alluring temptation to postulate a panoply of special subjective properties (typically called qualia) to which you (alone) have access. Those are fine items for our manifest image, but they must be "bracketed," as the phenomenologists say, when we turn to scientific explanation. Failure to appreciate this leads to an inflated list of things that need to be explained, featuring, preeminently, a Hard Problem that is nothing

---

100   Operationalism is the proposal by some logical positivists back in the 1920s that we don't know what a term means unless we can define an operation that we can use to determine when it applies to something. Some have declared that the Turing Test is to be taken as an *operationalist definition* of intelligence. The "operationalist sleight of hand" that Searle warns against is the claim that we really can't claim to know what consciousness is until we figure out how we can learn about the consciousness of others. Searle's alternative is itself a pretty clear case of operationalism: *If I want to know what consciousness is, my measurement operation is simple: I just look inside and whatever I see—that's consciousness!* It works for him, of course, but not for others.

but an artifact of the failure to recognize that evolution has given us a gift that sacrifices literal truth for utility.

Imagine asking for some advice and being advised, "Use your pancreas!" or "Use your liver!" You would have no idea what action to take. And when a teacher urges you to "use your brain" you'd be utterly stymied if you didn't interpret this as the directive to "use your *mind*," that thinking thing with which you are so intimately acquainted that it is hardly distinguishable from you, yourself. No wonder we are reluctant to see it as illusory; if *it* is illusory, so are *we*!

If we, our *selves*, were all "just" part of each other's user-illusions, wouldn't that imply that, really, life has no meaning? No. The manifest image that has been cobbled together by genetic evolutionary processes over billions of years, and by cultural evolutionary processes over thousands of years, is an extremely sophisticated system of helpful metaphorical renderings of the underlying reality uncovered in the scientific image. It is a user-illusion that we are so adept at using that we take it to be unvarnished reality, when in fact it has many coats of intervening interpretive varnish on it. The manifest image composes our *Umwelt*, the world we live in for almost all human purposes—aside from science. We learn about reality via the categories of colors, sounds, aromas, solid objects, sunsets and rainbows, people and their intentions, promises, threats, and assurances, institutions, and artifacts. We view our prospects, make our decisions, plan our lives, commit our futures in its terms, and this is what makes the manifest image *matter*—to us. It's life or death for us, and what else could make it matter more? Our own reflections on all this are necessarily couched in terms of meanings, or contents, the only readily usable "access" we have to what goes on between our ears and behind our eyes.

If Searle has stressed for years the importance of adopting a *first*-person point of view (what do *I* see? what is it like to be *me*?), a philosopher who early appreciated the "antiseptic virtue" of adopting the *third*-person point of view (what does *it* want? what are *they* conscious of?) is Jonathan Bennett, whose little monograph, *Rationality*

(1964), set out to study human rationality indirectly by studying the (non)rationality of bees! By insisting on adopting the third-person point of view, and starting with a humble-but-oh-so-competent creature, Bennett minimizes the temptation to be taken in by the unavoidable practice of *identification by content* that is the hallmark of introspective methods.

This is what I mean: if you want to talk about your own mental states, you *must* identify them by their content: "Which idea? My idea of HORSE. Which sensation? My sensation of *white*." How else? There is no way you can identify your own mental states "from the inside" as, for instance, *concept J47* or *color-sensation 294*. By taking for granted the content of your mental states, by picking them out *by their content*, you sweep under the rug all the problems of indeterminacy or vagueness of content. Reading your own mind is too easy; reading the mind of a honeybee places the problems front and center. We won't have a complete science of consciousness until we can align our manifest-image identifications of mental states by their contents with scientific-image identifications of the subpersonal information structures and events that are causally responsible for generating the details of the user-illusion we take ourselves to operate in.

Here is another source of the staying power of the Cartesian point of view. By presupposing that we normal folks are rational and hence have understanding (not just competence), we tacitly endorse our everyday use of the intentional stance as not just practical and valuable but as also *the plain truth* about human minds. This puts us in distinguished company: we are intelligent designers, rather like the Intelligent Designer who designed us. We wouldn't want to give up that honor, would we? And so we normally give both ourselves and our fellow human beings more credit for the authorship of our creations, and more blame for our misdeeds, than would be warranted by an unvarnished view of the causation involved.

Besides—and here comes a big payoff—the Cartesian point of view fits nicely, it seems, with traditional ideas of free will and moral responsibility. I recognized the penetration of this hunch some years ago when I tried to uncover the submerged grounds for resistance to any version of the account of consciousness sketched here, and I discovered that many cognitive scientists—not only laypeople—were reluctant even to *consider* such doctrines. After I had laid to rest a number of their objections, they would often eventually let the cat out of the bag: "But what about free will? Wouldn't a completely materialistic account of consciousness show that we can't be morally responsible?" No, it wouldn't, and here, in a nutshell, is why (I've had my say about that question in two books and many articles, so on this occasion I will be brief). The traditional view of free will, as a personal power somehow isolated from physical causation, is both incoherent and unnecessary as a grounds for moral responsibility and meaning. The scientists and philosophers who declare free will a fiction or illusion are right; it is part of the user-illusion of the manifest image. That puts it in the same category with colors, opportunities, dollars, promises, and love (to take a few valuable examples from a large set of affordances). If free will is an illusion then so are they, and for the same reason. This is not an illusion we should want to dismantle or erase; it's where we live, and we couldn't live the way we do without it. But when these scientists and philosophers go on to claim that their "discovery" of this (benign) illusion has important implications for the law, for whether or not we are responsible for our actions and creations, their arguments evaporate. Yes, we should shed the cruel trappings of retributivism, which holds people *absolutely* responsible (in the eyes of God) for their deeds; we should secure in its place a sane, practical, defensible system of morality and justice that still punishes when punishment is called for, but with a profoundly different framing or attitude. One can get a sense of this by asking yourself: If—because free will is an illusion—no one is ever responsible for what they do, should we abolish yellow and red cards in soccer, the penalty box in ice hockey, and all the other penalty systems in sports?

The phenomena of free will and moral responsibility, worthy items in the ontology of the human manifest image, survive robustly once we strip off some of the accrued magic of tradition and reground them in scientific reality. Regardless of whether I am right in my claim that the phenomena of free will and responsibility, properly reformed and understood, are defensible elements in our most serious ontology, we need to recognize how the fear that these important features of everyday life are doomed generates a powerful undercurrent of resistance that distorts the imaginations of people trying to figure out what human consciousness is.

Nicholas Humphrey is the writer who has most forcefully drawn attention to the unargued prejudice people tend to have in favor of "spiritual" accounts in his *Soul Searching: Human Nature and Supernatural Belief* (1995). As he shows, people tend to treat belief in the supernatural as not only excusable but also morally praiseworthy. Credulity is next to godliness. The human mind, many think, is the last bastion of what is sacred in this world, and to explain it would be to destroy it, so to be safe, we had better declare consciousness conveniently out of bounds to science. And as we have seen, there is tremendous emotional resistance to any considerations that might seem to cast doubt on the presumption that nonhuman animals— the grizzly bear, the puppy, the dolphin—have minds that are, if not *just* like ours, at least enough like it to provide them some moral standing, perhaps not moral responsibility but at least the right not to be ill treated.

For all these reasons, resisting the force of Cartesian gravity takes some strenuous exercises of the imagination, and we need to avoid overshooting in our resistance as well. We have to set aside some intuitions that seem *almost* indubitable and take seriously some suggestions that seem, at first, paradoxical. That's difficult, but science has shown us again and again how to do it. Even school children can shed pre-Copernican and pre-Galilean intuitions without flinching, and by the time they are teenagers they can get comfortable replacing some of their Newtonian intuitions with Einsteinian intuitions. Getting comfortable with quantum physics is still a work in progress—for me, I confess, in spite of much mental calisthenics.

Easier (for me, in any case) is embracing Darwin's strange inversion of reasoning, and Turing's, and Hume's. By offering a sketch of the *causes* of Cartesian gravity, I have tried to help the unpersuaded find a vantage point from which they can diagnose their own failures of imagination and overcome them.

Human consciousness is unlike all other varieties of animal consciousness in that it is a product in large part of cultural evolution, which installs a bounty of words and many other thinking tools in our brains, creating thereby a cognitive architecture unlike the "bottom-up" minds of animals. By supplying our minds with systems of representations, this architecture furnishes each of us with a perspective—a user-illusion—from which we have a limited, biased access to the workings of our brains, which we involuntarily misinterpret as a rendering (spread on the external world or on a private screen or stage) of both the world's external properties (colors, aromas, sounds, . . .) and many of our own internal responses (expectations satisfied, desires identified, etc.). The incessant torrent of self-probing and reflection that we engage in during waking life is what permits us, alone, to comprehend our competences and many of the reasons for the way the world is. Thanks to this infestation of culturally evolved symbiont information structures, our brains are empowered to be intelligent designers, of artifacts and of our own lives.

15

The Age of Post-Intelligent Design

What are the limits of our comprehension?

If the brain were so simple we could understand it, we would be so simple we couldn't.

—Emerson M. Pugh, *The Biological Origin of Human Values*

Human comprehension has been steadily growing since prehistoric times. For forty millennia and more, we have been living in the age of intelligent design—crafting pots, tools, weapons, clothes, dwellings and vehicles; composing music and poetry; creating art; inventing and refining agricultural practices; and organizing armies, with a mixture of dutiful obedience to tradition, heedless and opportunistic improvisation, and knowing, intentional, systematic R&D, irregularly punctuated with moments of "inspired" genius. We applaud intelligent design in all arenas, and aspire from infancy to achieve recognition for our creations. Among the artifacts we have created is the concept of God, the Intelligent Designer, in our own image. That's how much we value the intelligent designers in our societies.

We recognize the value of these fruits of our labors, and our laws and traditions have been designed to create an artificial

environment in which we can preserve and enhance our accumulated wealth. It is a perfectly *real* environment, not a merely *virtual*
world, but it is no less an artifact, and we call it *civilization*. We are
well aware that our species is no more immune to extinction than
any other, and that we might all expire in a plague or technological
catastrophe, or—only slightly less dire—we might destroy civilization and revert to the "state of nature," as Hobbes called it, where
life is nasty, brutish, and short. But has it ever occurred to us that
this age of comprehending heroes might simply expire while *Homo
sapiens* went right on reproducing, outliving its name, the *knowing*
hominin? There are some unsettling signs that we are becoming
overcivilized, thanks to the ingenuity of all the labor-saving inventions on which we have become so dependent, and are entering the
age of post-intelligent design.

The epigraph to this chapter, Pugh's clever reflection on the
audacious project of using our brains to understand our brains, has
been attributed, in different versions, to many authors, and it may
well have been independently reinvented many times. A variant species includes one of my favorite George Carlin one-liners:

> For years and years and years, I thought *my brain* was the most
> important organ of my body, until one day I thought, hmm.
> *Look who's telling me* that!

Is there an important truth lurking here, or is this just another
way Cartesian gravity has of diverting us from the quest to understand human consciousness? Noam Chomsky (1975) has proposed a
distinction that has drawn a lot of attention and converted a few dis-
ciples: on the one hand there are *problems*, which we can solve, and
on the other hand there are *mysteries*, which we can't. Science and
technology have solved many *problems* about matter, energy, gravity,
electricity, photosynthesis, DNA, and the causes of tides, tubercu-
losis, inflation, and climate change, for instance. Progress is being
made on thousands of other problems. But no matter how advanced
our scientific problem-solving becomes, there are problems that are

beyond human comprehension altogether, which we might better call mysteries. Consciousness tops Chomsky's list of mysteries, along with free will. Some thinkers—now known as *mysterians*—have been eager to take this unargued claim on his authority and run with it. There may possibly be mysteries systematically beyond the ken of humanity now and forever, but the argument in favor of this disheartening conclusion put forward by Chomsky and the other mysterians, while superficially appealing, is not persuasive. Here is a rendering of the Argument from Cognitive Closure, drawing on various versions:

> It is an undeniable fact of biology that our brains are strictly limited, like the brains of all other creatures. From our *relatively* Olympian vantage point, we can see that fish are clever in their way but obviously not equipped to understand plate tectonics, while dogs draw a blank when it comes to the concept of democracy. Every brain must suffer from *cognitive closure* (McGinn 1990) with regard to a host of issues that are simply beyond it, unimaginable and unfathomable. We don't have a miraculous *res cogitans* between our ears, but just lots of brain tissue subject to the laws of physics and biology.

So far, so good. I have no objection to anything in this beginning, which articulates uncontroversial facts about the physical world. But then it goes on:

> It would be profoundly unbiological—wouldn't it?—to suppose that our human brains were somehow exempt from these natural limits. Such delusions of grandeur are obsolete relics from our prescientific past.

This would be compelling if it weren't for the equally obvious biological fact that human brains have become equipped with add-ons, thinking tools by the thousands, that multiply our brains' cognitive powers by many orders of magnitude. Language, as we have seen, is the key invention, and it expands our individual cognitive powers by providing a medium for uniting them with all the cognitive powers of every clever human being who has ever thought. The smartest chimpanzee never gets to compare notes with other chimpanzees in her group, let alone the millions of chimpanzees who have gone before.

The key weakness of the Argument from Cognitive Closure is the systematic elusiveness of good examples of mysteries. As soon as you frame a question that you claim we will never be able to answer, you set in motion the very process that might well prove you wrong: you raise a topic of investigation. While your question may get off on the wrong foot, this fact about it is likely to be uncovered by the process of trying to answer it. The reflexive curiosity of philosophy—going "meta" about every question asked—is almost a guarantee that there will be something approximating exhaustive search—sometimes no better than random, sometimes brilliantly directed—of variations on the question that might prove more perspicuous. Asking better and better questions is the key to refining our search for solutions to our "mysteries," and this refinement is utterly beyond the powers of any languageless creature. "What is democracy?" A dog will never know the answer, to be sure, but it will never even understand the question. We can understand the questions, which radically changes our quest, turning unimaginable mysteries into problems worth trying to solve.

Perhaps in consideration of this practically limitless power of language to extend our grasp, Chomsky has recently moderated his position (2014). While there is a "conceptual distinction" between problems and mysteries, "we accept the best explanations science can give us" even when we can't imagine how they work. "It doesn't matter what we can conceive any more. We've given up on that." In other words, thanks to language, and the tools of science it makes possible, we can have a good scientific theory of some perplexing phenomenon, a theory worth endorsing, while not *really* understanding it. That is, we could be justified in accepting it, even betting our lives on the truth of its predictions, while not understanding how or why it works. Whether or not this revision would appeal to the mysterians, it is still an interesting idea. But could it be true?

Downloading thousands of culturally acquired thinking tools may permit us to magnify our powers dramatically, but doesn't that just postpone cognitive closure? How much schooling can an individual mind/brain absorb? Here we notice an ambiguity in the mysterian surmise. Is it the claim that there are mysteries that *no single human mind* can comprehend or that there are mysteries that are beyond the *pooled* comprehension of whole civilizations? The idea of distributed comprehension—the idea that *we as a group* might understand something that none of us individually could fully understand—strikes some people as preposterous, so loyal are they to the ideal of the do-it-yourself intelligent designer, the genius who has it all figured out. This is a *motif* with many familiar variations. A painting by the *studio* of Rembrandt is less valuable, less a masterpiece, than a painting by Rembrandt himself. Novels almost always have solo novelists (the hardworking editors who reshape the penultimate draft often don't even get recognized), and when creative teams succeed—Gilbert and Sullivan, Rodgers and Hammerstein—they almost always involve a division of labor: one does the lyrics and one does the music, for instance. But coauthored works of nonfiction have been common for centuries, and, in the sciences today, there are fields in which a single-authored paper is a rarity.

One of the founding documents of cognitive science, *Plans and the Structure of Behavior* (1960) was written by George Miller, Eugene Galanter, and Karl Pribram. Its introduction of the idea of a TOTE unit (Test-Operate-Test-Exit) was an early formalization of the idea of feedback loops, and it played an important role in the transition from behaviorism to cognitive modeling. For all its early influence, it is seldom read these days, and a joke once commonly heard was that Miller wrote it, Galanter formalized it, and Pribram believed it. The very idea that such a division of labor might be possible—and successful—was risible at the time, but no longer. Science is full of collaborations in which theoreticians—who understand the math—and experimentalists and fieldworkers—who rely on the theoreticians without mastering the math—work together to create multiple-author works in which many of the details are only partially understood by each author. Other combinations of specialized understanding flourish as well.

So let's imagine a multi-author, multi-volume book, *The Scientific Theory of Consciousness*, that comes to be uncontroversially accepted by the scientific community. The volumes become, if you like, the standard textbooks on human consciousness, used in courses across neuroscience, psychology, philosophy, and other fields where consciousness is an important phenomenon—but although some intrepid souls claim to have read through the whole boxed set, nobody claims to have mastered all the levels of explanation. Would it count as vindicating Chomsky's mysterianism—consciousness is still a mystery, since no single theorist can really conceive of it—or as knocking over yet another of the mysterians' candidates for unfathomable mysteries?

We have learned, as civilization has progressed, that a division of labor makes many things possible. A single person, or family, can make a simple house or canoe, and a small community can raise a barn or a stockade, but it takes hundreds of workers with dozens of different talents to build a cathedral or a clipper ship. Today peer-reviewed papers with hundreds of coauthors issue from CERN and other bastions of Big Science. Often none of the team members can claim to have more than a bird's-eye-view comprehension of the whole endeavor, and we have reached a point where even the most brilliant solo thinkers are often clearly dependent on their colleagues for expert feedback and confirmation.

Consider Andrew Wiles, the brilliant Princeton mathematician who in 1995 proved Fermat's Last Theorem, a towering achievement in the history of mathematics. A close look at the process he went through, including the false starts and unnoticed gaps in the first version of his proof, demonstrates that this triumph was actually the work of many minds, a community of communicating experts, both collaborating and competing with each other for the glory, and without the many layers of achieved and battle-tested mathematics on which Wiles's proof depended, it would have been impossible for Wiles or anyone else to judge that the theorem had, in fact, been proven.[101] If you are a lone wolf mathematician and think you have proved Fermat's Last Theorem, you have to consider the disjunction: *Either I have just proved Fermat's Last Theorem or I am going mad*, and since history shows that many brilliant mathematicians have been deluded in thinking they had succeeded, you have to take the second alternative seriously. Only the formal concession and subsequent congratulations of your colleagues could or should put that anxiety to rest.

Even artists, poets, and musicians, treasured for their individuality and "divine afflatus," do best when they have an intimate working knowledge and understanding of the works of their predecessors. The twentieth-century rebels who made something of a fetish of defying "the canon," attempting to create works of hyper-originality, are either fading into oblivion or proving that the staying power of their creations is due to more appreciation of their traditions than they were willing to admit. The painter Philip Guston once eloquently acknowledged his indirect dependence on all he had extracted and digested from the intelligent design of others:

> I believe it was John Cage who once told me, "When you start working, everybody is in your studio—the past, your friends, enemies, the art world, and above all, your own ideas—all are there. But as you continue painting, they start leaving, one by one, and you are left completely alone. Then, if you're lucky, even you leave." (2011, p. 30)

What kind of limits are there on the brains we were born with? For now we can note that whether the limits are practical or absolute, we have found, and largely perfected, a work-around that

---

101   Simon Singh, "The Whole Story," http://simonsingh.net/books/fermats
-last-theorem/the-whole-story/, is by far the most accessible account I have found, an edited version of an essay published in *Prometheus* magazine.

postpones the confrontation with our frailty: collaboration, both systematic and informal. Groups can do things, and (arguably) understand things, that individuals cannot, and much of our power derives from that discovery. It is possible to resist this idea of group comprehension, but only—so far as I can see—by elevating comprehension to a mystical pinnacle that has little or nothing to do with the comprehension we rely on, in ourselves and others, to solve our problems and create our masterpieces. This blunts the edge of the mysterian argument. By ignoring the power of collaborative understanding, it raises an obsolete issue, viewing comprehension as an all-or-nothing blessing, which it seldom, if ever, is.

Descartes, in his day, was very concerned to secure *perfect* comprehension for his "clear and distinct" ideas, and for this, he argued, he needed to prove the existence of a benign, all-powerful, nondeceiving God. His thought-experimental hypothesis was that there might otherwise be an evil demon bent on deceiving him about his most confidently held convictions, and this "possibility in principle" fixed his method—and tied his hands quite securely. For Descartes, only the kind of certainty we reserve for dead-obvious mathematical truths ($2 + 2 = 4$, a plane triangle has three straight sides and interior angles adding up to two right angles) was good enough to count as *real* knowledge, and only the crystalline comprehension we can have of the individual steps of a maximally simple proof could count as *perfect* comprehension. Where Descartes relied on God as the guarantor of his proofs, today we rely on the improbability of multiple thinkers arriving, by different routes, at the *same* wrong result. (Note that this is an application of the principle that dictated taking at least three chronometers on your sailing ship, so that when they began to disagree on what time it was, the odd one out was very probably wrong.) We tend to overlook the importance of the fact that we have voluminous experience of many people independently coming up with the same answer to multiplication and division questions, for instance, but if that were not our experience, no amount of analytic reflection on the intrinsic necessity of mathematics—or the existence of a benign God—would convince us to trust our calculations. Is arithmetic a sound system of calculation? *Probably*—so very probably that you can cheerfully bet your life on it.

"Look Ma, no hands!"

Civilization advances by extending the number of important operations we can perform without thinking about them.
—Alfred North Whitehead

What I cannot create, I do not understand.
—Richard Feynman

I have argued that the basic, bottom-up, clueless R&D done by natural selection has gradually created cranes—labor-saving products that make design work more effective—which have opened up Design Space for further cranes, in an accelerating zoom into the age of intelligent design, where top-down, reflective, reason-formulating, systematic, foresighted R&D can flourish. This process has succeeded in changing the balance of selective forces that shape us and all other organisms and in creating highly predictive theories that retrospectively explain the very processes of their own creation. This cascade of cranes is not a miracle, not a gift from God, but a natural product of the fundamental evolutionary process, along with the other fruits of the Tree of Life.

To review, over several thousand years we human beings have come to appreciate the powers of individual minds. Building on the instinctive habits of all living things, we distinguish food from poison, and, like other locomoting organisms, we are extra sensitive to animacy (guided movement) in other moving things, and more particularly to the beliefs and desires (information and goals) that guide those movements, tracking as best we can *who knows what* and *who wants what*, in order to guide our own efforts at hide and seek. This native bias is the genetic basis for the intentional stance, our practice of treating each other as rational agents guided by largely true beliefs and largely well-ordered desires. Our uninterrupted interest in these issues has generated the *folk psychology* that we rely on to make sense of one another. We use it to predict and explain not just the repetitive behaviors we observe in our neighbors and ourselves, and the "forced moves" that anyone would be stupid not to execute, but even many of the strokes of "insight" that are the mark of "genius." That is, our expectations are very frequently confirmed, which cements our allegiance to the intentional stance, and when our expectations are confounded, we tend to fall back on "explanations" of our failure that are at best inspired guesswork and at worst misleading mythmaking.

We encourage our children to be curious and creative, and we self-consciously identify the ruts and boundaries in our own thinking processes so that we can try to overcome them. The minds we prize most are the minds that are neither too predictable (boring, unchallenging) nor too chaotic. *Practice makes perfect*, and we have invented games that encourage us to rehearse our mind-moves, such as chess and Go and poker, as well as prosthetic devices—telescopes, maps, calculators, clocks, and thousands of others—that permit us to apply our mind-moves in ever more artificial and sophisticated environments. In every sphere of inquiry and design, we have highly organized populations of experts collaborating to create and perfect theories and other artifacts, and we have adopted traditions and market mechanisms to provide the time, energy, and materials for these projects. We are the intelligent designers living in a world intelligently designed for intelligent designers by our ancestors. And now, after centuries of dreaming about this prospect, we have begun designing and producing artifacts that can design and produce artifacts (that can design and produce artifacts . . .).

Many hands make light work. That's another adage that is as applicable to mind-work as to barn-raising, but we are now discovering that *hands-off* design work is often not only lighter and easier, but, thanks to the artifacts we have recently designed, more—in a word—competent.

Nanotechnology, the new and burgeoning field of chemistry and materials science that is beginning to construct artifacts atom by atom, has featured the brilliant and patient handiwork of pioneers who have developed sophisticated tools for manipulating (moving, cutting, isolating, immobilizing, etc.) bits of matter at the nanometer scale (a nanometer is one-billionth of a meter). Like GOFAI before it, nanotechnology began as top-down intelligent design, a brilliant method for *hand making* large inventories of "miracle drugs, "smart materials," and other nanorobots. It has had triumphs and will surely have many more, especially with the new nanotool of CRISPR at its disposal (see, for a brief nontechnical introduction, Specter 2015). Like PCR (polymerase chain reaction), the technique that revolutionized gene sequencing, CRISPR, which permits genes to be edited and spliced together more or less ad lib, replaces highly sophisticated and laborious techniques, a labor-saving invention that reduces the time and effort required by orders of magnitude. Jennifer Doudna of UC Berkeley and Emmanuelle Charpentier, now of the Max Planck Institute, are two of the supremely intelligent designers of this new crane.

These techniques, like those developed by Pixar and other computer-animation companies, create push-button automated processes that replace thousands of days of *brilliant drudgery* (not an oxymoron—extremely talented people doing extremely repetitive but demanding work). When Walt Disney Productions released *Snow White and the Seven Dwarfs* in 1937, it astonished the world with its lifelike animations, the fruits of the labors of hundreds of talented animators, working in highly organized teams to solve the problems of getting lifelike action, complete with all the jiggle and bounce of reality, onto thousands of cels, or frames, of film. Those heroic artistic sweatshops are historical relics now; the talents one needed to be that kind of frame-by-frame animator are now largely obsolete, and the same is true about the talents of early molecular biologists who ingeniously isolated gene fragments and patiently coaxed them to divulge their sequences, one codon at a time. Similar tales of the automation of heretofore tedious intellectual tasks could be told about other fields, from astronomy to textual analysis. In general these tasks amount to gathering, sorting, and refining data on a large scale, giving the human data-interpreter more free time to reflect on the results. (I will never forget the time I spent a day in the laboratory of a promising young neuroscientist gathering data from experiments on macaques with chronically implanted electrodes in their brains. Late in the day I asked him a question about his views on a theoretical controversy then boiling about the role of activity in various brain areas on modulating consciousness; he sighed and replied, "I don't have time to think! I'm too busy running experiments.") The new techniques that minimize the brilliant drudgery are amazingly competent, but they are still tools—not robotic colleagues—utterly dependent on the decisions and intentions of intelligent tool users and directors—lab directors and studio directors.

Today, however, we are beginning to appreciate, and exploit, the truth of Orgel's Second Rule: Evolution is cleverer than you are. The bottom-up, tireless algorithms of natural selection (and their close cousins) are being harnessed by intelligent designers in many fields to do the dirty work of massive searches, finding precious needles in all kinds of haystacks.

Some of this exploration involves actual biological natural selection in the laboratory. For instance, Frances Arnold, at Caltech, does award-winning protein engineering, creating novel proteins by breeding them, in effect. She has devised systems for generating huge populations of variant genes—DNA recipes for proteins—and then testing the resulting proteins for talents never before found in Nature.

> We are developing new tools for protein engineering and using them to create new and improved catalysts for carbon fixation, sugar release from renewable polymers such as cellulose, and biosynthesis of fuels and chemicals. (Arnold 2013)

What she recognized was that since the space of *possible* proteins is Vastly greater than the space of *existing* proteins, there are almost certainly traversable paths of gradual evolution that have never yet been explored to destinations that would provide us with wonder drugs, wonder tissues, and wonder catalysts, a host of nanorobots that can do our bidding once we find them. When she was a graduate student, a senior scientist warned her that there were no known proteins that had anything like the properties she was hoping to obtain. "That's because there's never been selection for them" was her intrepid reply.

> Consequently, these enzymes may open up whole new regions of "chemical space" that could not be explored in previous medicinal chemistry efforts. (Arnold 2013)

Frances Arnold has created a technology for generating novel proteins—long sequences of amino acids that, when linked together, fold into evolved nanorobots with remarkable powers. A strikingly different technology developed by David Cope, emeritus professor of music at University of California at Santa Cruz, uses a computer program to generate novel music—long sequences of notes and chords that, when linked together, yield musical compositions with remarkable powers: imitation Bach, imitation Brahms, Wagner, Scott Joplin, and even musical comedy songs (Cope and Hofstadter 2001). How "original" are the thousands of compositions churned out by Cope's EMI (Experiments in Musical Intelligence)? Well, they are clearly derivative and involve heavy borrowing from the great composers whose styles they mimic, but they are nonetheless not mere copies, and not mere copies with a few random mutations; they are much better than that. They involve taking in and digesting the works of the masters and extracting from that computational process the core, the gist, the style of that composer, and then composing novel pieces in that style, a very sophisticated musical feat. (Try it, if you are a musician, and see: compose a piano piece that is pure Chopin or Mozart—or Count Basie or Erroll Garner. Simple parody or caricature is not that hard, especially of a jazz pianist as mannered as Erroll Garner, but composing good music requires significant musical insight and talent—in a human composer.)

Experiments in Musical Intelligence, designed and improved by

Cope over more than three decades, has produced many well-constructed piano pieces, songs, symphonies, and other compositions, all untouched by Cope's editorial hand, except for the final aesthetic judgment of which of the bounty most deserve to be heard. I arranged for a nice test of EMI—one of many that have been conducted over the years—at the Montreal Bach Festival in December of 2015, where I gave a talk summarizing some of the main points of this book, and closing with a performance, by Ukrainian pianist Serhiy Salov, of four short piano pieces. I told the audience of over 300 Bach lovers that at least one was by Bach and at least one was by EMI, and after they were played, the audience voted (with eyes closed). Two EMI inventions were declared genuine Bach by dozens in the audience—maybe not a majority in either case, but close—and when I asked those who had got them all right to stand, only about a dozen rose to a round of applause.

Cope, like Arnold, sets the goals and decides when to declare victory but otherwise is hands off. These very different research projects are thus variations on Darwin's theme of *methodical selection*, in which the selective force of natural selection is focused through the nervous system of a discerning, purposeful, foresighted agent. But the heavy lifting is left to the inexorable pattern-finding powers of the algorithms of natural selection, in cascades of uncomprehending generate-and-test cycles that gradually refine the search process.

Since natural selection is a substrate-neutral family of algorithms that can occur in any medium with a few simple properties, evolution *in silico* (simulated in a computer program) is sometimes faster and cheaper than evolution *in vivo*, and can be applied to almost any question or problem you formulate. Pedro Domingos's recent book *The Master Algorithm* (2015) is a lively and authoritative survey of all the new varieties of Darwinian and—shall we say—*Darwinesque* systems of "machine learning" or "deep learning." Domingos simplifies the stampede by identifying five "tribes of machine learning": symbolists (the descendants of GOFAI); connectionists (the descendants of McCulloch and Pitts's logical neurons—see chapter 6, p. 110); evolutionaries (John Holland's genetic algorithms and their offspring); Bayesians (those who have devised practical algorithms for achieving the competences of hierarchical networks of Bayesian expectation-generators); and analogizers (the descendants of the nearest-neighbor algorithm invented by Fix and Hodges [1951]). In different ways, all five echo the pattern of natural selection. Obviously, being computer based, they all are ultimately composed of Turing's simplest comprehension-free competences (conditional branching and arithmetic), and except perhaps for the creations of the symbolists, they are bottom-up, needle-in-haystack-finding repetitive churnings that gradually, with great reliability, home in on good (or good enough) answers to daunting problems.

John Holland, the beloved and recently deceased mentor of dozens of brilliant cognitive scientists and computer scientists at the Santa Fe Institute and the University of Michigan, invented genetic algorithms, where the parallels with evolution by natural selection are obvious (and delicious to Darwinians): there is the generation of huge populations of variant codings, which are each given the opportunity to make progress on solving a problem, with the winners of this environmental test getting to reproduce (complete with a sort of sex, and "crossover" like the random gene-mixing that occurs in the creation of our sperm and ova). Over many generations, the competence of the initially randomly concocted strings of computer code is multiplied and refined. Genetic algorithms have been used to design the fascinating evolved virtual creatures of Karl Sims (see the many websites devoted to this serious playground of imagination) and also such no-nonsense engineering triumphs as circuit boards and computer programs. Domingos notes (p. 133) that in 2005, a patent was issued for a genetically designed factory-optimization system (General Leslie Groves, they are closing in on you). Architects have begun using genetic algorithms to optimize the functional properties of buildings—for instance their strength, safety, use of materials, and use of light and energy.

In scientific research, machine learning is being harnessed to solve, by brute force, problems that are simply beyond human analysis. It is noteworthy that the late Richard Feynman, brilliant theoretical physicist that he was, spent many of his last days exploring the use of supercomputers to solve problems in physics that defied his wizardry with equations. And he lived to see his maxim rendered more or less obsolete. While it may still be true that what you cannot create you cannot understand, creating something is no longer the guarantee of understanding that it used to be. It is now possible to make—very indirectly—things that do what we want them to do but which we really cannot understand. This is sometimes called black box science. You buy the latest high-tech black box, feed in your raw data, and out comes the analysis; the graphs are ready to print and publish, yet you couldn't explain in detail how it works, repair it if it broke, and *it is not clear that anybody else could either.* This possibility was always staring us in the face, of course; things we "make by hand" (boats, bridges, engines, symphonies) we can (more or less) control as we construct, understanding each step along the way. Things we "make the old-fashioned way" (children, grandchildren, . . .) defy our comprehension because of our obliviousness to the details of the processes that create them. Today, we are generating brain-children, and brain-grandchildren, and brain-great-grandchildren that depend on processes we cannot follow in detail, even when we can prove that the results are trustworthy.

The use of computers in research has generated several quite distinct kinds of challenges to Feynman's maxim. Some mathematical proofs executed (in part or entirely) by computer are simply too long for a single human mathematician to check each step, which has been, for good reason, the standard of acceptance for several thousand years. What should give? A famous case is the computer-assisted proof in 1976 of the four-color theorem first discussed by Möbius in the 1840s: Any map of contiguous areas separated by shared boundaries can be colored in with just four colors such that the same color never appears on both sides of a boundary. After many failed proofs by some of the world's cleverest mathematicians, Kenneth Appel and Wolfgang Haken harnessed a computer to deal with and dismiss the nearly 2,000 different possibilities that had to be ruled out, as they themselves had proven. For some years, their proof was not generally accepted because it seemed to involve a humanly uncheckable series of steps by the computer, but the wide consensus among mathematicians today is that this is a proven theorem. (And alternative proofs have since been constructed, also using computers.) This was an "intuitive" result: nobody had managed to produce a clear counterexample in spite of person-centuries of trying, so most mathematicians figured it was true long before it was proven. But there are also counterintuitive theorems that have been proven with the help of computers. In chess, for instance, the fifty-move rule, which declared any game a draw that proceeded for fifty moves without a capture or pawn move, was long viewed by experts as more than generous, but it was upset by the discovery—thanks to computer analysis—of some mating nets (winning series of moves by one side that cannot be escaped once entered), that involve no captures or pawn moves and exceed fifty moves by hundreds of moves. After some experimentation with revising the number, it was officially decided by FIDE, the international governing body of the game, to keep the fifty-move rule, since it was a possibility in principle that would never come up in serious (human) play.

The computer programs that analyze chess positions, like those that prove mathematical propositions, are traditional, top-down, intelligently designed programs. The programs Domingos is mainly concerned with are strikingly different. As he puts it, "We can think of machine learning as the inverse of programming, in the same way that the square root is the inverse of the square, or integration is the inverse of differentiation" (p. 7). Yet another strange inversion of reasoning, or better, another *instance* of the basic Darwinian inversion: competence without comprehension. The "central hypothesis" of Domingos's book is beyond audacious:

All knowledge—past, present, and future—can be derived from data by a single, universal learning algorithm.

I call this learner the Master Algorithm. If such an algorithm is possible, inventing it would be one of the greatest scientific achievements of all time. In fact, the Master Algorithm is the last thing we'll ever have to invent because, once we let it loose, it will go on to invent everything else that can be invented. All we need to do is provide it with enough of the right kind of data, and it will discover the corresponding knowledge. (p. 25)

It isn't clear if he really means it, since he soon backpedals:

OK, some say, machine learning can find statistical regularities in data, but it will never discover anything deep, like Newton's laws. It arguably hasn't yet, but I bet it will. (p. 39)

A wager, then, not a hypothesis he thinks he can secure by reasoned argument at this time in his book. In any case it's useful to have his articulation of this extreme prospect, since no doubt many people have half-formed nightmares about just such eventualities, and it will help shine some skeptical light on them. We can begin with the claim Domingos responds to with his wager. Can machine learning ever advance beyond finding "statistical regularities"? Domingos bets that it will, but what is the basis for his optimism?

The structure of an intelligent agent

We have seen how Bayesian networks are excellent at teasing out the statistical regularities that matter to the organism—its affordances. Animal brains, equipped by natural selection with such networks, can guide the bodies they inhabit with impressive adroitness, but by themselves have scant ability to *adopt novel perspectives*. That, I have argued, requires an infestation of memes, cognitive competences (habits, ways) designed elsewhere and installed in brains, habits that profoundly change the cognitive architecture of those brains, turning them into minds, in effect. So far, the only animals whose brains are thus equipped are *Homo sapiens*.

Just as the eukaryotic cell came into existence in a relatively sudden instance of *technology transfer*, in which two independent legacies of R&D were united in a single stroke of symbiosis to create a big leap forward, the human mind, the *comprehending* mind, is—and had to be—a product of symbiosis, uniting the fruits of two largely independent legacies of R&D. We start, I have argued, with animal brains that have been, to a considerable extent, redesigned to be excellent bases for thinking tools designed elsewhere—memes. And chief among them, words. We *acquire* most of our words unconsciously, in this sense: we were not aware of learning seven new words a day when we were young, and for most words—words that aren't explicitly introduced to us—we only gradually home in on their meanings thanks to unconscious processes that find the patterns in our early experience of these words. Once we *have* the words, we can begin *using* them, but without necessarily noticing what we are doing. (For every word in your vocabulary, there was a debutante token, the first time you used it either in a public speech act or an internal monologue or musing. How often have you been aware of doing that with the new words that have entered your working vocabulary in, say, the last decade? Ever?) Once words become our familiar tools, not mere sounds associated with contexts, we can start using them to create new perspectives on everything we encounter.

So far, there are few signs of this sort of phenomenon emerging in the swiftly growing competences of deep-learning machines. As Domingos stresses, learning machines are (very intelligently) designed to avail themselves of Darwinesque, bottom-up processes of self-redesign. For IBM's Watson, the program that beat champion contestants Ken Jennings and Brad Rutter in the *Jeopardy* television quiz program in 2011, the words it was competent to string together into winning answers were not *thinking tools* but just nodes located in a multidimensional space of other nodes, not so much memes as fossil traces of human memes, preserving stupendous amounts of information about human beliefs and practices without themselves being active participants in those practices. Not yet, but maybe someday. In short, Watson doesn't yet think thoughts using the words about which it has so much statistical information. Watson can answer questions (actually, thanks to *Jeopardy*'s odd convention, Watson can compose questions to which the *Jeopardy* clues are the answers: *Jeopardy*: "The capital of Illinois," contestant: "What is Springfield?"), but this is not having a conversation.

It is the capacity to self-monitor, to subject the brain's patterns of reaction to yet another round (or two or three or seven rounds) of pattern discernment, that gives minds their breakthrough powers.[102] In the current environment of machine learning, it is the human users, like Frances Arnold in her protein workshop and David Cope with his Experiments in Musical Intelligence, the designers and other operators of the machines, who are in position to play those roles, evaluating, adjusting, criticizing, tweaking, and discarding the dubious results that often emerge. They are the critics whose quality control activities provide the further selective forces that could "in principle" raise these systems into comprehension, promoting them from tools into colleagues, but that's a giant step or series of giant steps. From this perspective we can see more clearly that our meme-infested minds harbor *users*, critics of the raw deliverances of our animal brains without which we would be as naïve as other mammals, who are wily on their own turf but clueless in the face of serious novelty.

Curiosity killed the cat, according to one meme, and animal curiosity, driven bottom-up by the presence of novelty, is an important high-risk, high-payoff feature in many species, but only human beings have the capacity for controlled, systematic, foresighted, hypothesis-testing curiosity, a feature of the users that emerge in each brain, users who can exploit their brains' vast capacity for uncovering statistical regularities. The user-illusion of conscious-

---

102 Watson, notably, does have some specialized self-monitoring layers: it must assess its "confidence" in each candidate answer and can also adjust its threshold of confidence, taking more or less risk in answering. I don't want to sell Watson short; it is a multilayered, multitalented system.

ness plays the same role in each of us that the human-computer interfaces of Watson and other deep-learning systems play; they provide something like a showcase for talents, a "marketplace of ideas" in which real-time evaluation and competition can enhance the speed and resolution of quality control.

So human *thinking*—as Darwin recognized in the phenomenon he called *methodical selection*—can speed up natural selection by focusing selective power through the perceptual and *motivational* systems of domesticators. Frances Arnold isn't just *farming* her proteins; she is doing intensive, directed *breeding* of the new proteins. That should alert us to the prospect that our marvelous minds are not immune to fads and fancies that bias our self-redesign efforts in bizarre and even self-defeating ways. Like the preposterous plumage encouraged into existence by methodical pigeon fanciers, and the pathetic disabilities patiently engineered into various "toy" dog varieties, human beings can—often with the help of eager accomplices—shape their minds into grotesque artifacts that render them helpless or worse.

This suggests—but certainly does not prove—that without us machine *users* to interpret the results, critically and insightfully, deep-learning machines may grow in competence, surpassing animal brains (including ours) by orders of magnitude in the bottom-up task of finding statistical regularities, but never achieve (our kind of) comprehension. "So what?" some might respond. "The computer kind of bottom-up comprehension will eventually submerge the human kind, overpowering it with the sheer size and speed of its learning." The latest breakthrough in AI, AlphaGo, the deep-learning program that has recently beaten Lee Seedol, regarded by many as the best human player of Go in the world, supports this expectation in one regard if not in others. I noted that Frances Arnold and David Cope each play a key quality-control role in the generation processes they preside over, as critics whose scientific or aesthetic judgments decide which avenues to pursue further. They are, you might say, piloting the exploration machines they have designed through Design Space. But AlphaGo itself does something similar, according to published reports: its way of improving its play is to play thousands of Go games against itself, making minor exploratory mutations in them all, *evaluating* which are (probably) progress, and using those evaluations to adjust the further rounds of practice games. It is just another level of generate and test, in a game that could hardly be more abstract and insulated from real-world noise and its attendant concerns, but AlphaGo is learning to make "intuitive" judgments about situations that have few of the hard-edged landmarks that computer programs excel at sorting through. With the self-driving car almost ready for mass adoption—a wildly optimistic prospect that not many took seriously only a few years ago—will the self-driving scientific exploration vehicle be far behind?

So practical, scientific, and aesthetic judgment may soon be off-loaded or outsourced to artificial agents. If Susan Blackmore is right, this abdication or alienation of human judgment is already being pioneered in the digital world of popular music and Internet memes—*tremes*, in her new terminology (see ch. 11, p. 237). There has been a surfeit of memes for centuries, with complaints dating to the earliest days of the printing press, and ever since then people have been willing to pay for filters, to strain out the time-wasting, mind-clogging, irritating memes one way or another. Don't try to read every poem by every poet; wait until some authoritative poet or critic puts forth a highly selective anthology. But which authority should you trust? Which meets your needs and tastes? You can subscribe to a literary journal that regularly reviews such volumes, along with the work of individual poets. But which literary journal should you trust? Check out their reputations as reported in still other journals you can buy. There is a living to be made catering to the expressed needs of individual meme seekers, and if business slacks off, you can try to create a new need which you can then service. All very familiar. But we are entering a new era where the filters and second-guessers and would-be trendsetters may not be people at all, but artificial agents. This will not suit everybody, as we will see in the next section. But that may not stop hierarchical layers of such differential replicators from burgeoning, and then we may indeed face the calamity encountered by the Sorcerer's Apprentice and the multiplying brooms.

In an IBM television advertisement, Watson, "in conversation" with Bob Dylan, says that it can "read 800 million pages a second." Google Translate, another icon among learning machines, has swept aside the GOFAI systems that were top-down attempts to "parse" and interpret (and thereby understand, in at least a pale version of human comprehension) human language; Google Translate is an astonishingly swift, good—though still far from perfect—translator between languages, but it is entirely parasitic on the corpus of translation that has already been done by human bilinguals (and by volunteer bilingual informants who are invited to assist on the website). Prospecting for patterns, sifting through millions of passages that have already been well translated (well enough to be found online), Google Translate settles into a likely (probable) acceptable translation *without any actual comprehension at all.*

This is a contentious claim that requires some careful unpacking. There is a joke about an Englishman who says, "The French call it a *couteau,* the Italians call it a *cotello,* the Germans call it a *Messer,* and we English call it a *knife*—which, after all, is *what it is*!" The self-satisfied insularity of the Englishman is attached to something he knows—what a knife is—that has no counterpart (it seems) in the "knowledge" of Google Translate. In the jargon of cognitive science, the Englishman's knowledge of the meaning of "knife" (and "*couteau*" and the other synonymous terms) is *grounded* in nonlinguistic knowledge, acquaintance, familiarity with knives, with cutting and sharpening, the heft and feel of a carving knife, the utility of a pen knife, and so forth. The Englishman has, with respect to the word *knife,* what you probably do not have with respect to the English word *snath,* even if you know that the Germans call it a *Sensenwurf.* But hang on. Google Translate no doubt has a rich body of data about the contexts in which "knife" appears, a neighborhood that features "cut," "sharp," "weapon" but also "wield," "hold," "thrust," "stab," "carve," "whittle," "drop," and "bread," "butter," "meat" and "pocket," "sharpen," "edge," and many more terms, with their own neighborhoods. Doesn't all this refined and digested information about linguistic contexts amount to a *sort* of grounding of the word "knife" after all? Isn't it, in fact, the only sort of grounding most of us have for technical terms such as "messenger RNA" and "Higgs boson"? It does guide the translation process down ever more appropriate channels. If you rely on Google Translate to be your bilingual interpreter, it will hardly ever let you down. Doesn't that demonstrate a serious degree of comprehension? Many will say NO! But if we are to keep this adamant denial from being a mere ritualistic defiance of the machine, there had better be something the *real* comprehender can *do* with his or her (or its) comprehension that is beyond the powers of Google Translate.

Maybe this will do the trick: It is one thing to *translate* a term paper from English to French, and another thing to *grade* that term paper. That won't do, because Thomas Landauer's pioneer development of "latent semantic analysis" (see, e.g., Littman et al. 1998) has already created a computer program that does precisely that (Rehder et al. 1998). A professor sets an essay question on an exam, and writes an A+ answer to the question, which is then given to both the computer program and a human teaching assistant as an example of what a good essay on the topic should be. (The A+ answer is *not* shown to the examination takers, of course.) Then the program and the teaching assistant grade all the student answers, and the program's grades are in closer agreement to the professor's judgments than the grades submitted by the teaching assistant, who is presumably a budding expert in the field. This is unnerving, to say the least; here is a computer program that doesn't understand English, let alone the subject matter of the course, but simply (!) on the basis of sophisticated statistical properties exhibited by the professor's model essay evaluates student answers to the same questions with high reliability. Assessment competence without comprehension! (Landauer has acknowledged that *in principle* a student could contrive an essay that was total nonsense but that had all the right statistical properties, but any student who could do that would deserve an A+ in any case!)

Then how about the task of simply *having a sensible conversation with a human being*? This is the classic Turing Test, and it really can separate the wheat from the chaff, the sheep from the goats, quite definitively. Watson may beat Ken Jennings and Brad Rutter, two human champions in the TV game *Jeopardy*, but that isn't free-range conversation, and the advertisements in which Watson chats with Jennings or Dylan or a young cancer survivor (played by an actress) are scripted, not extemporaneous. A real, open-ended conversation between two speaking agents is, as Descartes (1637) observed in his remarkably prescient imagining of a speaking automaton, a spectacular exhibition of great—if not *infinite*, as Descartes ventured to say—cognitive skills. Why? Because ordinary human conversation is conducted in a space of possibilities governed by Gricean free-floating rationales! I may not expressly *intend* that you *recognize* my intention to get you to *believe* that what I am saying is true (or that it is irony, or kidding, or transparent exaggeration), but if you are not up to that kind of recognition, and if you are also not up to originating speech acts having similar free-floating rationales that explain your own responses and challenges, you will not be a convincing, or engaging, conversationalist. Grice's iterated layers of cognition may not accurately represent real-time features underlying a *performance*, but they do specify a *competence*.

A participant in a high-powered conversation has to be able to recognize patterns in its own verbal actions and reactions, to formulate hypothetical scenarios, to "get" jokes, call bluffs, change the subject when it gets tedious, explain its earlier speech acts when queried, and so forth. All this requires—if it is not magic—the representation of all the discriminations that must be somehow *noticed* in order to provide the settings for the mental and ultimately verbal actions taken. For instance, if you don't or can't notice (in some minimal, possibly subliminal sense) that I'm joking, you can't go along with the gag, except by accident. Such noticing is not simply a matter of your animal brain making a discrimination; it is rather some kind of heightened influence that not only retrospectively distinguishes what is noticed from its competitors at the time but also,

just as importantly, contributes to the creation of a *noticer*, a relatively long-lasting "executive," not a place in the brain but a sort of
political coalition that can be in control over the subsequent competitions for some period of time. Such differences in aftermath
("And then what happens?") can be striking.

Imagine being asked to complete partial words (the "word stem
completion paradigm") and being confronted with

sta _____

or

fri _____

What occurred to you? Did you think *start* or *stable* or *station*, for
instance, and *frisk*, *fried*, *friend*, or *frigid*? Suppose that, a few seconds before you get the word stem to complete, an answer word is
flashed very briefly on the screen, thus:

staple

followed a second later by *sta*_____. The temptation to answer
"staple" would be huge, of course. But suppose the experimenters
said at the outset of the experiment: "If you've just seen a word
flash, *don't* use it as the answer!" Then, not surprisingly, you can
overcome the urge and say something different most of the time,
maybe *stake* or *starlight*. You are unlikely to say "staple" because you
can follow the exclusion policy recommended by the experimenter.
But that's only if you notice (or are conscious of) the flashed word.
If the word is flashed for only 50msec and followed by a "mask"—a
patterned screen—for 500msec, you are *more* likely to say "staple"
in spite of trying to follow the instruction (Debner and Jacoby
1994). Notice how clean the design of this experiment is: two
groups of subjects, one group told to *use* the "priming" word if it's
a good answer, and the other group told *not* to use the "priming"
word if it's a good answer. Both groups get primes that are either

50msec or 500msec long, followed by a mask. The mask doesn't mask the 500msec-long primes—subjects notice them, can report them, are conscious of them, and either use them or refrain from using them, as requested. But the mask does mask the 50msec-long primes—subjects claim not to have seen any priming word at all (this is a standard "backward masking" phenomenon). In both kinds of cases, short and long, there is discrimination by the brain of the prime, as shown by the fact that in the exclusion condition, the short-duration primes *raise* the probability of being the answer given, while the long-duration primes *lower* that probability. Dehaene and Naccache (2001) note "the impossibility for subjects [i.e., executives] to strategically use the unconscious information."

My claim, then, is that deep learning (so far) *discriminates* but doesn't *notice*. That is, the flood of data that a system takes in does not have relevance for the system except as more "food" to "digest." Being bedridden, not having to fend for itself, it has no goals beyond increasing its store of well-indexed information. Beyond the capacity we share with Watson and other deep learning machines to acquire know-how that depends on statistical regularities that we extract from experience, there is the capacity to *decide* what to search for and *why*, given one's current aims. It is the absence of *practical* reason, of intelligence harnessed to pursue diverse and shifting and self-generated ends, that (currently) distinguishes the truly impressive Watson from ordinary sane people. If and when Watson ever reaches the level of sophistication where it can enter fully into the human practice of reason-giving and reason-evaluating, it will cease to be merely a tool and become a colleague. And then Watson, not just Watson's creators and maintainers, would be eligible for being considered *responsible* for its actions.

The way in which deep-learning machines are dependent on human understanding deserves further scrutiny. In chapter 8, (pp. 157–160), we considered Deacon's bold criticism of traditional AI: would-be mind designers who abstract away from the requirements of energy capture and self-protection thereby restrict their search to parasitic systems, always dependent on human maintenance—they are tools, not colleagues. Now we can see that the kind of comprehension AI systems are currently exhibiting—and it is becoming breathtakingly competitive with the best human comprehension—is also parasitic, strictly dependent on the huge legacy of human comprehension that it can tap. Google Translate would be nowhere without the millions of good translations by bilinguals that it draws upon, and Watson's inhumanly compendious factual knowledge is likewise dependent on all those millions of pages it sucks off the Internet every day. To adapt once again Newton's famous remark, these programs stand on the shoulders of giants, helping themselves to all the cleverness on display in the earlier products of intelligent design.

This is nicely illustrated by a problem I set for my students when Watson beat Jennings and Rutter on *Jeopardy*. I gave them the assignment of coming up with questions that they thought would stump Watson but be easy for Jennings or Rutter (or any normal human being). (It's notable that on *Jeopardy*, the rules had to be adjusted in Watson's favor. For instance, the problems set to Watson were all verbal, with no vision or hearing required.) The likely stumpers (in my opinion) involve imagination in one way or another:

> The happy word you could spell on the ground using a cane, a hula hoop, and a slingshot.
>
> Ans. What is joy?
>
> Make a small animal huge by changing one letter in its name.
>
> Ans. What is mouse to moose?
>
> The numeral, between 0 and 9, that would make a good shape for a hot tub and adjacent swimming pool.
>
> Ans. What is 8?

I have better examples, but I wouldn't publish them—or put them on the Internet—since then Watson would probably sweep them up and keep them on hand for a future contest! Watson doesn't need an imagination when it can poach on the imaginations of others. Note that in this regard Watson is deeply Darwinian: neither Watson nor natural selection depend on foresight or imagination because they are driven by processes that relentlessly and without comprehension extract information—statistical patterns that can guide design improvements—from what has already happened. They are both blind to types of events that haven't happened in the scope of their selection processes. Of course if there really is nothing new under the sun, this is no limitation, but human imagination, the capacity we have to envision realities that are not accessible to us by simple hill climbing from where we currently are, does seem to be a major game-changer, permitting us to *create*, by *foresighted design*, opportunities and, ultimately, enterprises and artifacts that could not otherwise arise. A conscious human mind is not a miracle, not a violation of the principles of natural selection, but a novel extension of them, a new crane that adjusts evolutionary biologist Stuart Kauffman's concept of the *adjacent possible*: many more places in Design Space are adjacent *to us* because we have evolved the ability to think about them and either seek them or shun them. The unanswered question for Domingos and other exponents of deep learning is whether learning a sufficiently detailed and dynamic *theory* of agents with imagination and reason-giving capabilities would enable a system (a computer program, a Master Algorithm) to generate and exploit the abilities of such agents, that is to say, to generate all the morally relevant powers of a person.[103]

My view is (still) that deep learning will not give us—in the next fifty years—anything like the "superhuman intelligence" that has attracted so much alarmed attention recently (Bostrom 2014; earlier invocations are Moravec 1988; Kurzweil 2005; and Chalmers 2010; see also the annual Edge world question 2015; and Katchadourian

---

103   Spike Jonze's science fiction film, *Her* (2013), starring Joaquin Phoenix and the voice of Scarlett Johansson as the Siri-like virtual person on his cell phone with whom he falls in love, is one of the two best speculative explorations of this question to date, along with Alex Garland's *Ex Machina* (2015).

2015). The accelerating growth of competence in AI, advancing under the banner of deep learning, has surprised even many professionals in the field, not just long-time commentators and critics such as myself. There is a long tradition of hype in AI, going back to the earliest days, and many of us have a well-developed habit of discounting the latest "revolutionary breakthrough" by, say, 70% or more, but when such high-tech mavens as Elon Musk and such world-class scientists as Sir Martin Rees and Stephen Hawking start ringing alarm bells about how AI could soon lead to a cataclysmic dissolution of human civilization in one way or another, it is time to rein in one's habits and reexamine one's suspicions. Having done so, my verdict is unchanged but more tentative than it used to be. I have always affirmed that "strong AI" is "possible in principle"— but I viewed it as a negligible practical possibility, because it would cost too much and not give us anything we really needed. Domingos and others have shown me that there may be feasible pathways (technically and economically feasible) that I had underestimated, but I still think the task is orders of magnitude larger and more difficult than the cheerleaders have claimed, for the reasons presented in this chapter, and in chapter 8 (the example of Newyorkabot, p. 164).

So I am not worried about humanity creating a race of super-intelligent agents destined to enslave us, but that does not mean I am not worried. I see other, less dramatic, but much more likely, scenarios in the immediate future that are cause for concern and call for immediate action.

What will happen to *us*?

Artifacts already exist—and many more are under development— with competences so far superior to any human competence that they will usurp our authority as experts, an authority that has been unquestioned since the dawn of the age of intelligent design. And when we cede hegemony to these artifacts, it will be for very good reasons, both practical and moral. Already it would be *criminally negligent* for me to embark with passengers on a transatlantic sailboat cruise without equipping the boat with several GPS systems. Celestial navigation by sextant, compass, chronometer, and Nautical Almanac is as quaint a vestige of obsolete competence as sharpening a scythe or driving a team of oxen. Those who delight in such skills can indulge in them, using the Internet to find one another, and we celestial navigators can prudently bring our old-fashioned gear along, and practice with it, on the off chance that we will need a backup system. But we have no right to jeopardize lives by shunning the available high-tech gadgets.

We all still learn the multiplication table up to 12x12 and how to use it for larger numbers (don't we?), and we can do long division problems with pencil and paper, but few know how to execute an algorithm for extracting a square root. So what? Don't waste your effort and brain cells on tasks you can order by pressing a few keys or just asking Google or Siri. The standard response to the worriers is that when educating our children, we do need to teach them the *principles* of all the methods we ourselves are still adept at using, and for this comprehension a certain minimal level of actual experience with the methods is practically valuable, but we can (probably) get the principles to sink in without subjecting our children to old-fashioned drudgery. This seems to make good sense, but how far does it generalize?

Consider medical education. Watson is just one of many computer-based systems that are beginning to outperform the best diagnosticians and specialists on their own turf. Would *you* be willing to indulge your favorite doctor in her desire to be an old-fashioned "intuitive" reader of symptoms instead of relying on a computer-based system that had been proven to be a hundred times more reliable at finding rare, low-visibility diagnoses than any specialist? Your health insurance advisor will oblige you to submit to the tests, and conscientious doctors will see that they must squelch their yearnings to be diagnostic heroes and submit to the greater authority of the machines whose buttons they push. What does this imply about how to train doctors? Will we be encouraged to jettison huge chunks of traditional medical education—anatomy, physiology, bio-chemistry—along with the ability to do long division and read a map? *Use it or lose it* is the rule of thumb cited at this point, and it has many positive instances. Can your children read road maps as easily as you do or have they become dependent on GPS to guide them? How concerned should we be that we are dumbing ourselves down by our growing reliance on intelligent machines?

So far, there is a fairly sharp boundary between machines that enhance our "peripheral" intellectual powers (of perception, algo-rithmic calculation, and memory) and machines that at least pur-port to replace our "central" intellectual powers of comprehension (including imagination), planning, and decision-making. Hand cal-culators; GPS systems; Pixar's computer graphics systems for inter-polating frames, calculating shadows, adjusting textures and so forth; and PCR and CRISPR in genetics are all quite clearly on the peripheral side of the boundary, even though they accomplish tasks that required substantial expertise not so long ago. We can expect that boundary to shrink, routinizing more and more cognitive tasks, which will be fine *so long as we know where the boundary currently is*. The real danger, I think, is not that machines more intelligent than we are will usurp our role as captains of our destinies, but that we will *over*-estimate the comprehension of our latest thinking tools, prematurely ceding authority to them far beyond their competence.

There are ways we can reinforce the boundary, even as we allow it to shrink, by making it salient to everyone. There are bound to be innovations that encroach on this line, and if recent history is our guide, we should expect each new advance to be oversold. There are antidotes, which we should go to some lengths to provide. We know that people are quick to adopt the intentional stance toward anything that impresses them as at all clever, and since the default assumption of the intentional stance is rationality (or comprehen-sion), positive steps should be taken to *show* people how to temper their credulity when interacting with an anthropomorphic system. First, we should expose and ridicule all gratuitous anthropomor-phism in systems, the cute, ever-more-human voices, the perky (but canned) asides. *When you are interacting with a computer, you should know you are interacting with a computer.* Systems that deliberately conceal their shortcuts and gaps of incompetence should be deemed fraudulent, and their creators should go to jail for committing the crime of creating or using an artificial intelligence that impersonates a human being.

We should encourage the development of a tradition of hyper-modesty, with all advertising duly accompanied by an obligatory list of all known limits, shortcomings, untested gaps, and other sources of cognitive illusion (the way we now oblige pharmaceutical companies to recite comically long lists of known side effects whenever they advertise a new drug on television). Contests to expose the limits of comprehension, along the lines of the Turing Test, might be a good innovation, encouraging people to take pride in their ability to suss out the fraudulence in a machine the same way they take pride in recognizing a con artist. Who can find the quickest, surest way of exposing the limits of this intelligent tool? (Curiously, the tolerance and politeness we encourage our children to adopt when dealing with strangers now has the unwanted effect of making them gullible users of the crowds of verbalizing nonagents they encounter. They must learn that they should be aggressive and impolitely inquisitive in dealing with newly encountered "assistants.")

We should hope that new cognitive prostheses will continue to be designed to be parasitic, to be tools, not collaborators. Their only "innate" goal, set up by their creators, should be to respond, constructively and transparently, to the demands of the user. A cause for concern is that as learning machines become more competent at figuring out what we, their users, probably intend, they may be designed to conceal their "helpful" extrapolations from us. We already know the frustration of unwanted automatic "correction" of what are deemed typographical errors by spell-checkers, and many of us disable these features since their capacity to misconstrue our intentions is still too high for most purposes. That is just the first layer of semi-comprehension we have to deal with.

There are already stress lines developing around current developments that call for comment. Google has a program for enhancing their search engine by automatically figuring out what it suspects the user *really* meant by entering the input symbol string. (http:// googleblog.blogspot.com/2010/01/helping-computers-understand-language.html) This would no doubt be useful for many purposes, but not for all. As Douglas Hofstadter has noted, in an open letter to a former student, then at Google working on this project:

> It worries me and in fact deeply upsets me that Google is trying to undermine things that I depend on on a daily basis, all the time.
>
> When I put something in quotes in a Google search, I always mean it to be taken literally, and for good reason. For example (just one type of example among many), as a careful writer, I am constantly trying to figure out the best way of saying something in one language or another, and so I will very frequently check two possible phrasings against each other, in order to see whether one has a high frequency and the other a very low frequency. This is an extremely important way for me to find things out about phrasings. If Google, however, doesn't take my phrasings literally but feels free to substitute other words willy-nilly inside what I wrote, then I am being royally misled if I get a high count for a certain phrase. This is very upsetting to me.
>
> I want machines to be reliably mechanical, not to be constantly slipping away from what I ask them to do. Supposed "intelligence" in machines may at times be useful, but it may also be extremely unuseful and in fact harmful, and in my experience, the artificial intelligence (here I use the word "artificial" in the sense of "fake," "non-genuine") that these days is put into one technological device after another is virtually always a huge turnoff to me.
>
> I am thus not delighted by what your group is doing, but in fact greatly troubled by it. It is just one more attempt to make mechanical devices not reliable as such. You ask Google to do

X, presuming that it will do precisely X, but in fact it does Y instead, where Y is what it "thinks" you meant. To me, this kind of attempt to read my mind is fantastically annoying if not dangerous, because it almost never is correct or even in the right ballpark. I want machines to remain reliably mechanical, so that I know for sure what I am dealing with. I don't want them to try to "outsmart" me, because all they will do in the end is mislead and confuse me. This is a very elementary point, and yet it seems to be being totally ignored at Google (or at least in your group). I think it is a very big mistake. (Personal correspondence, 2010, with Abhijit Mahabal)

At the very least, such systems should (1) prominently announce when they are trying to be "mind readers" not merely "mechanical" and (2) offer users the option of turning off the unwanted "comprehension" in the same way you can turn off your all-too-enterprising spell-checker. A "strict liability" law might provide a much needed design incentive: anyone who *uses* an AI system to make decisions that impact people's lives and welfare, like users of other dangerous and powerful equipment, must be trained (and bonded, perhaps) and held to higher standards of accountability, so that it is always in their interests to be extra-scrupulously skeptical and probing in their interactions, lest they be taken in by their own devices. This would then indirectly encourage designers of such systems to make them particularly transparent and modest, since users would shun systems that could lead them down the primrose path to malpractice suits.

There is another policy that can help keep the abdication of our cognitive responsibilities in check. Consider technology for "making us stronger": on the one hand, there is the bulldozer route, and on the other hand, the Nautilus machine route. The first lets you do prodigious feats while still being a 98-pound weakling; the second makes you strong enough to do great things on your own. Most of the software that has enhanced our cognitive powers has been of the bulldozer variety, from telescopes and microscopes to genomesequencers and the new products of deep learning. Could there also be Nautilus-type software for bulking up the comprehension powers of individuals? Indeed there could be, and back in 1985, George Smith and I, along with programmers Steve Barney and Steve Cohen, founded the Curricular Software Studio at Tufts with the aim of creating "imagination prostheses," software that would furnish and discipline students' minds, opening up notorious pedagogical bottlenecks, allowing students to develop fluent, dynamic, accurate models in their imagination of complex phenomena, such as population genetics, stratigraphy (interpreting the geological history of layers of rock), statistics, and the computer itself. The goal was to make systems that, once mastered, could be set aside, because the users had internalized the principles and achieved the level of comprehension that comes from intensive exploration. Perhaps it is now time for much larger projects designed to help people think creatively and accurately about the many complex phenomena confronting us, so that they can be independently intelligent, comprehending users of the epistemological prostheses under development, not just passive and uncritical beneficiaries of whatever technological gifts they are given.

———

We have now looked at a few of the innovations that have led us to relinquish the *mastery of creation* that has long been a hallmark of understanding in our species. More are waiting in the wings. We have been motivated for several millennia by the idea expressed in Feynman's dictum, "What I cannot create, I do not understand." But recently our ingenuity has created a slippery slope: we find ourselves indirectly making things that we only partially understand, and they in turn may create things we don't understand at all. Since some of these things have wonderful powers, we may begin to doubt the value—or at least the preeminent value—of understanding. "Comprehension is so *passé*, so *vieux jeux*, so old-fashioned! Who needs understanding when we can all be the beneficiaries of artifacts that save us that arduous effort?"

Is there a good reply to this? We need something more than tradition if we want to defend the idea that comprehension is either intrinsically good—a good in itself, independently of all the benefits it indirectly provides—or practically necessary if we are to continue living the kinds of lives that matter to us. Philosophers, like me, can be expected to recoil in dismay from such a future. As Socrates famously said, "the unexamined life is not worth living," and ever since Socrates we have taken it as self-evident that achieving an ever greater understanding of *everything* is our highest professional goal, if not our highest goal absolutely. But as another philosopher, the late Kurt Baier, once added, "the over-examined life is nothing to write home about either." Most people are content to be the beneficiaries of technology and medicine, scientific fact-finding and artistic creation without much of a clue about how all this "magic" has been created. Would it be so terrible to embrace the *over*-civilized life and trust our artifacts to be good stewards of our well-being?

I myself have been unable to concoct a persuasive argument for the alluring conclusion that comprehension is "intrinsically" valuable—though I find comprehension to be one of life's greatest thrills—but I think a good case can be made for preserving and enhancing human comprehension *and* for protecting it from the artifactual varieties of comprehension now under development in deep learning, for deeply *practical* reasons. Artifacts can break, and if few people understand them well enough either to repair them or substitute other ways of accomplishing their tasks, we could find ourselves and all we hold dear in dire straits. Many have noted that for some of our high-tech artifacts, the supply of repair persons is dwindling or nonexistent. A new combination color printer and scanner costs less than repairing your broken one. Discard it and start fresh. Operating systems for personal computers follow a similar version of the same policy: when your software breaks or gets corrupted, don't bother trying to diagnose and fix the error, unmutating the mutation that has crept in somehow; reboot, and fresh new versions of your favorite programs will be pulled up from safe storage in memory to replace the copies that have become defective. But how far can this process go?

Consider a typical case of uncomprehending reliance on technology. A smoothly running automobile is one of life's delights; it enables you to get where you need to get, on time, with great reliability, and for the most part, you get there in style, with music playing, air conditioning keeping you comfortable, and GPS guiding your path. We tend to take cars for granted in the developed world, treating them as one of life's constants, a resource that is always available. We plan our life's projects with the assumption that of course a car will be part of our environment. But when your car breaks down, your life is seriously disrupted. Unless you are a serious car buff with technical training you must acknowledge your dependence on a web of tow-truck operators, mechanics, car dealers, and more. At some point, you decide to trade in your increasingly unreliable car and start afresh with a brand new model. Life goes on, with hardly a ripple.

But what about the huge system that makes this all possible: the highways, the oil refineries, the automakers, the insurance companies, the banks, the stock market, the government? Our civilization has been running smoothly—with some serious disruptions—for thousands of years, growing in complexity and power, Could it break down? Yes, it could, and to whom could we then turn to help us get back on the road? You can't buy a new civilization if yours collapses, so we had better keep the civilization we have running in good repair. Who, though, are the reliable mechanics? The politicians, the judges, the bankers, the industrialists, the journalists, the professors—the leaders of our society, in short—are much more like the average motorist than you might like to think: doing their local bit to steer their part of the whole contraption, while blissfully ignorant of the complexities on which the whole system depends. According to the economist and evolutionary thinker Paul Seabright (2010), the optimistic tunnel vision with which they operate is not a deplorable and correctable flaw in the system but an enabling condition. This distribution of partial comprehension is not optional. The edifices of social construction that shape our lives in so many regards depend on our myopic confidence that their structure is sound and needs no attention from us.

At one point Seabright compares our civilization with a termite castle. Both are artifacts, marvels of ingenious design piled on ingenious design, towering over the supporting terrain, the work of vastly many individuals acting in concert. Both are thus by-products of the evolutionary processes that created and shaped those individuals, and in both cases, the design innovations that account for the remarkable resilience and efficiency observable were not the brainchildren of individuals, but happy outcomes of the largely unwitting, myopic endeavors of those individuals, over many generations. But there are profound differences as well. Human *cooperation* is a delicate and remarkable phenomenon, quite unlike the almost mindless cooperation of termites, and indeed quite unprecedented in the natural world, a unique feature with a unique ancestry in evolution. It depends, as we have seen, on our ability to engage each other within the "space of reasons," as Wilfrid Sellars put it. Cooperation depends, Seabright argues, on trust, a sort of almost invisible social glue that makes possible both great and terrible projects, and this trust is not, in fact, a "natural instinct" hard-wired by evolution into our brains. It is much too recent for that.[104] Trust is a byproduct of social conditions that are at once its enabling condition and its most important product. We have bootstrapped ourselves into the heady altitudes of modern civilization, and our natural

---

104 Seabright points out that no band of chimpanzees or bonobos could tolerate the company of strangers—proximity to conspecifics who are not family or group members—that we experience with equanimity virtually every day, a profound difference. The (relative) calm with which many ungulate species can crowd together at a watering hole is not trust; it is instinctual indifference to familiar nonpredators, more like our attitude toward trees and bushes than our attitude toward other human beings in the landscape. Trust is a cultural phenomenon, as I observed in chapter 7.

emotions and other instinctual responses do not always serve our new circumstances.

Civilization is a work in progress, and we abandon our attempt to understand it at our peril. Think of the termite castle. We human observers can appreciate its excellence and its complexity in ways that are quite beyond the nervous systems of its inhabitants. We can also aspire to achieving a similarly Olympian perspective on our own artifactual world, a feat only human beings could imagine. If we don't succeed, we risk dismantling our precious creations in spite of our best intentions. Evolution in two realms, genetic and cultural, has created in us the capacity to know ourselves. But in spite of several millennia of ever-expanding intelligent design, we still are just staying afloat in a flood of puzzles and problems, many of them created by our own efforts of comprehension, and there are dangers that could cut short our quest before we—or our descendants—can satisfy our ravenous curiosity.

Home at last

This completes our journey from bacteria to Bach and back. It has been a long and complicated trek through difficult terrain, encountering regions seldom traveled by philosophers, and other regions beset by philosophers and typically shunned by scientists. I have invited you to take on board some distinctly counterintuitive ideas and tried to show you how they illuminate the journey. I would now like to provide a summary of the chief landmarks and remind you of why I found them necessary waypoints on the path.

We began with the problem of the mind and Descartes's potent polarization of the issues. On one side, the sciences of matter and motion and energy and their support, thanks to evolution, of life; on the other side, the intimately familiar but at the same time utterly mysterious and private phenomena of consciousness. How can this dualist wound be healed? The first step in solving this problem, I argued, is Darwin's *strange inversion of reasoning*, the revolutionary insight that all the design in the biosphere can be, must ultimately be, the product of blind, uncomprehending, purposeless processes of natural selection. No longer do we have to see Mind as the Cause of everything else.

Evolution by natural selection can mindlessly uncover the *reasons without reasoners*, the free-floating rationales that explain why the parts of living things are arranged as they are, answering both questions: *How come?* and *What for?* Darwin provided the first great instance of *competence without comprehension* in the process of natural selection itself. Then *Turing's strange inversion of reasoning* provided an example, and a workbench for exploring the possibilities, of another variety of competence without comprehension: computers, which unlike the human agents for which they were named, do not have to understand the techniques they exploit so competently. There is so much that can be accomplished by competence with scant comprehension—think of termite castles and stotting antelopes—that we are faced with a new puzzle: What is comprehension for, and how could a human mind like Bach's or Gaudí's arise? Looking more closely at how computers are designed to use information to accomplish tasks heretofore reserved for comprehending human thinkers helped clarify the distinction between "bottom-up" design processes exhibited by termites—and by natural selection itself—and "top-down" intelligent design processes. This led to the idea of *information as design worth stealing*, or buying or copying in any case. Shannon's excellent theory of information clarifies the basic idea—*a difference that makes a difference*—and provides it with a sound theoretical home, and ways of measuring information, but we need to look further afield to see why such differences are so valuable, so worth measuring in the first place.

The various processes of Darwinian evolution are not all the same, and some are "more Darwinian" than other processes that are just as real, and just as important in their own niches, so it is important to be a *Darwinian about Darwinism*. Godfrey-Smith's Darwinian Spaces is a good thinking tool for helping us plot not only the relations between the way different species evolve but also the way evolution itself evolves, with some lineages exhibiting de-Darwinization over time.

Returning to the puzzle about how brains made of billions of neurons without any top-down control system could ever develop into human-style minds, we explored the prospect of decentralized, distributed control by neurons equipped to fend for themselves, including as one possibility *feral neurons*, released from their previous role as docile, domesticated servants under the selection pressure created by a new environmental feature: cultural invaders. *Words striving to reproduce*, and other memes, would provoke adaptations, such as revisions in brain structure in coevolutionary response. Once cultural transmission was secured as the chief behavioral innovation of our species, it not only triggered important changes in neural architecture but also added novelty to the environment—in the form of thousands of Gibsonian affordances—that enriched the ontologies of human beings and provided in turn further selection pressure in favor of adaptations—thinking tools—for keeping track of all these new opportunities. *Cultural evolution itself evolved* away from undirected or "random" searches toward more effective design processes, foresighted and purposeful and dependent on the comprehension of agents: intelligent designers. For human comprehension, a huge array of thinking tools is required. Cultural evolution de-Darwinized itself with its own fruits.

This vantage point lets us see the manifest image, in Wilfrid Sellars's useful terminology, as a special kind of artifact, partly genetically designed and partly culturally designed, a particularly effective *user-illusion* for helping time-pressured organisms move adroitly through life, availing themselves of (over)simplifications that create an image of th*e world we live in* that is somewhat in tension with the scientific image to which we must revert in order to explain the emergence of the manifest image. Here we encounter yet another revolutionary *inversion of reasoning*, in David Hume's account of our knowledge of causation. We can then see human *consciousness as a user-illusion*, not rendered in the Cartesian Theater (which does not exist) but constituted by the representational activities of the brain coupled with the appropriate reactions to those activities ("and then what happens?").

This closes the gap, the Cartesian wound, but only a sketch of this all-important unification is clear at this time. The sketch has enough detail, however, to reveal that human minds, however intelligent and comprehending, are not the most powerful imaginable cognitive systems, and our intelligent designers have now made dramatic progress in creating machine learning systems that use bottom-up processes to demonstrate once again the truth of Orgel's Second Rule: Evolution is cleverer than you are. Once we appreciate the universality of the Darwinian perspective, we realize that our current state, both individually and as societies, is both imperfect and impermanent. We may well someday return the planet to our bacterial cousins and their modest, bottom-up styles of design improvement. Or we may continue to thrive, in an environment we have created with the help of artifacts that do most of the heavy cognitive lifting their own way, in an age of post-intelligent design. There is not just coevolution between memes and genes; there is codependence between our minds' top-down reasoning abilities and the bottom-up uncomprehending talents of our animal brains. And if our future follows the trajectory of our past—something that is partly in our control—our artificial intelligences will continue to be dependent on us even as we become more warily dependent on them.

Appendix

The Background

The ideas and arguments in this book are the foreground of a set of arguments I have been building for half a century. I have tried to write the book so that the reader needs no knowledge of this earlier work of mine, and I have included only a few references to it. Readers, especially if not yet persuaded, may wonder if I have further support for the claims I advance here, and the answer is yes. Some readers may also want to know the history of the arguments so that they can trace the trajectory through time of the various claims, which have evolved. This appendix provides references to all the supporting work, along with some of the most useful critical work that has been provoked by them.

But before turning to these background works, let me keep the promise in fn. 91, p. 296 about Peter Carruthers and put in the foreground some deeply relevant work I have discovered but not yet found time to digest properly to incorporate into this book. In addition to Carruthers's *The Centered Mind*, and his earlier work, I would like to recommend Murray Shanahan, *Embodiment and the Inner Life*, 2010, Radu Bogdan, *Our Own Minds: Sociocultural Grounds for Self-Consciousness*, 2010, and *Mindvaults: Sociocultural Grounds for Pretending and Imagining*, 2013, Andy Clark, *Surfing Uncertainty*, 2015, and Olivier Morin, *How Traditions Live and Die*, 2016. In an ideal world I would have made the time to include all these in my preparation and no doubt my book would be better for it, but at some point one has to draw the blinds and concentrate on writing. I hope to comment on all of these (and more) in the near future, probably on my website.

Chapter 1

12 the terms *romantic* and *killjoy*. I first analyzed the tug of war between *romantics* and *killjoys* with regard to animal minds in "Intentional Systems in Cog-

nitive Ethology: The Panglossian Paradigm Defended," a Target Article in *Behavioral and Brain Sciences* (1983). That essay also initiated my critique of Gould and Lewontin's notorious "Panglossian paradigm" essay, which I developed over more than a dozen years into an increasingly detailed attack on the misdirection found in Gould's work. See, for example, my chapter on Gould in *Darwin's Dangerous Idea* (hereafter *DDI*) (1995) and the subsequent exchanges with Gould in the *New York Review of Books* (1997, 1997b), and Dennett, "Darwinian Fundamentalism" (1997).

14   a cliff over which you push your opponents. I described this cliff in "Current Issues in the Philosophy of Mind" (1978b).

18   trying to *answer* rhetorical questions. On rhetorical questions as telltale weak points in arguments, see *Intuition Pumps* (hereafter *IP*) (2013).

19   inflating it into something deep and metaphysical. Humphrey and Dennett, "Speaking for Our Selves," 1989, is reprinted in Dennett, *Brainchildren* (1998).

Chapter 2

26   an intelligent designer from another galaxy. See Dennett, "Passing the Buck to Biology," in *Behavioral and Brain Sciences* (1980).

29   The reverse-engineering perspective is ubiquitous in biology. Biology as a species of engineering is treated at length in *DDI* (1995), chapter 8.

Chapter 3

36   The design stance was introduced in "Intentional Systems" in 1971 and has figured prominently in my work ever since. See especially *Brainstorms* (1978), "Three Kinds of Intentional Psychology" (1981a), "Making Sense of Ourselves," (1981b), and *The Intentional Stance* (1987).

42   vestigial. For discussion of some examples of structures in artifacts that have lost their reasons for existing, see *DDI* (1995, p. 199).

44   cycles. The claim that cycles—not just reproductive cycles—can raise the probability that something new will happen was the theme of my essay in Edge.org's 2011 question: "What scientific concept would improve everybody's cognitive toolkit?" then published (2012c) in John Brockman, ed., *This Will Make You Smarter.*

49   Prime Mammal: I discuss David Sanford's argument for the nonexistence of mammals in *DDI* (1995) and again in *IP* (2013).

50   free-floating rationales. I introduced the term "free-floating rationales" in my "Intentional Systems in Cognitive Ethology" (1983), and have elaborated on it in *DDI* (1995), *IP* (2013), and elsewhere.

Chapter 4

53 This chapter is drawn with many revisions from my 2009 PNAS article, "Darwin's 'Strange Inversion of Reasoning'" and from "Turing's 'Strange Inversion of Reasoning'" (2013d).

Chapter 5

77 Penrose's argument against AI—and, in effect, against natural selection!—has been analyzed at length by me in "Murmurs in the Cathedral," my review of Penrose's book, in *Times Literary Supplement* (1989), and *DDI* (1995, ch. 15), and *Brainchildren* (1998), especially the chapters in part II, on AI and AL.

80 optimality. For more on optimality assumptions in evolutionary biology see "Intentional Systems in Cognitive Ethology," 1983, and *DDI* (1995), and my work on the design stance, the intentional stance, and the physical stance cited above for chapter 3.

Chapter 6

112 The example of Jacques and Sherlock is adapted from "Three Kinds of Intentional Psychology" (1981) and also appears in *IP* (2013). In the earlier versions I was explicitly discussing the *beliefs* of an intentional system and using the example to resist the slide to the tempting conclusion that there had to be a "language of thought" in which all beliefs are somehow encoded, a theme that has had many outings in my work. See also, *The Intentional Stance* (1987) and Westbury and Dennett, "Mining the Past to Construct the Future: Memory and Belief as Forms of Knowledge" (1998). Today I am generalizing the point by talking about information. The *information* about the viscosity of the medium through which birds fly that is *embodied* in the design of the birds' wings (and their controls) does not have to be *encoded* in that design. As I have revised and enhanced the concept of the intentional stance, I have (I hope) clarified and extended this theme: the utility of adopting an informational/intentional perspective does *not* depend on the existence of a system of representation to encode that valuable commodity, semantic information.

114 transparency. For the dangers of transparency, see *IP* (2013) and Dennett and Roy (2015).

117 The importance of asking *cui bono?* was first discussed in *DDI* (1995).

121 Good Tricks are defined and discussed in *DDI* (1995) along with "just-so stories."

127 For a different perspective on internal trust, see McKay and Dennett, "The Evolution of Misbelief," *Behavioral and Brain Sciences* (2009).

128 See "Real Patterns" (1991b) for an extended analysis of the relationship between ontology and discernible patterns.

135 The example of a wagon with spoked wheels was discussed in both *Consciousness Explained* (hereafter *CE*) (1991, p. 204) and *DDI* (1995, p. 348).

Chapter 7

138 Whenever all three factors are present. This is the Darwinian "algorithm" I discuss in detail in Dennett (1995).

147 Hutterites. There is a critical discussion of Sober and Wilson's analysis of the Hutterites in "Some Varieties of Greedy Ethical Reductionism," in *DDI* (1995, pp. 467–481).

Chapter 8

155 parallel architectures. Paul Churchland and I have had a long-running debate over the powers of parallel architectures. As my most recent installment's title makes clear, our differences are diminishing: "Two Steps Closer on Consciousness" (2006d). See also my discussion in *Freedom Evolves* (2003, pp. 106ff).

157 See my review of Deacon's book: "Aching Voids and Making Voids" (2013b).

162 Tecumseh Fitch. The previous six paragraphs are drawn, with revisions, from my contribution to the 2008 Edge.org question: What have you changed your mind about and why?

165 How do brains pick up affordances? This section benefited from discussions with David Baßler and is drawn with revisions from "Why and How Does Consciousness Seem the Way It Seems?" and "How Our Belief in Qualia Evolved, and Why We Care So Much—A Reply to David H. Baßler," *Open MIND*, Metzinger 2015 and from "Expecting Ourselves to Expect: The Bayesian Brain as a Projector," my *BBS* commentary (2013c) on Clark (2013).

166 competence models versus performance models. Related (but better) distinctions have been influential in cognitive science: David Marr and Tommy Poggio (1976) proposed three levels: computational, algorithmic, and implementation; Allen Newell (1982) proposed the knowledge level, symbol level, hardware level; my three "stances," intentional, design, and physical, were proposed earlier (1971) and were somewhat more general. Each ver-

sion of what is basically the same point has its drawbacks: Marr and Poggio's term "computational" is a misleading term for the specs-level; Newell's term "symbol level" expresses a GOFAI bias, and my "intentional" stance imposes a controversial and typically misunderstood bit of philosophical jargon on cognitive science.

170   Wonder tissue is first mentioned in print in 1984, in "Cognitive Wheels" (1984b).

Chapter 9

177   Avital and Jablonka. For more on Avital and Jablonka's book, see my review in *Journal of Evolutionary Biology* (2002b).

186   "down with democracy." The vicissitudes of "deciphering" the brain's activities as a way of attributing beliefs has been a theme of mine since 1975. See "Brain Writing and Mind Reading," reprinted in *Brainstorms* (1978).

Chapter 10

219   novel variations or wider contexts. My interest in cognitive ethology and the problems of attributing higher order intentional states to nonhuman animals goes back to my 1983 *BBS* Target Article, "Intentional Systems in Cognitive Ethology: The 'Panglossian Paradigm' Defended" and has been pursued off and on ever since. See, for example, "Cognitive Ethology: Hunting for Bargains or a Wild Goose Chase?" (1989), "Do Animals Have Beliefs?" (1995b), and "Animal Consciousness: What Matters and Why" (1995b).

Chapter 11

222   Real patterns are analyzed in Dennett (1991b).

229   thinko. The concept of thinkos was introduced in "From Typo to Thinko: When Evolution Graduated to Semantic Norms" (2006c).

Chapter 12

253   Humphrey. I reply to Nicholas Humphrey, "Cave Art, Autism, and the Evolution of the Human Mind," in *Cambridge Archaeological Journal* (1998b).

254   apes with brains being manipulated. See the chapters on memes in *CE* (1991), *DDI* (1995), and "The Evolution of Culture" (2001b).

254 On the lancet fluke in the ant climbing the blade of grass, see "The Evolution of Culture" (2001b) and *Breaking the Spell: Religion as a Natural Phenomenon* (2006).

265 vervet monkey alarm calls. My *BBS* article on cognitive ethology (1983), discussed the difficulties of distinguishing anecdote from scientific evidence of animal intelligence. "Beliefs about beliefs," (1978d), my commentary on Premack and Woodruff's *BBS* Target Article, "Does the Chimpanzee Have a Theory of Mind?" (1978), was followed by a wide variety of studies, in the field and the laboratory, of signs of higher-order intentionality (such as the false belief tasks) in animals and children. See, for instance, Whiten and Byrne (1988) *Machiavellian Intelligence* and *Machiavellian Intelligence, II: Extensions and Evaluations* (1997), as well as more recent work by Tomasello, Call, Povinelli, and many others. In *Kinds of Minds* 1996, I discussed the distinction between having beliefs about beliefs and thinking about thinking, and reviewed some of this literature. See also my discussion in "*The Selfish Gene* as a Philosophical Essay" (2006b).

277 the design of the LAD by natural selection. In "Passing the Buck to Biology" (*BBS*, 1980) I offered what I thought was a friendly amendment to Chomsky's Target Article and was astonished by his intransigent reply. I had thought the whole point of the LAD was to claim that Darwinian, rather than Skinnerian, or Popperian, or Gregorian R&D was mainly responsible for designing grammar, but I was mistaken.

Chapter 13

294 an order which is there. I discuss the passage from Anscombe in "Real Patterns" (1991b, p. 43).

295 McGeer (2004). See Griffin and Dennett (2008) on scaffolding in the context of autism.

301 "I can do meta." I quote Hofstadter's quip, "Anything you can do I can do meta" in "Hofstadter's Quest" (1995b). Hofstadter claims that I am the author of the quip, but it doesn't matter who came up with it first. In 2004, Edge.com *What's Your Law?* Charles Simonyi promulgated Simonyi's Law, "Anything that can be done could be done 'meta.'" Good ideas are Good Tricks; they tend to get reinvented frequently.

301 JVM and virtual machines. For a more detailed introduction to virtual machines and their relation to hardware and software, see *IP* (2013, ch. 4, "An Interlude about Computers") for something of a primer.

317 Oscar Wilde. The example of Oscar Wilde laboriously generating his witticisms was first discussed in "Get Real" (1994).

318 There are lengthy discussions of the Hard Problem and the "real magic" of

consciousness in *Sweet Dreams: Philosophical Obstacles to a Science of Consciousness* (2005) and *IP* (2013).

319   The Spamlet and Spakesheare thought experiment first appeared in my Presidential Address to the Eastern Division of the American Philosophical Association, published as "In Darwin's Wake, Where Am I?" (Proc. APA 2000).

331   the extended mind. I discuss the value of using marks in the environment in *Kinds of Minds* (1996) and "Learning and Labeling" (1993).

Chapter 14

336   "Martian" scientists. The point about consciousness being evident to Martian scientists was first developed in *Sweet Dreams* (2003).

337   "what is it like?" See the amusing and insightful reflections by Hofstadter in Hofstadter and Dennett, *The Mind's I* (1981), pp. 403–414.

342   von Neumann. For a discussion of game theory and privacy, see *Freedom Evolves* (2003).

343   linguistic communication. The two paragraphs at the close of the section have been drawn, with many revisions, from "My Body Has a Mind of Its Own" (2007c) and "The Evolution of 'Why'" (2010), which offer further discussion of the sources of this position in the work of McFarland (1989, 1989b), Drescher (1991), Brandom (1994), Haugeland (1998), and Millikan (2000b). See also my review (2009b) of McFarland's 2009 book *Guilty Robots, Happy Dogs*.

344   The coinage of "selfy" modeled on "sexy" is discussed in *CE* (1991).

345   Wegner. There is an extended discussion of Wegner's view of free will as illusory in *Freedom Evolves* (2003).

351   heterophenomenology. These paragraphs are largely based on "Heterophenomenology Reconsidered" (2007b), where there is a much fuller discussion; see especially the response to Schwitzgebel (2007) and, in fn. 6, to Siewert (2007), whose thinking helped me reframe these ideas. Heterophenomenology (1982, 1991, 2003c, 2005, 2007c) lets us keep track of the affordances, the "things" in a person's "notional world" without endorsing that ontology as part of our scientific ontology.

352   fn. 99, on mental imagery research: Shepard and Metzler (1971) was the brilliant trigger for this explosion of work, with Pylyshyn (1973) and Kosslyn (1975) setting the table for the subsequent controversies (see, e.g., Kosslyn 1980, Kosslyn et al. 1979, Kosslyn et al. 2001, Pylyshyn 2002). For my discussions of the mental imagery controversies, see *Content and Consciousness* (1969, "The Nature of Images and the Introspective Trap"), *Brainstorms* (1978, chs.

9 and 10), *CE* (1991), and "Does Your Brain Use the Images in It, and If So, How?" (commentary on Pylyshyn 2002) in *Behavioral and Brain Sciences* 2002.

354   Hume's strange inversion. The paragraphs on Hume's strange inversion are drawn, with revisions, from my commentary on Clark, *BBS*, 2013. An earlier presentation of some of the ideas can be found in Hurley, Dennett, and Adams, *Inside Jokes* (2011).

363   Hard Question. For an extended discussion of the Hard Question, see *CE* (1991), and for an example of asking and beginning to answer the Hard Question, see "The Friar's Fringe of Consciousness" (2015b).

363   can't have it both ways. I can offer intuition pumps to render my claim at least entertainable by those who find it frankly incomprehensible at first. See especially "The Tuned Deck," in "Explaining the 'Magic' of Consciousness" (2002c) (from which some material in the previous paragraphs is drawn), *Sweet Dreams* (2005), and *IP* (2013).

367   For more on Bennett, and this identification by content, see "Jonathan Bennett's *Rationality*" (forthcoming). See also Azzouni (2013) on how meanings dwell in our phenomenology in spite of not existing.

368   Free will. Sam Harris (2012) and others (e.g., Coyne 2012) have recently argued that science shows that there is no free will, and that this has revolutionary consequences (unspecified) for our policies of morality and law, crime, punishment, and reward. They overlook the fact that the manifest image is where we live and is what matters. Color, from the perspective of the scientific image, is an illusion, but it still matters to us and we wisely take the time and effort to surround ourselves with color schemes we prize. It is as if Harris et al. were arguing that money is an illusion, so we should abandon it. I don't see them giving away all their illusory money, nor do I see them ducking responsibility for their intentional actions, so it isn't clear what they think they are showing. For more on free will, see my books *Elbow Room: The Varieties of Free Will Worth Wanting* (1984 [rev. ed. 2015]) and *Freedom Evolves* (2003) and "Erasmus: Sometimes a Spin Doctor Is Right," my Erasmus Lecture (2012b); on Harris: "Reflections on *Free Will*" (2014b) and "Daniel Dennett on Free Will Worth Wanting," in the podcast *Philosophy Bites Again* (2014).

Chapter 15

371   Pugh. Quoted by George E. Pugh, *The Biological Origin of Human Values* (1978), p. 154. In a footnote, the author writes that this quote was from his father, Emerson Pugh, around 1938. The quote is also widely attributed to Lyall Watson, born in 1939. http://todayinsci.com/P/Pugh_Emerson /PughEmerson-Quotations.htm.

376 multi-volume book on consciousness. In "Belief in Belief," chapter 8 in *Breaking the Spell* (2006), I discuss the division of labor between scientists in different fields who rely on each other's expertise ("they do the understanding and we do the believing!") and point out that in theology, the point is often made that *nobody* understands the terms of the discussion.

383 evolved nanorobots. Evolving robots on the macroscale have also achieved some impressive results in very simplified domains, and I have often spoken about the work in evolutionary robotics by Inman Harvey and Phil Husbands at Sussex (e.g., Harvey et al. 1997), but I have not discussed it in print.

383 David Cope. Cope's *Virtual Music* (2001) includes essays by Douglas Hofstadter, composers, musicologists, and me: "Collision Detection, Muselot, and Scribble: Some Reflections on Creativity." The essays are filled with arresting observations and examples, and the volume includes many musical scores and comes with a CD.

384 substrate-neutral. See *DDI* (1995) on substrate neutrality.

385 analogizers. See also Douglas Hofstadter's many works on the importance of analogy finding, especially *Metamagical Themas* (1985), *Fluid Concepts and Creative Analogies* (1995), and *Surfaces and Essences: Analogy as the Fuel and Fire of Thinking*, coauthored with Emmanuel Sander (first published in French as *L'Analogie. Cœur de la pensée*; published in English in the United States in April 2013).

385 Architects. For some examples of the use of genetic algorithms in architecture, see Sullivan-Fedock (2011), Asadi et al. (2014), Yu et al. (2014), and for optimizing electoral redistricting, see Chou et al., (2012).

390 self-monitor. I discuss this design strategy in an unpublished paper, "A Route to Intelligence: Oversimplify and Self-monitor" (1984c) that can be found on my website: http://ase.tufts.edu/cogstud/dennett/recent.html.

395 Turing Test. For an analysis and defense of the Turing Test as a test of genuine comprehension, see my "Can Machines Think?" (1985), reprinted in *Brainchildren*, with two postscripts (1985) and (1997); "Fast Thinking" in *The Intentional Stance* (1987); and especially "The Chinese Room," in *IP* (2013), where I discuss examples of the cognitive layering that must go into some exchanges in a conversation (pp. 326–327).

396 Debner and Jacoby (1994). For more on such experiments, see my "Are We Explaining Consciousness Yet?" (2001c) and also Dehaene and Naccache (2001), Smith and Merikle (1999), discussed in Merikle et al. (2001).

399 *theory* of agents with imagination. I discuss the prospects of such a powerful *theory* or *model* of an intelligent agent, and point out a key ambiguity in the original Turing Test, in an interview with Jimmy So about the implications of *Her*, in "Can Robots Fall in Love" (2013), *The Daily Beast*,

http://www.thedailybeast.com/articles/2013/12/31/can-robots-fall-in-love-and-why-would-they.html.

400 a negligible practical possibility. When explaining why I thought strong AI was possible in principle but practically impossible, I have often compared it to the task of making a robotic bird that weighed no more than a robin, could catch insects on the fly, and land on a twig. No cosmic mystery, I averred, in such a bird, but the engineering required to bring it to reality would cost more than a dozen Manhattan Projects, and to what end? We can learn all we need to know about the principles of flight, and even bird flight, by making simpler models on which to test our theories, at a tiny fraction of the cost. People have recently confronted me with news items about the latest miniature drones, smaller than robins, as small as flying insects, and asked me if I wanted to reconsider my analogy. No, because they are not autonomous self-controlled robots but electronic marionettes, and besides, they can't catch flies or land on twigs. Maybe someday they will be able to do these things, but only if DARPA wastes billions of dollars. I have discussed the Singularity in "The Mystery of David Chalmers" (2012) and "The Singularity—An Urban Legend?" (2015).

402 prematurely ceding authority. I enlarge upon this concern in "Information, Technology, and the Virtues of Ignorance" (*Daedalus*, 1986), reprinted in *Brainchildren* (1998) and in "The Singularity—An Urban Legend?" (2015).

408 car breaks down. These paragraphs are drawn with revisions from my foreword to the second edition of Seabright, *The Company of Strangers*, 2010.

REFERENCES

Alain, Chartier. (1908) 1956. *Propos d'un Normand 1906–1914*. Parais: Gallimard.

Allen, Colin, Mark Bekoff, and George Lauder, eds. 1998. *Nature's Purposes: Analyses of Function and Design in Biology*. Cambridge, Mass.: MIT Press.

Anscombe, G. E. M. 1957. *Intention*. Oxford: Blackwell.

Arnold, Frances. 2013. Frances Arnold Research Group. http://www.che.caltech.edu/groups/fha/Projects3b.htm.

Asadia, Ehsan, Manuel Gameiro da Silva, Carlos Henggeler Antunes, Luís Dias, and Leon Glicksman. 2014. "Multi-Objective Optimization for Building Retrofit: A Model Using Genetic Algorithm and Artificial Neural Network and an Application." *Energy and Buildings* 81: 444–456.

Avital, Eytan, and Eva Jablonka. 2000. *Animal Traditions: Behavioural Inheritance in Evolution*. Cambridge: Cambridge University Press.

Azzouni, Jody. 2013. *Semantic Perception: How the Illusion of a Common Language Arises and Persists*. New York: Oxford University Press.

Bailey, Ida E., Felicity Muth, Kate Morgan, Simone L. Meddle, and Susan D. Healy. 2015. "Birds Build Camouflaged Nests." *The Auk* 132 (1): 11–15.

Baron-Cohen, Simon. 2003. *The Essential Difference: Male and Female Brains and the Truth about Autism*. New York: Basic Books.

Bateson, Gregory. 1973. *Steps to an Ecology of Mind: Collected Essays in Anthropology, Psychiatry, Evolution, and Epistemology*. St Albans, Australia: Paladin.

Baum, Eric B. 2004. *What Is Thought?* Cambridge, Mass.: MIT Press.

Behe, Michael. 1996. *Darwin's Black Box: The Biochemical Challenge to Evolution*. New York: Free Press.

Bennett, Jonathan Francis. 1964. *Rationality: An Essay Towards an Analysis*. London: Routledge.

Bennett, Jonathan. 1976. *Linguistic Behaviour.* London: Cambridge University Press.

Berger S. L., T. Kouzarides, R. Shiekhattar, and A. Shilatifard. 2009. "An Operational Definition of Epigenetics." *Genes Dev.* 23 (7): 781–783.

Bernstein, Leonard. 1959. "Why Don't You Run Upstairs and Write a Nice Gershwin Tune?" In *The Joy of Music*, 52–62. New York: Simon and Schuster.

Beverley, R. M. 1868. *The Darwinian Theory of the Transmutation of Species Examined.* (Published anonymously "By a Graduate of the University of Cambridge.") London: Nisbet (quoted in a review, *Athenaeum* 2102 [Feb. 8]: 217).

Bickerton, Derek. 2009. *Adam's Tongue: How Humans Made Language, How Language Made Humans.* New York: Hill and Wang.

———. 2014. *More Than Nature Needs: Language, Mind, and Evolution.* Cambridge, Mass.: Harvard University Press.

Blackmore, Susan. 1999. *The Meme Machine.* New York: Oxford University Press.

———. 2008. "Memes Shape Brains Shape Memes." *Behavioral and Brain Sciences* 31: 513.

———. 2010. "Dangerous Memes, or What the Pandorans Let Loose." In *Cosmos and Culture: Cultural Evolution in a Cosmic Context*, edited by Steven Dick and Mark Lupisella, 297–318. NASA SP-2009-4802.

Bogdan, Radu J. 2013. *Mindvaults: Sociocultural Grounds for Pretending and Imagining.* Cambridge, Mass.: MIT Press.

Borges, J. L. 1962. *Labyrinths: Selected Stories and other Writings.* New York: New Directions.

Bostrom, Nick. 2014. *Superintelligence: Paths, Dangers, Strategies.* New York: Oxford University Press.

Boyd, Robert, and Peter J. Richerson. 1985. *Culture and the Evolutionary Process.* Chicago: University of Chicago Press.

———. 2005. *The Origin and Evolution of Cultures.* Oxford: Oxford University Press. http://site.ebrary.com/id/10233633.

Boyd, Robert, P. Richerson, and J. Henrich. 2011. "The Cultural Niche: Why Social Learning Is Essential for Human Adaptation." *PNAS* 108 (suppl. 2): 10918–10925.

Brandom, Robert. 1994. *Making It Explicit: Reasoning, Representing, and Discursive commitment.* Cambridge, Mass.: Harvard University Press.

Bray, Dennis. 2009. *Wetware: A Computer in Every Living Cell.* New Haven: Yale University Press.

Brenowitz E.A., D. J. Perkel, and L. Osterhout. 2010. "Language and Birdsong: Introduction to the Special Issue." *Brain and Language* 115 (1): 1–2.

Brockman, John, ed. 2011. *This Will Make You Smarter.* New York: Harper Torchbook.

Bromberger, Sylvain. 2011. "What Are Words? Comments on Kaplan (1990), on Hawthorne and Lepore, and on the Issue." *Journal of Philosophy* 108 (9): 486–503.

Burt, Austin, and Robert Trivers. 2008. *Genes in Conflict: The Biology of Selfish Genetic Elements*. Cambridge, Mass.: Belknap Press.

Butterworth, G. 1991. "The Ontogeny and Phylogeny of Joint Visual Attention." In *Natural Theories of Mind*, edited by A. Whiten, 223–232. Oxford: Basil Blackwell.

Byrne, Richard W., and Andrew Whiten. 1988. *Machiavellian Intelligence*. Oxford: Clarendon Press.

Carruthers, Peter. 2015. *The Centered Mind: What the Science of Working Memory Shows Us about the Nature of Human Thought*. Oxford: Oxford University Press.

Cavalli-Sforza, L. L., and Marcus W. Feldman. 1981. *Cultural Transmission and Evolution: A Quantitative Approach*. Princeton, N.J.: Princeton University Press.

Chalmers, David. 1996. *The Conscious Mind: In Search of a Fundamental Theory*. New York: Oxford University Press.

———. 2010. "The Singularity: A Philosophical Analysis." *Journal of Consciousness Studies* 17 (9–10): 7–65.

Cheney, Dorothy L., and Robert M. Seyfarth. 1980. "Vocal Recognition in Free-Ranging Vervet Monkeys." *Animal Behaviour* 28 (2): 362–367.

———. 1990. "Attending to Behaviour versus Attending to Knowledge: Examining Monkeys' Attribution of Mental States." *Animal Behaviour* 40 (4): 742–753.

Chomsky, Noam. 1965. *Aspects of the Theory of Syntax*. Cambridge: MIT Press.

———. 1975. *Reflections on Language*. New York: Pantheon Books.

———. 1995. *The Minimalist Program*. Cambridge, Mass.: MIT Press.

———. 2000. *New Horizons in the Study of Language and Mind*. Cambridge: Cambridge University Press.

———. 2000b. *On Nature and Language*. New York: Cambridge University Press.

———. 2014. "Mysteries and Problems." October 18, https://www.youtube.com/watch?v=G8G2QUK_1Wg.

Chou, C., S. Kimbrough, J. Sullivan-Fedock, C. J. Woodard, and F. H. Murphy. 2012. "Using Interactive Evolutionary Computation (IEC) with Validated Surrogate Fitness Functions for Redistricting." Presented at the Genetic and Evolutionary Computation, ACM. Philadelphia.

Christiansen, Morten H., and Nick Chater. 2008. "Language as Shaped by the Brain." *Behavioral and Brain Sciences* 31 (5): 489–509.

Christiansen, Morten H., and Simon Kirby. 2003. *Language Evolution*. Oxford: Oxford University Press.

Churchland, Paul M. 1979. *Scientific Realism and the Plasticity of Mind*. Cambridge: Cambridge University Press.

Claidière N., K. Smith, S. Kirby, and J. Fagot. 2014. "Cultural Evolution of Systematically Structured Behaviour in a Non-Human Primate." *Proceedings of the Royal Society: Biological Sciences*. 281 (1797). doi: 10.1098/rspb.2014.1541.

Clark, Andy. 2013. "Whatever Next? Predictive Brains, Situated Agents, and the Future of Cognitive Science." *Behavioral and Brain Sciences* 36 (3): 181–204.

———. 2015. *Surfing Uncertainty: Prediction, Action, and the Embodied Mind*, New York: Oxford University Press.

Clark, Andy, and David Chalmers. 1998. "The Extended Mind." *Analysis* 58 (1): 7–19.

Cloud, Daniel. 2014. *The Domestication of Language: Cultural Evolution and the Uniqueness of the Human Animal*. New York: Columbia University Press.

Colgate, Stirling A., and Hans Ziock. 2011. "A Definition of Information, the Arrow of Information, and Its Relationship to Life." *Complexity* 16 (5): 54–62.

Cope, David, and Douglas R. Hofstadter. 2001. *Virtual Music: Computer Synthesis of Musical Style*. Cambridge, Mass.: MIT Press.

Coppinger, Raymond, and Lorna Coppinger. 2001. *Dogs: A Startling New Understanding of Canine Origin, Behavior, and Evolution*. New York: Scribner.

Corballis, Michael C. 2003. "From Mouth to Hand: Gesture, Speech, and the Evolution of Right-Handedness." *Behavioral and Brain Sciences* 26 (2): 199–208.

———. 2009. "The Evolution of Language." *Annals of the New York Academy of Sciences* 1156 (1): 19–43.

Coyne, Jerry. 2012. "You Don't Have Free Will." *Chronicle of Higher Education*, March 18.

Crick, Francis. 1994. *The Astonishing Hypothesis: The Scientific Search for the Soul*. New York: Scribner.

Cronin, Helena. 1992. *The Ant and the Peacock: Altruism and Sexual Selection from Darwin to Today*. Cambridge: Cambridge University Press.

Crystal, Jonathon D., and Allison L. Foote. 2009. "Metacognition in Animals." *Comparative Cognition and Behavior Review* 4: 1–16.

Darwin, Charles. 1859. *On the Origin of Species*. Washington Square: New York University Press.

———. (1862) 1984. *The Various Contrivances by Which Orchids Are Fertilised by Insects*. Chicago: University of Chicago Press.

———. 1868. *The Variation of Animals and Plants under Domestication*. New York: Orange Judd.

———. 1871. *The Descent of Man and Selection in Relation to Sex*. London: J. Murray.

Darwin, Charles, and Frederick Burkhardt. 1997. *The Correspondence of Charles Darwin*. Vol. 10. Cambridge: Cambridge University Press.

Dawkins, Richard. 1976. *The Selfish Gene*. Oxford: Oxford University Press. 1989 Rev. ed.

———. 1986. *The Blind Watchmaker*. New York: Norton.

———. 2004. *The Ancestor's Tale: A Pilgrimage to the Dawn of Evolution*. Boston: Houghton Mifflin.

———. 2004b. "Extended Phenotype—But Not *Too* Extended. A Reply to Laland, Turner and Jablonka." *Biology and Philosophy* 19: 377–396.

Dawkins, Richard, and John R. Krebs. 1978. "Animal Signals: Information or Manipulation." *Behavioural Ecology: An Evolutionary Approach,* 2: 282–309.

de Boer, Bart, and W. Tecumseh Fitch. 2010. "Computer Models of Vocal Tract Evolution: An Overview and Critique." *Adaptive Behaviour* 18 (1): 36–47.

Deacon, Terrence William. 1997. *The Symbolic Species: The Co-Evolution of Language and the Brain.* New York: W. W. Norton.

———. 2011. *Incomplete Nature: How Mind Emerged from Matter.* New York: W. W. Norton.

Debner, J. A., and L. L. Jacoby. 1994. "Unconscious Perception: Attention, Awareness, and Control." *Journal of Experimental Psychology. Learning, Memory, and Cognition* 20 (2): 304–317.

Defoe, Daniel. 1883. *The Life and Strange Surprising Adventures of Robinson Crusoe of York, Mariner: As Related by Himself.* London: E. Stock.

Dehaene, S., and L. Naccache. 2001. "Towards a Cognitive Neuroscience of Consciousness: Basic Evidence and a Workspace Framework." *COGNITION* 79 (1–2): 1–37.

Dennett, Daniel C. *Content and Consciousness.* 1969. London and New York: Routledge and Kegan Paul, and Humanities Press.

———. 1971. "Intentional Systems." *Journal of Philosophy* 68 (4): 87–106.

———. 1978. *Brainstorms: Philosophical Essays on Mind and Psychology.* Montgomery, Vt.: Bradford Books.

———. 1978b. "Current Issues in Philosophy of Mind." *American Philosophical Quarterly* 15 (4): 249–261.

———. 1978c. "Why Not the Whole Iguana?" *Behavioral and Brain Sciences* 1 (1): 103–104.

———. 1978d. "Beliefs about Beliefs." Commentary on Premack and Woodruff. *Behavioral and Brain Sciences* 1 (4) 568–570.

———. 1980. "Passing the Buck to Biology." *Behavioral and Brain Sciences* 19.

———. 1981. "Three Kinds of Intentional Psychology." In *Reduction, Time and Reality,* edited by R. Healey, 37–60. Cambridge: Cambridge University Press.

———. 1982. "How to Study Consciousness Empirically: Or Nothing Comes to Mind." *Synthese* 53: 159–180.

———. 1983. "Intentional Systems in Cognitive Ethology: The 'Panglossian Paradigm' Defended"; and "Taking the Intentional Stance Seriously." *Behavioral and Brain Sciences* 6 (3): 343–390.

———. 1984. *Elbow Room: The Varieties of Free Will Worth Wanting.* Cambridge, Mass.: MIT Press.

———. 1984b. "Cognitive Wheels: The Frame Problem of AI." In *Minds, Machines and Evolution,* edited by C. Hookway, 129–151. Cambridge: Cambridge University Press 1984.

————. 1984c. "A Route to Intelligence: Oversimplify and Self-monitor." Available at http://ase.tufts.edu/cogstud/papers/oversimplify.pdf.

————. 1987. *The Intentional Stance.* Cambridge, Mass.: MIT Press.

————. 1989. "Cognitive Ethology: Hunting for Bargains or a Wild Goose Chase?" In *Goals, No-Goals and Own Goals,* edited by D. Noble and A. Montefiore, 101–116. Oxford: Oxford University Press.

————. 1991. *Consciousness Explained.* Boston: Little, Brown.

————. 1991b. "Real Patterns." *Journal of Philosophy* 88 (1): 27–51.

————. 1993. "Learning and Labeling." Commentary on A. Clark and A. Karmiloff-Smith, "The Cognizer's Innards." *Mind and Language* 8 (4): 540–547.

————. 1994. "Get Real." *Philosophical Topics* 22 (1): 505–568.

————. 1995. *Darwin's Dangerous Idea.* New York: Simon and Schuster.

————. 1995b. "Do Animals Have Beliefs?" In *Comparative Approaches to Cognitive Sciences,* edited by Herbert Roitblat and Jean-Arcady Meyer, 111–118. Cambridge, Mass.: MIT Press.

————. 1995c. "Hofstadter's Quest: A Tale of Cognitive Pursuit." *Complexity* 1 (6): 9–12.

————. 1995d. "Animal Consciousness—What Matters and Why." *Social Research* 62 (3): 691–710.

————. 1996. *Kinds of Minds: Toward an Understanding of Consciousness.* New York: Basic Books.

————. 1997. "Darwinian Fundamentalism: An Exchange." *New York Review,* 64.

————. 1998. *Brainchildren: Essays on Designing Minds.* Cambridge, Mass.: MIT Press.

————. 1998b. "Reply to Nicholas Humphrey, Cave Art, Autism, and the Evolution of the Human Mind." *Cambridge Archeological Journal* 8 (2): 184–85.

————. 1999. Review of *Having Thought: Essays in the Metaphysics of Mind,* by John Haugeland, *Journal of Philosophy* 96 (8): 430–5.

————. 2001. "In Darwin's Wake, Where Am I?" *Proceedings and Addresses of the American Philosophical Association.* 75 (2): 11–30.

————. 2001b. "The Evolution of Culture." *The Monist* 84 (3): 305–324.

————. 2001c. "Are We Explaining Consciousness Yet?" *Cognition* 79: 221–237.

————. 2002. "Does Your Brain Use the Images in It, and If So, How?" *Behavioral and Brain Sciences* 25 (2): 189–190.

————. 2002b. "Tarbutniks Rule." Review of *Animal Traditions: Behavioural Inheritance in Evolution,* 2000 by Eytan Avital and Eva Jablonka, *Journal of Evolutionary Biology* 15 (2): 332–334.

————. 2002c. "Explaining the 'Magic' of Consciousness. Exploring Consciousness, Humanities, Natural Science, Religion. *Proceedings of the International Symposium,* Milano, November 19–20, 2001 (published in December 2002, Fondazione Carlo Erba), 47–58.

———. 2003. *Freedom Evolves*. New York: Viking.

———. 2003b. "The Baldwin Effect: A Crane, Not a Skyhook." In *Evolution and Learning: The Baldwin Effect Reconsidered*, edited by Bruce H. Weber and David J. Depew. Cambridge, Mass.: MIT Press, 60–79.

———. 2003c. "Who's on First? Heterophenomenology Explained." *Journal of Consciousness Studies* 10 (9–10): 19–30.

———. 2005. *Sweet Dreams: Philosophical Obstacles to a Science of Consciousness*. Cambridge, Mass.: Bradford Book/MIT Press.

———. 2006. *Breaking the Spell: Religion as a Natural Phenomenon*. New York: Viking.

———. 2006b. "The Selfish Gene as a Philosophical Essay." In *Richard Dawkins: How a Scientist Changed the Way We Think*, edited by A. Grafen and M. Ridley, 101–115. Oxford: Oxford University Press.

———. 2006c. "From Typo to Thinko: When Evolution Graduated to Semantic Norms." In *Evolution and Culture*, edited by S. Levinson and P. Jaisson 133–145. Cambridge, Mass.: MIT Press.

———. 2006d. "Two Steps Closer on Consciousness." In *Paul Churchland*, edited by Brian Keeley, 193–209. New York: Cambridge University Press.

———. 2007. "Instead of a Review." *Artificial Intelligence* 171 (18): 1110–1113.

———. 2007b. "Heterophenomenology Reconsidered." *Phenomenology and the Cognitive Sciences* 6 (1–2): 247–270.

———. 2007c. "My Body Has a Mind of Its Own." In *Distributed Cognition and the Will: Individual Volition and Social Context*, edited by D. Ross, D. Spurrett, H. Kincaid, and G. L. Stephens, 93–100. Cambridge, Mass.: MIT Press.

———. 2008. "Competition in the Brain." In *What Have You Changed Your Mind About?*, edited by John Brockman, 37–42. New York: HarperCollins.

———. 2009. "Darwin's 'Strange Inversion of Reasoning.'" *Proceedings of the National Academy of Sciences of the United States of America* 106: 10061–10065.

———. 2009b. "What Is It Like to be a Robot?" Review of *Guilty Robots, Happy Dogs*, by David McFarland. *BioScience* 59 (8): 707–709.

———. 2010. "The Evolution of Why?" In *Reading Brandom: On Making It Explicit*, edited by B. Weiss and J. Wanderer, 48–62. New York: Routledge.

———. 2012. "The Mystery of David Chalmers." *Journal of Consciousness Studies* 19 (1–2): 86–95.

———. 2012b. *Erasmus: Sometimes a Spin Doctor Is Right*. Amsterdam: Praemium Erasmianum Foundation.

———. 2012c. "Cycles." In *This Will Make You Smarter*, edited by J. Brockman and Edge.org, 110–119. New York: Harper Torchbook.

———. 2013. *Intuition Pumps and Other Tools for Thinking*. New York: W. W. Norton.

———. 2013b. "Aching Voids and Making Voids." Review of *Incomplete Nature: How*

Mind Emerged from Matter, by Terrence W. Deacon, *Quarterly Review of Biology* 88 (4): 321–324.

―――. 2013c. "Expecting Ourselves to Expect: The Bayesian Brain as a Projector." *Behavioral and Brain Sciences* 36 (3): 209–210.

―――. 2013d. "Turing's 'Strange Inversion of Reasoning.'" In *Alan Turing: His Work and Impact*, edited by S. Barry Cooper and J. van Leeuwen, 569–573. Amsterdam: Elsevier.

―――. 2014. "Daniel Dennett on Free Will Worth Wanting." In *Philosophy Bites Again*, edited by D. Edmonds and N. Warburton, 125–133. New York: Oxford University Press.

―――. 2014b. "Reflections on Free Will." Review of *Free Will*, by Sam Harris. Naturalism.org.

―――. 2015. "The Singularity—An Urban Legend?" In *What to Think about Machines That Think*, edited by John Brockman, 85–88. New York: HarperCollins.

―――. 2015b. "The Friar's Fringe of Consciousness." In *Structures in the Mind: Essays on Language, Music, and Cognition in Honor of Ray Jackendoff*, edited by Ida Toivonen, Piroska Csuri, and Emile van der Zee, 371–378. Cambridge, Mass.: MIT Press.

―――. 2015c. "Why and How Does Consciousness Seem the Way It Seems?" In *Open MIND*, edited by T. Metzinger and J. M. Windt. Frankfurt and Main: MIND Group. doi: 10.15502/9783958570245.

―――. 2015d. "How Our Belief in Qualia Evolved, and Why We Care So Much— A Reply to David H. Baßler." In *Open MIND*, edited by T. Metzinger and J. M. Windt. Frankfurt: MIND Group. doi: 10.15502/9783958570665.

―――. Forthcoming. "Jonathan Bennett's *Rationality*." In *Ten Neglected Classics*, edited by Eric Schliesser.

Dennett, Daniel C., and Ryan T. McKay. 2006. "A Continuum of Mindfulness." *Behavioral and Brain Sciences* 29: 353–354.

Descartes, René. (1637) 1956. *Discourse on Method*. New York: Liberal Arts Press.

―――. 1641. *Meditations on First Philosophy*. Paris: Michel Soly.

Diamond, Jared. 1978. "The Tasmanians: The Longest Isolation, the Simplest Technology." *Nature* 273: 185–186.

Diesendruck, Gil, and Lori Markson. 2001. "Children's Avoidance of Lexical Overlap: A Pragmatic Account." *Developmental Psychology* 37 (5): 630–641.

Domingos, Pedro. 2015. *The Master Algorithm: How the Quest for the Ultimate Learning Machine Will Remake Our World*. New York: Basic Books.

Drescher, Gary L. 1991. *Made-up Minds: A Constructivist Approach to Artificial Intelligence*. Cambridge, Mass.: MIT Press.

Dyson, Freeman J. 1988. *Infinite in All Directions: Gifford Lectures Given at Aberdeen, Scotland, April–November 1985*. New York: Harper and Row.

Edelman, Gerald M. 1989. *The Remembered Present: A Biological Theory of Consciousness*. New York: Basic Books.

Eigen, Manfred. 1992. *Steps Towards Life*. Oxford: Oxford University Press.

Eldredge, Niles. 1983. "A la recherche du docteur Pangloss." *Behavioral and Brain Sciences* 6 (3): 361–362.

Eliasmith, Chris. 2013. *How to Build a Brain: A Neural Architecture for Biological Cognition*. New York: Oxford University Press.

Emery, N. J. 2000. "The Eyes Have It: The Neuroethology, Function and Evolution of Social Gaze." *Neuroscience & Biobehavioral Reviews* 24: 581–604.

Everett, Daniel L. 2004. "Coherent Fieldwork." In *Linguistics Today*, edited by Piet van Sterkenberg, 141–162. Amsterdam: John Benjamins.

Fisher, D. 1975. "Swimming and Burrowing in *Limulus anti Mesolimulus*." *Fossils and Strata* 4: 281–290.

Fitch, W. Tecumseh. 2008. "Nano-Intentionality: A Defense of Intrinsic Intentionality." *Biology & Philosophy* 23 (2): 157–177.

———. 2010. *The Evolution of Language*. Cambridge: Cambridge University Press. http://dx.doi.org/10.1017/CBO9780511817779.

Fitch, W. T., L. Huber, and T. Bugnyar. 2010. "Social Cognition and the Evolution of Language: Constructing Cognitive Phylogenies." *Neuron* 65 (6): 795–814.

FitzGibbon, C. D., and J. H. Fanshawe. 1988. "Stotting in Thomson's Gazelles: An Honest Signal of Condition." *Behavioral Ecology and Sociobiology* 23 (2): 69–74.

Floridi, Luciano. 2010. *Information: A Very Short Introduction*. Oxford: Oxford University Press.

Fodor, Jerry, 1998. "Review of Steven Pinker's *How the Mind Works*, and Henry Plotkin's *Evolution in Mind*." *London Review of Books*. Reprinted in Fodor, *In Critical Condition*. Cambridge, Mass.: Bradford Book/MIT Press.

———. 2008. *LOT 2: The Language of Thought Revisited*. Oxford: Clarendon Press.

Francis, Richard C. 2004. *Why Men Won't Ask for Directions: The Seductions of Sociobiology*. Princeton, N.J.: Princeton University Press.

Frischen, Alexandra, Andrew P. Bayliss, and Steven P. Tipper. 2007. "Gaze cueing of Attention: Visual Attention, Social Cognition, and Individual Differences." *Psychological Bulletin* 133(4): 694–724.

Friston, Karl, Michael Levin, Biswa Sengupta, and Giovanni Pezzulo. 2015. "Knowing One's Place: A Free-Energy Approach to Pattern Regulation." *Journal of the Royal Society Interface*, 12: 20141383.

Frith, Chris D. 2012. "The Role of Metacognition in Human Social Interactions." *Philosophical Transactions of the Royal Society B: Biological Sciences* 367 (1599): 2213–2223.

Gelman, Andrew. 2008. "Objections to Bayesian Statistics." *Bayesian Anal.* 3 (3): 445–449.

Gibson, James J. 1966. "The Problem of Temporal Order in Stimulation and Perception." *Journal of Psychology* 62 (2): 141–149.

———. 1979. *The Ecological Approach to Visual Perception*. Boston: Houghton Mifflin.

Godfrey-Smith, Peter. 2003. "Postscript on the Baldwin Effect and Niche Construction." In *Evolution and Learning: The Baldwin Effect Reconsidered*, edited by Bruce H. Weber and David J. Depew, 210–223. Cambridge, Mass.: MIT Press.

———. 2007. "Conditions for Evolution by Natural Selection." *Journal of Philosophy* 104: 489–516.

———. 2009. *Darwinian Populations and Natural Selection*. Oxford: Oxford University Press.

Gorniak, Peter, and Deb Roy. 2006. "Perceived Affordances as a Substrate for Linguistic Concepts." *MIT Media Lab*. See also Gorniak's MIT dissertation, "The Affordance-based Concept."

Gould, Stephen Jay. 1989. *Wonderful Life: The Burgess Shale and the Nature of History*. New York: W. W. Norton.

———. 1991. *Bully for Brontosaurus: Reflections in Natural History*. New York: W. W. Norton.

———. 1997. "Darwinian Fundamentalism." Part I of review of *Darwin's Dangerous Idea, New York Review of Books*, June 12.

———. 1997b. "Evolution: The Pleasures of Pluralism." Part II of review of *Darwin's Dangerous Idea*, June 26.

Gould, Stephen Jay, and Richard C. Lewontin. 1979. "The Spandrels of San Marco and the Panglossian Paradigm: A Critique of the Adaptationist Programme." Proceedings of the Royal Society of London, the Evolution of Adaptation by Natural Selection (Sept. 21), Series B, *Biological Sciences* 205 (1161): 581–598.

Graziano, Michael S. A. 2013. *Consciousness and the Social Brain*. Oxford, New York: Oxford University Press.

Grice, H. Paul. 1957. "Meaning." *The Philosophical Review* 66: 377–388.

———. 1968. "Utterer's Meaning, Sentence Meaning, and Word Meaning." *Foundations of Language*, 4. Reprinted as ch. 6 in Grice 1989, 117–137.

———. 1969. "Utterer's Meaning and Intentions." *Philosophical Review*, 78. Reprinted as ch. 5 in Grice 1989, 86–116.

———. 1972. *Intention and Uncertainty*. London: Oxford University Press.

———. 1989. *Studies in the Way of Words*. The 1967 William James Lectures at Harvard University. Cambridge, Mass.: Harvard University Press.

Griffin, Donald R., and Carolyn A. Ristau. 1991. "Aspects of the Cognitive Ethology of an Injury-Feigning Bird, the Piping Plover." In *Cognitive Ethology: The Minds of Other Animals: Essays In Honor of Donald R. Griffin*. Hillsdale, N.J.: L. Erlbaum Associates.

Griffin, R., and Dennett, D. C. 2008. "What Does The Study of Autism Tell Us about the Craft of Folk Psychology?" In *Social Cognition: Development, Neuroscience, and Autism*, edited by T. Striano and V. Reid, 254–280. Malden, Mass.: Wiley-Blackwell.

Griffiths, Paul, 1995. "The Cronin Controversy." *Brit. J. Phil. Sci.* 46: 122–138.

———. 2008. "Molecular and Developmental Biology." In *The Blackwell Guide to the Philosophy of Science*, edited by Peter Machamer and Michael Silverstein, 252–271. Oxford: Blackwell.

Guston, Philip. 2011. *Philip Guston: Collected Writings, Lectures, and Conversations*, edited by Clark Coolidge. Berkeley, Los Angeles, London: University of California Press.

Haig, David. 1997. "The Social Gene." *Behavioural Ecology: An Evolutionary Approach*, edited by John R. Krebs and Nicholas Davies, 284–304. Oxford: Blackwell Science.

———. 2008. "Conflicting Messages: Genomic Imprinting and Internal Communication." In *Sociobiology of Communication: An Interdisciplinary Perspective*, edited by Patrizia D'Ettorre and David P. Hughes, 209–223. Oxford: Oxford University Press.

Halitschke, Rayko, Johan A. Stenberg, Danny Kessler, André Kessler, and Ian T. Baldwin. 2008. "Shared Signals—'Alarm Calls' from Plants Increase Apparency to Herbivores and Their Enemies in Nature." *Ecology Letters* 11 (1): 24–34.

Hansell, M. H. 2000. *Bird Nests and Construction Behaviour*. Cambridge: Cambridge University Press.

———. 2005. *Animal Architecture*. Oxford: Oxford University Press.

———. 2007. *Built by Animals*. Oxford: Oxford University Press.

Hardy, Alister, 1960. "Was Man More Aquatic in the Past?" *The New Scientist*, 642–645.

Hardy, Thomas. 1960. *Selected Poems of Thomas Hardy*. London: Macmillan.

Harris, Sam. 2012. *Free Will*. New York: Free Press.

Harvey, I., P. Husbands, D. Cliff, A. Thompson, and N. Jakobi. 1997. "Evolutionary Robotics: the Sussex Approach." *Robotics and Autonomous Systems* 20 (2–4): 205–224.

Haugeland, John. 1985. *Artificial Intelligence: The Very Idea*. Cambridge, Mass.: MIT Press.

———. 1998. *Having Thought: Essays in the Metaphysics of Mind*. Cambridge, Mass.: Harvard University Press.

Hauser, Marc D. 1996. *The Evolution of Communication*. Cambridge, Mass.: MIT Press.

Hauser, Marc D., Noam Chomsky, and W. Tecumseh Fitch. 2002. "The Faculty of Language: What Is It, Who Has It, and How Did It Evolve?" *Science* 298 (5598): 1569–1579.

Heeks, R. J. 2011. "Discovery Writing and the So-called Forster Quote." April 13. https://rjheeks.wordpress.com/2011/04/13/discovery-writing-and-the-so-called-forster-quote/.

Henrich J. 2004. "Demography and Cultural Evolution: Why Adaptive Cultural Processes Produced Maladaptive Losses in Tasmania." *American Antiquity* 69 (2): 197–221.

———. 2015. *The Secret of Our Success.* Princeton, N.J.: Princeton University Press.

Hewes, Gordon Winant. 1973. *The Origin of Man.* Minneapolis: Burgess.

Hinton, Geoffrey E. 2007. "Learning Multiple Layers of Representation." *Trends in Cognitive Sciences* 11 (10): 428–434.

Hofstadter, Douglas. 1979. *Gödel, Escher, Bach: An Eternal Golden Braid.* New York: Basic Books.

———. 1981. "Reflections." In *The Mind's I,* edited by Hofstadter and Dennett. 403–404.

———. 1982. "Can Creativity Be Mechanized?" *Scientific American* 247: 20–29.

———. 1982b. "Who Shoves Whom Around Inside the Careenium? Or What Is the Meaning of the Word 'I'?" *Synthese* 53 (2): 189–218.

———. 1985. *Metamagical Themas: Questing for the Essence of Mind and Pattern.* New York: Basic Books.

———. 2007. *I Am a Strange Loop.* New York: Basic Books.

Hofstadter Douglas, and Daniel Dennett, eds. 1981. *The Mind's I: Fantasies and Reflections on Self and Soul.* New York: Basic Books and Hassocks, Sussex: Harvester.

Hohwy, Jakob. 2012. "Attention and Conscious Perception in the Hypothesis Testing Brain." *Frontiers in Psychology* 3 (96): 1–14.

———. 2013. *The Predictive Mind.* New York: Oxford University Press.

Huebner, Bryce, and Daniel Dennett. 2009. "Banishing 'I' and 'We' from Accounts of Metacognition." Response to Peter Carruthers 2008. "How We Know Our Own Minds: The Relationship Between Mindreading and Metacognition." *Behavioral and Brain Sciences* 32: 121–182.

Hughlings Jackson, J. 1915. "Hughlings Jackson on Aphasia and Kindred Affections of Speech." *Brain* 38: 1–190.

Hume, David. 1739. *A Treatise of Human Nature.* London: John Noon.

Humphrey, Nicholas K. 1976. "The Social Function of Intellect." *Growing Points in Ethology*: 303–317.

———. 1995. *Soul Searching: Human Nature and Supernatural Belief.* London: Chatto and Windus.

———. 1996. *Leaps of Faith: Science, Miracles, and the Search for Supernatural Consolation.* New York: Basic Books.

———. 1998. "Cave Art, Autism, and the Evolution of the Human Mind." *Cambridge Archeological Journal* 8 (2): 184–185.

———. 2000. *How to Solve the Mind-Body Problem.* Thorverton, UK: Imprint Academic.

————. 2006. *Seeing Red: A Study in Consciousness.* Cambridge, Mass.: Harvard University Press.

————. 2009. "The Colour Currency of Nature." In *Colour for Architecture Today,* edited by Tom Porter and Byron Mikellides, 912. London: Taylor and Francis.

————. 2011. *Soul Dust: The Magic of Consciousness.* Princeton, N.J.: Princeton University Press.

Humphrey, Nicholas K. and Daniel Dennett. 1989. "Speaking for Ourselves: An Assessment of Multiple Personality-Disorder." *Raritan—A Quarterly Review* 9 (1): 68–98.

Hurford, James R. 2014. *The Origins of Language: A Slim Guide.* New York: Oxford University Press.

Hurley, Matthew M., D. C. Dennett, and Reginald B. Adams. 2011. *Inside Jokes: Using Humor to Reverse-Engineer the Mind.* Cambridge, Mass.: MIT Press.

Jackendoff, Ray. 1994. *Patterns in the Mind.* New York: Basic Books.

————. 1996. "How Language Helps Us Think." *Pragmatics and Cognition* 4 (1): 1–34.

————. 2002. *Foundations of Language: Brain, Meaning, Grammar, Evolution.* New York: Oxford University Press.

————. 2007. *Language, Consciousness, Culture: Essays on Mental Structure.* Cambridge, Mass.: MIT Press.

————. 2007b. "Linguistics in Cognitive Science: The State of the Art." *Linguistic Review* 24: 347–401.

Jackendoff, Ray, Neil Cohn, and Bill Griffith. 2012. *A User's Guide to Thought and Meaning.* New York: Oxford University Press.

Jakobi, Nick. 1997. "Evolutionary Robotics and the Radical Envelope-of-Noise Hypothesis." *Adaptive Behavior* 6: 325–367.

Jolly, Alison. 1966. *Lemur Behavior: A Madagascar Field Study.* Chicago: University of Chicago Press.

Kameda, Tatsuya, and Daisuke Nakanishi. 2002. "Cost-benefit Analysis of Social/Cultural Learning in a Nonstationary Uncertain Environment: An Evolutionary Simulation and an Experiment with Human Subjects." *Evolution and Human Behavior* 23 (5): 373–393.

Kaminski, J. 2009. "Dogs (*Canis familiaris*) are Adapted to Receive Human Communication." In *Neurobiology of "Umwelt:" How Living Beings Perceive the World,* edited by A. Berthoz and Y. Christen, 103–107. Berlin: Springer Verlag.

Kaminski, J., J. Brauer, J. Call, and M. Tomasello. 2009. "Domestic Dogs Are Sensitive to a Human's Perspective." *Behaviour* 146: 979–998.

Kanwisher N., et al. 1997. "The Fusiform Face Area: A Module in Human Extrastriate Cortex Specialized for Face Perception." *Journal of Neuroscience* 17 (11): 4302–4311.

Kanwisher, N. and D. Dilks. 2013. "The Functional Organization of the Ventral Visual Pathway in Humans." In *The New Visual Neurosciences*, edited by L. Chalupa and J. Werner. Cambridge, Mass.: MIT Press.

Kaplan, David. "Words." 1990. *Proceedings of the Aristotelian Society, Supplementary Volumes*: 93–119.

Katchadourian, Raffi. 2015. "The Doomsday Invention: Will Artificial Intelligence Bring Us Utopia or Destruction?" *New Yorker*, November 23, 64–79.

Kauffman, Stuart. 2003. "The Adjacent Possible." Edge.org, November 9, https://edge.org/conversation/stuart_a_kauffman-the-adjacent-possible.

Keller, Helen. 1908. *The World I Live In*. New York: Century.

Kessler, M. A., and B. T. Werner. 2003. "Self-Organization of Sorted Patterned Ground." *Science* 299 (5605): 380–383.

Kobayashi, Yutaka, and Norio Yamamura. 2003. "Evolution of Signal Emission by Non-infested Plants Growing Near Infested Plants to Avoid Future Risk." *Journal of Theoretical Biology* 223: 489–503.

Kosslyn, Stephen Michael. 1975. "Information Representation in Visual Images." *Cognitive Psychology* 7 (3): 341–370.

———. 1980. *Image and Mind*. Cambridge, Mass.: Harvard University Press.

Kosslyn, Stephen M., et al. 1979. "On the Demystification of Mental Imagery." *Behavioral and Brain Sciences* 2 (4): 535–548.

Kosslyn, S. M., et al. 2001. "The Neural Foundations of Imagery." *Nature Reviews Neuroscience* 2: 635–642.

Kurzweil, Ray. 2005. *The Singularity Is Near: When Humans Transcend Biology*. New York: Viking.

Laland, Kevin, J. Odling-Smee, and Marcus W. Feldman. 2000. "Group Selection: A Niche Construction Perspective." *Journal of Consciousness Studies* 7 (1): 221–225.

Landauer, Thomas K., Peter W. Foltz, and Darrell Laham. 1998. "An Introduction to Latent Semantic Analysis." *Discourse Processes* 25 (2–3): 259–84.

Lane, Nick. 2015. *The Vital Question: Why Is Life the Way It Is?* London: Profile.

Levin, M., 2014. "Molecular Bioelectricity: How Endogenous Voltage Potentials Control Cell Behavior and Instruct Pattern Regulation in Vivo." *Molecular Biology of the Cell* 25: 3835–3850.

Levine, Joseph. 1983. "Materialism and Qualia: The Explanatory Gap." *Pacific Philosophical Quarterly* 64: 354–361.

Levitin, Daniel J. 1994. "Absolute Memory for Musical Pitch: Evidence from the Production of Learned Melodies." *Perception & Psychophysics* 56 (4): 414–423.

Levitin, Daniel J., and Perry R. Cook. 1996. "Memory for Musical Tempo: Additional Evidence That Auditory Memory Is Absolute." *Perception & Psychophysics* 58 (6): 927–935.

Lewis, S. M., and C. K. Cratsley. 2008. "Flash Signal Evolution, Mate Choice and Predation in Fireflies." *Annual Review of Entomology* 53: 293–321.

Lieberman, Matthew D. 2013. *Social: Why Our Brains Are Wired to Connect.* New York: Crown.

Littman, Michael L., Susan T. Dumais, and Thomas K. Landauer. 1998. "Automatic Cross-Language Information Retrieval Using Latent Semantic Indexing." In *Cross-Language Information Retrieval*, 51–62. New York: Springer.

Lycan, William G. 1987. *Consciousness.* Cambridge, Mass.: MIT Press.

MacCready, P. 1999. "An Ambivalent Luddite at a Technological Feast." *Designfax*, August.

MacKay, D. M. 1968. "Electroencephalogram Potentials Evoked by Accelerated Visual Motion." *Nature* 217: 677–678.

Markkula, G. 2015. "Answering Questions about Consciousness by Modeling Perception as Covert Behavior." *Frontiers in Psychology* 6: 803–815.

Marr, D. and T. Poggio. 1976. "From Understanding Computation to Understanding Neural Circuitry." *Artificial Intelligence Laboratory. A.I. Memo.* Cambridge, Mass.: MIT.

Marx, Karl. (1861) 1942. Letter to Lasalle, London, January 16, 1861. *Gesamtausgabe.* International Publishers.

Mayer, Greg. 2009. "Steps toward the Origin of Life." Jerry Coyne's blog, https://whyevolutionistrue.wordpress.com/2009/05/15/steps-toward-the-origin-of-life/.

McClelland, Jay, and Joan Bybee. 2007. "Gradience of Gradience: A Reply to Jackendoff." *Linguistic Review* 24: 437–455

McCulloch, Warren S., and Walter Pitts. 1943. "A Logical Calculus of the Ideas Imminent in Nervous Activity." *Bulletin of Mathematical Biophysics* 5: 115–133.

McFarland, David. 1989. *Problems of Animal Behaviour.* Harlow, Essex, UK: Longman Scientific and Technical.

———. 1989b. "Goals, No-Goals and Own Goals." In *Goals, No-Goals and Own Goals: A Debate on Goal-Directed and Intentional Behaviour*, edited by Alan Montefiore and Denis Noble, 39–57. London: Unwin Hyman.

McGeer, V. 2004. "Autistic Self-awareness." *Philosophy, Psychiatry & Psychology* 11: 235–25l.

McGinn, Colin. 1991. *The Problem of Consciousness: Essays towards a Resolution.* Cambridge, Mass.: Blackwell.

———. 1999. *The Mysterious Flame: Conscious Minds in a Material World.* New York: Basic Books.

McKay, Ryan T., and Daniel C. Dennett. 2009. "The Evolution of Misbelief." *Behavioral and Brain Sciences* 32 (6): 493.

Mercier, Hugo, and Dan Sperber. 2011. "Why Do Humans Reason? Arguments for an Argumentative Theory." *Behavioral and Brain Sciences* 34: 57–111.

Merikle, Philip M., Daniel Smilek, and John D. Eastwood. 2001, "Perception without Awareness: Perspectives from Cognitive Psychology." *Cognition* 79 (1/2): 115–134.

Miller, Geoffrey F. 2000. *The Mating Mind: How Sexual Choice Shaped the Evolution of Human Nature.* New York: Doubleday.

Miller, George A., Eugene Galanter, and Karl H. Pribram. 1960. *Plans and the Structure of Behavior.* New York: Henry Holt.

Miller, Melissa B., and Bonnie L. Bassler. 2001. "Quorum Sensing in Bacteria." *Annual Reviews in Microbiology* 55 (1): 165–199.

Millikan, Ruth Garrett. 1984. *Language, Thought, and Other Biological Categories: New Foundations for Realism.* Cambridge, Mass.: MIT Press.

———. 1993. *White Queen Psychology and Other Essays for Alice.* Cambridge, Mass.: MIT Press.

———. 2000. *On Clear and Confused Ideas: An Essay about Substance Concepts.* Cambridge: Cambridge University Press.

———. 2000b. "Naturalizing Intentionality." In *Philosophy of Mind, Proceedings of the Twentieth World Congress of Philosophy,* edited by Bernard Elevitch, vol. 9, 83–90. Philosophy Documentation Center.

———. 2004. *Varieties of Meaning.* The 2002 Jean Nicod Lectures. Cambridge, Mass.: MIT Press.

———. 2005. *Language: A Biological Model.* Oxford: Clarendon Press.

———. Forthcoming. *Unicepts, Language, and Natural Information.*

Minsky, Marvin. 1985. *The Society of Mind.* New York: Simon and Schuster.

Misasi J., and N. J. Sullivan. 2014. "Camouflage and Misdirection: The Full-on Assault of Ebola Virus Disease." *Cell* 159 (3): 477–486.

Moravec, Hans P. 1988. *Mind Children: The Future of Robot and Human Intelligence.* Cambridge, Mass.: Harvard University Press.

Mordvintsev, A., C. Olah, and M. Tyka. 2015. "Inceptionism: Going Deeper into Neural Networks." Google Research Blog. Retrieved June 20.

Morgan, Elaine. 1982. *The Aquatic Ape.* New York: Stein and Day.

———. 1997. *The Aquatic Ape Hypothesis.* London: Souvenir Press.

Morin, Olivier. 2016. *How Traditions Live and Die: Foundations for Human Action.* New York: Oxford University Press.

Nagel, Thomas. 1974. "What Is It Like to Be a Bat?" *Philosophical Review* 83 (4): 435–450.

Newell, Allen. 1992. "The Knowledge Level." *Artificial Intelligence* 18 (1): 87–127.

Nimchinsky, E.A., E. Gilissen, J. M. Allman., D. P. Perl, J. M. Erwin, and P.R. Hof. 1999. "A Neuronalmorphologic Type Unique to Humans and Great Apes." *Proc Natl Acad Sci.* 96 (9): 5268–5273.

Nørretranders, Tor. 1998. *The User Illusion: Cutting Consciousness Down to Size.* New York: Viking.

Peirce, Charles S. 1906. *Collected Papers of Charles Sanders Peirce*, edited by Charles Hartshorne and Paul Weiss. Cambridge, Mass.: Harvard University Press.

Penrose, Roger. 1989. *The Emperor's New Mind: Concerning Computers, Minds, and the Laws of Physics.* Oxford: Oxford University Press.

Pinker, Steven. 1997. *How the Mind Works.* New York: W. W. Norton.

———. 2003. "Language as an Adaptation to the Cognitive Niche." *Studies in the Evolution of Language* 3: 16–37.

———. 2009. "Commentary on Daniel Dennett." Mind, Brain, and Behavior Lecture, Harvard University, April 23. https://www.youtube.com/watch?v=3H8i5x-jcew.

Pinker, Steven, and Ray Jackendoff. 2005. "The Faculty of Language: What's Special about It?" *Cognition* 95 (2): 201–236.

Powner, M. W., B. Gerland, and J. D. Sutherland. 2009. "Synthesis of Activated Pyrimidine Ribonucleotides in Prebiotically Plausible Conditions." *Nature* 459 (7244): 239–242.

Pugh, George Edgin. 1978. *The Biological Origin of Human Values.* New York: Basic Books.

Pylyshyn, Zenon. 1973. "What the Mind's Eye Tells the Mind's Brain: A Critique of Mental Imagery." *Psychological Bulletin* 80: 1–24.

———. 2002. "Mental Imagery: In Search of a Theory." *Behavioral and Brain Sciences* 25 (2): 157–182.

Quine, W. V. 1951. "Two Dogmas of Empiricism." *Philosophical Review* 60: 20–43.

Rehder, M. F., Michael B. Schreiner, W. Wolfe, Darrell Laham, Thomas K. Landauer, and Walter Kintsch. 1998. "Using Latent Semantic Analysis to Assess Knowledge: Some Technical Considerations." *Discourse Processes* 25 (2/3): 337–354.

Rendell L., R. Boyd, D. Cownden, M. Enquist, K. Eriksson, M. W. Feldman, L. Fogarty, S. Ghirlanda, T. Lillicrap, and K. N. Laland. 2010. "Why Copy Others? Insights from the Social Learning Strategies Tournament." *Science* 328 (5975): 208–213.

Richard, Mark. Forthcoming. *Meanings as Species.*

Richerson, P. J., and R. Boyd. 2004. *Not by Genes Alone.* Chicago: University of Chicago Press.

Ridley, Matt. 2010. *The Rational Optimist.* New York: Harper Collins.

Ristau, Carolyn A. 1983. "Language, Cognition, and Awareness in Animals?" *Annals of the New York Academy of Sciences* 406 (1): 170–186.

Rogers, D. S., and P. R. Ehrlich. 2008. "Natural Selection and Cultural Rates of Change." *Proceedings of the National Academy of Sciences of the United States of America* 105 (9): 3416–3420.

Rosenberg, Alexander. 2011. *The Atheist's Guide to Reality: Enjoying Life without Illusions.* New York: W. W. Norton.

Roy, Deb. 2011. "The Birth of a Word." TED talk, http://www.ted.com/talks /deb_roy_the_birth_of_a_word0.

Sanford, David H. 1975. "Infinity and Vagueness." *Philosophical Review* 84 (4): 520–535.

Scanlon, Thomas. 2014. *Being Realistic about Reasons.* New York: Oxford University Press.

Schönborn, Christoph. 2005. "Finding Design in Nature." *New York Times,* July 7.

Schwitzgebel, Eric. 2007. "No Unchallengeable Epistemic Authority, of Any Sort, Regarding Our Own Conscious Experience—Contra Dennett?" *Phenomenology and the Cognitive Sciences* 6 (1–2): 1–2.

Seabright, Paul. 2010. *The Company of Strangers: A Natural History of Economic Life.* Rev. ed. Princeton, N.J.: Princeton University Press.

Searle, John. R. 1980. "Minds, Brains, and Programs." *Behavioral and Brain Sciences* 3 (3): 417–457.

———. 1992. *The Rediscovery of the Mind.* Cambridge, Mass.: MIT Press.

Selfridge, Oliver G. 1958. "Pandemonium: A Paradigm for Learning in Mechanisation of Thought Processes." In *Proceedings of a Symposium Held at the National Physical Laboratory,* 513–526.

Sellars, Wilfrid. 1962. *Science, Perception, and Reality.* London: Routledge and Paul.

Seung, H. S. 2003. "Learning in Spiking Neural Networks by Reinforcement of Stochastic Synaptic Transmission." *Neuron* 40 (6): 1063–73.

Seyfarth, Robert, and Dorothy Cheney. 1990. "The Assessment by Vervet Monkeys of Their Own and Another Species' Alarm Calls." *Animal Behaviour* 40 (4): 754–764.

Seyfarth, Robert, Dorothy Cheney, and Peter Marler. 1980. "Vervet Monkey Alarm Calls: Semantic Communication in a Free-Ranging Primate." *Animal Behaviour* 28 (4): 1070–1094.

Shanahan, Murray. 2010. *Embodiment and the Inner Life.* New York: Oxford University Press.

Shannon, Claude Elwood. 1948. "A Mathematical Theory of Communication." *Bell System Technical Journal* 27 (3).

Shannon, Claude Elwood, and Warren Weaver. 1949. *The Mathematical Theory of Communication.* Urbana: University of Illinois Press.

Shepard, Roger N., and Jacqueline Metzler. 1971. "Mental Rotation of Three Dimensional Objects." *Science* 171 (3972): 701–703.

Shepard, Roger N., and Lynn A. Cooper. 1982. *Mental Images and Their Transformations.* Cambridge, Mass.: MIT Press.

Siegel, Lee. 1991. *Net of Magic: Wonders and Deceptions in India.* Chicago: University of Chicago Press.

Siewert, Charles. 2007. "In Favor of (Plain) Phenomenology." *Phenomenology and the Cognitive Sciences* 6 (1–2): 201–220.

Simmons, K. E. L. 1952. "The Nature of the Predator-Reactions of Breeding Birds." *Behaviour* 4: 161–171.

Simon, Herbert A. 1969. *The Sciences of the Artificial.* Cambridge, Mass.: MIT Press.

Skutch, Alexander F. 1976. *Parent Birds and Their Young.* Austin: University of Texas Press.

Smith, Brian Cantwell. 1985. "The Limits of Correctness in Computers." Symposium on Unintentional Nuclear War, Fifth Congress of the International Physicians for the Prevention of Nuclear War, Budapest, Hungary, June 28–July 1.

Smith, S. D., and P. M. Merikle. 1999. "Assessing the Duration of Memory for Information Perceived without Awareness." Poster presented at the 3rd Annual Meeting of the Association for the Scientific Study of Consciousness, Canada.

Smith, Stevie. 1957. *Not Waving but Drowning; Poems.* London: A. Deutsch.

Sober, Elliott, and David Sloan Wilson. 1995. "Some Varieties of Greedy Ethical Reductionism." In *DDI*, 467–481.

Sontag, Susan. 1977. *On Photography.* New York: Farrar, Straus and Giroux.

Specter, Michael. 2015. "The Gene Hackers: The Promise of CRISPR Technology." *New Yorker,* Nov. 16, 52.

Sperber, Dan, ed. 2000. *Metarepresentations: A Multidisciplinary Perspective.* Oxford: Oxford University Press.

Sperber, Dan, and Deirdre Wilson. 1986. *Relevance: Communication and Cognition.* Cambridge, Mass.: Harvard University Press.

Sterelny, Kim. 2003. *Thought in a Hostile World: The Evolution of Human Cognition.* Malden, Mass.: Blackwell.

———. 2012. *The Evolved Apprentice.* Cambridge, Mass.: MIT press.

Strawson, Galen. 2003. Review of *Freedom Evolves,* by Daniel Dennett. *New York Times Book Review,* March 2.

Strawson, Peter F. 1964. "Intention and Convention in Speech Acts." *Philosophical Review* 73 (Oct.): 439–460.

Sullivan-Fedock, J. 2011. "Increasing the Effectiveness of Energy Wind Harvesting with CFD Simulation-Driven Evolutionary Computation." Presented at the Seoul CTBUH 2011 World Conference. CTBUH: Seoul, South Korea.

Swiss, Jamy Ian. 2007. "How Magic Works." http://www.egconf.com/videos/how-magic-works.

Szostak, Jack. 2009. "Systems Chemistry on Early Earth." *Nature*, May 14, 171–172.

Tegla, Erno, Anna Gergely, Krisztina Kupan, Adam Miklo, and Jozsef Topa. 2012. "Dogs' Gaze Following Is Tuned to Human Communicative Signals." *Current Biology* 22: 209–212.

Thomas, Elizabeth Marshall. 1993. *The Hidden Life of Dogs*. Boston: Houghton Mifflin.

Thompson, D'Arcy Wentworth. 1917. *On Growth and Form*. Cambridge: Cambridge University Press.

Tinbergen, Niko. 1951. *The Study of Instinct*. Oxford: Clarendon Press.

———. 1959. *Social Behaviour in Animals, with Special Reference to Vertebrates*. London: Methuen.

———. 1961. *The Herring Gull's World; A Study of the Social Behaviour of Birds*. New York: Basic Books.

———. 1965. *Animal Behavior*. New York: Time.

Tomasello, Michael. 2014. *A Natural History of Human Thinking*. Cambridge: Harvard University Press.

Tononi G. 2008. "Consciousness as Integrated Information: A Provisional Manifesto." *Biological Bulletin* 215 (3): 216–42.

Trivers, Robert. 1985. *Social Evolution*. Menlo Park, Calif.: Benjamin/Cummings.

Turing, Alan M. 1936. "On Computable Numbers, with an Application to the Entscheidungs Problem." *Journal of Math* 58 (345–363): 5.

———. 1960. "Computing Machinery and Intelligence." *Mind*: 59: 433–460.

von Neumann, John, and Oskar Morgenstern. 1953 (©1944). *Theory of Games and Economic Behavior*. Princeton, N.J.: Princeton University Press.

von Uexküll, Jakob. 1934. "A Stroll through the Worlds of Animals and Men: A Picture Book of Invisible Worlds." In *Instinctive Behavior: The Development of a Modern Concept*, translated and edited by Claire H. Schiller. New York: International Universities Press.

Voorhees, B. 2000. "Dennett and the Deep Blue Sea." *J. Consc. Studies* 7: 53–69.

Walsh, Patrick T., Mike Hansell, Wendy D. Borello, and Susan D. Healy. 2011. "Individuality in Nest Building: Do Southern Masked Weaver (Ploceus velatus) Males Vary in Their Nest-building Behaviour?" *Behavioural Processes* 88 (1): 1–6.

Wegner, Daniel M. 2002. *The Illusion of Conscious Will*. Cambridge, Mass.: MIT Press.

Westbury, C., and D. C. Dennett. 2000. "Mining the Past to Construct the Future: Memory and Belief as Forms of Knowledge." In *Memory, Brain, and Belief*, edited by D. L. Schacter and E. Scarry, 11–32. Cambridge, Mass.: Harvard University Press.

Whiten, Andrew, and Richard W. Byrne. 1997. *Machiavellian Intelligence II: Extensions and Evaluations*. Cambridge: Cambridge University Press.

Wiener, Norbert. (1948) 1961. *Cybernetics: Or Control and Communication in the Animal and the Machine.* 2nd rev. ed. Paris/Cambridge, Mass.: Hermann and Cie/ MIT Press.

Wills, T., S. Soraci, R. Chechile, and H. Taylor. 2000. "'Aha' Effects in the Generation of Pictures." *Memory & Cognition* 28: 939–948.

Wilson, David Sloan. 2002. *Darwin's Cathedral: Evolution, Religion, and the Nature of Society.* Chicago: University of Chicago Press.

Wilson, Robert Anton. http://www.rawilson.com/sitnow.html.

Wimsatt, William, and Beardsley, Monroe. 1954. "The Intentional Fallacy." In *The Verbal Icon: Studies in the Meaning of Poetry.* Lexington: University of Kentucky Press.

Wrangham R., D. Cheney, R. Seyfarth, and E. Sarmiento. 2009. "Shallow-water Habitats as Sources of Fallback Foods for Hominins." *Am. J. Phys. Anthropol.* 140 (4): 630–642.

Wright, Robert. 2000. *NonZero: The Logic of Human Destiny.* New York: Pantheon Books.

Wyatt, Robert, and John A. Johnson. 2004. *The George Gershwin Reader.* New York: Oxford University Press.

Yu, Wei et al. 2015. "Application of Multi-Objective Genetic Algorithm to Optimize Energy Efficiency and Thermal Comfort in Building Design." *Energy and Buildings* 88: 135–143.

Zahavi, Amotz. 1975. "Mate Selection—A Selection for a Handicap." *Journal of Theoretical Biology* 59: 205–214.

INDEX

ABOUT THE AUTHOR

Daniel C. Dennett is University Professor and Co-director of the Center for Cognitive Studies at Tufts University. For over fifty years he has conducted research on human consciousness, contributing to advances in psychology, neuroscience, evolutionary biology, artificial intelligence, and robotics, as well as writing on such traditional philosophical topics as free will, ethics, epistemology, and metaphysics. In addition to his most famous books, *Consciousness Explained* (1991) and *Darwin's Dangerous Idea* (1995), he has published naturalistic accounts of religion, *Breaking the Spell* (2006), and humor, *Inside Jokes: Using Humor to Reverse-Engineer the Mind* (2011, co-authored with Matthew Hurley and Reginald Adams), and (with Linda LaScola) a study of closeted nonbelieving clergy, *Caught in the Pulpit* (2013). His *Intuition Pumps and Other Tools for Thinking* was published by Norton in 2013.

He lives with his wife, Susan, in Massachusetts and Maine. They have a daughter, a son, and five grandchildren.